ENGINEERING INTERVENTIONS IN SUSTAINABLE TRICKLE IRRIGATION

Water Requirements, Uniformity, Fertigation, and Crop Performance

Innovations and Challenges in Micro Irrigation

ENGINEERING INTERVENTIONS IN SUSTAINABLE TRICKLE IRRIGATION

Water Requirements, Uniformity, Fertigation, and Crop Performance

Edited by

Megh R. Goyal, PhD, PE
Basamma K. Aladakatti

APPLE
ACADEMIC
PRESS

Apple Academic Press Inc.
3333 Mistwell Crescent
Oakville, ON L6L 0A2 Canada

Apple Academic Press Inc.
9 Spinnaker Way
Waretown, NJ 08758 USA

ISBN 13: 978-1-77-463639-8 (pbk)
ISBN 13: 978-1-77-188601-7 (hbk)

Library and Archives Canada Cataloguing in Publication

Engineering interventions in sustainable trickle irrigation: water requirements, uniformity, fertigation, and crop performance / edited by Megh R. Goyal, PhD, PE Basamma K. Aladakatti.

(Innovations and challenges in micro irrigation; v. 8)
Includes bibliographical references and index.
Issued in print and electronic formats.
ISBN 978-1-77188-601-7 (hardcover).--ISBN 978-1-315-18424-1 (PDF)

1. Microirrigation. 2. Crops--Water requirements. 3. Crop improvement. I. Goyal, Megh Raj, editor II. Aladakatti, Basamma K., editor III. Series: Innovations and challenges in micro irrigation; v.8

S619.T74E54 2018   631.5'87   C2017-908082-2   C2017-908083-0

Library of Congress Cataloging-in-Publication Data

Names: Goyal, Megh Raj, editor. | Aladakatti, Basamma K., editor.
Title: Engineering interventions in sustainable trickle irrigation : water requirements, uniformity, fertiga-tion, and crop performance / editors: Megh R. Goyal, Basamma K. Aladakatti.
Description: Waretown, NJ : Apple Academic Press, 2018. | Series: Innovations and challenges in micro irrigation | Includes bibliographical references and index.
Identifiers: LCCN 2017060196 (print) | LCCN 2017061255 (ebook) | ISBN 9781315184241 (ebook) | ISBN 9781771886017 (hardcover : alk. paper)
Subjects: LCSH: Microirrigation. | Irrigation engineering.
Classification: LCC S619.T74 (ebook) | LCC S619.T74 E535 2018 (print) | DDC 628.1/6841--dc23
LC record available at https://lccn.loc.gov/2017060196

CONTENTS

LIST OF CONTRIBUTORS

A. Pannunzio
University of Buenos Aires, Av. San Martin 4500, Buenos Aires, Argentina. E-mail: pannunzio@agro.uba.ar

A. Selvaperumal
Department of Soil and Water Conservation Engineering, Agricultural Engineering College and Research Institute, Kumulur, Trichy, Tamil Nadu, India. E-mail: selvabtech.agri@gmail.com

Amol B. Khomne
Department of Soil and Water Conservation Engineering, College of Technology and Engineering, Maharana Pratap University of Agriculture and Technology, Udaipur 313001, Rajasthan, India. E-mail: khomneamol@gmail.com

Archana Irene
Department of Agronomy, Agricultural College and Research Institute, Tamil Nadu Agricultural University (TNAU), Coimbatore 625104, Tamil Nadu, India

Arun D. Bhagat
Department of Irrigation and Drainage Engineering, Dr. Annasaheb Shinde College of Agricultural Engineering (Dr. ASCAE), Mahatma Phule Agricultural University (MPKV), Rahuri 413722, India

Ashok Hugar
Department of Horticulture, College of Agriculture, University of Agricultural Sciences (UAS), Lingsugur Road, Raichur 584104, Karnataka, India

Ashok R. Mhaske
Agricultural Engineering Division, College of Agriculture at Nagpur, Dr. Panjabrao Deshmukh Krishi Vidyapeeth (State Government Agricultural University), Nagpur 440001, Maharashtra, India. E-mail: mhaskear@gmail.com

B. C. Mal
Former Vice-Chancellor, Chhattisgarh Swami Vivekananda Technical University, Bhilai, Chhattisgarh and Ex-Head of the Department of Agricultural and Food Engineering, Indian Institute of Technology Kharagpur 721302, West Bengal, India. E-mail: bmal@agfe.iitkgp.ernet.in

B. S. Polisgowdar
Department of Soil and Water Engineering, College of Agricultural Engineering, University of Agricultural Sciences, Raichur 584104, Karnataka, India. E-mail: polisgowdar61@yahoo.com

Basamma K. Aladakatti
Senior Research Fellow (Hydrology), Sujala-III Project, Department of Soil Science and Agril. Chemistry, College of Agriculture, University of Agricultural Sciences, Dharwad- 580005, India; Former PhD Student, Department of Soil and Water Engineering, College of Technology and Engineering, Maharana Pratap University of Agriculture and Technology, Udaipur 313001, India. E-mail: basammaka@gmail.com

Chandrika Patti
Department of Soil and Water Engineering, College of Agricultural Engineering, University of Agricultural Sciences (UAS), Lingsugur Road, Raichur 584104, Karnataka, India. E-mail: chandrikapatti@gmail.com

D. Tamilmani
Department of Soil and Water Conservation Engineering, Agricultural Engineering College and
Research Institute (AEC & RI), Tamil Nadu Agricultural University, Coimbatore 641003,
Tamil Nadu, India. E-mail: swc@tnau.ac.in

E. A. Holzapfel
University of Concepción, Casilla 537, Chillán, Chile

F. Bologna
Berries del Sol S.A., Ruta 28 km 5, Colonia Ayuí, Entre Ríos, Argentina

I. Muthuchamy
Department of Soil and Water Conservation Engineering, Agricultural Engineering College and
Research Institute, Kumulur, Trichy, Tamil Nadu, India

Jagadeesha Mulagund
Horticulture Department, Agricultural Engineering College and Research Institute (AEC & RI),
Tamil Nadu Agricultural University, Coimbatore 641003, Tamil Nadu, India.
E-mail: pmjaggu610@gmail.com

Jayakumar Manickam
Agronomist, Regional Coffee Research Station, Coffee Board, Chundale, Kerala, India.
E-mail: agrokumar2013@gmail.com

K. Nagarajan
Water Technology Centre, Tamil Nadu Agricultural University, Coimbatore 641003, Tamil Nadu,
India

K. Nitha
Department of Soil and Water Conservation Engineering, Agricultural Engineering College and
Research Institute (AEC & RI), Tamil Nadu Agricultural University, Coimbatore 641003,
Tamil Nadu, India

K. Shanmugasundaram
Department of Soil and Water Conservation Engineering, Agricultural Engineering College and
Research Institute (AEC & RI), Tamil Nadu Agricultural University, Kumulur 621712, Tamil Nadu,
India

Lala I. P. Ray
School of Natural Resource Management, College of Postgraduate Studies (Central Agricultural
University, Imphal), Umiam, Barapani 793103, Meghalaya, India. E-mail: lalaipray@ rediffmail.com

M. A. Patil
Department of Irrigation and Drainage Engineering, College of Technology & Engineering,
Maharana Pratap University of Agriculture and Technology, Udaipur 313001, India. E-mail: manal-
patil@gmail.com

M. B. Vinuta
Department of Soil and Water Conservation Engineering, Agricultural Engineering College and
Research Institute (AEC & RI), Tamil Nadu Agricultural University, Coimbatore 641003,
Tamil Nadu, India. E-mail: vinuta0413@yahoo.com

Mahesh Kothari
Department of Irrigation and Drainage Engineering, College of Technology & Engineering,
Maharana Pratap University of Agriculture and Technology, Udaipur 313001, India.
E-mail: drmahesh_kothari@yahoo.co.in

Mahesh Rajendran
Central Sericultural Research & Training Institute, Central Silk Board, Ministry of Textiles, Government of India, Post Berhampore 742101, Murshidabad, West Bengal, India. E-mail: maheagri@gmail.com

Mallikajuna S. Ayyanagowdar
Department of Soil and Water Engineering, College of Agricultural Engineering, University of Agricultural Sciences (UAS), Raichur 584101, India

Megh R. Goyal
University of Puerto Rico—Mayaguez Campus; Apple Academic Press, Inc., PO Box 86, Rincon, PR, 00677, USA. E-mail: goyalmegh@gmail.com

N. N. Firake
Department of Irrigation and Drainage Engineering, Dr. Annasaheb Shinde College of Agricultural Engineering (Dr. ASCAE), Mahatma Phule Agricultural University (MPKV), Rahuri 413722, India. E-mail: nn_firake7@rediffmail.com

P. K. Panigrahi
West Bengal and Fisheries Development Officer, Government of Odisha, India. E-mail: panigrahi.prasanta@gmail.com

P. K. Singh
Department of Soil and Water Conservation Engineering, College of Technology and Engineering, Maharana Pratap University of Agriculture and Technology, Udaipur 313001, Rajasthan, India. E-mail: pksingh35@yahoo.com

P. S. Kashyap
Department of SWCE, College of Technology, G. B. Pant University of Agriculture & Technology (GBPUA&T), Pantnagar 263145, Uttarakhand, India. E-mail: pskashyap@gmail.com

P. Texeira Soria
University of Buenos Aires, Av. San Martin 4500, Buenos Aires, Argentina

Pradip Dalavi
Department of Irrigation and Drainage Engineering, College of Technology & Engineering, Maharana Pratap University of Agriculture and Technology, Udaipur 313001, India

Priyanka Sharma
Department of Irrigation and Drainage Engineering, College of Technology & Engineering, Maharana Pratap University of Agriculture and Technology, Udaipur 313001, India. E-mail: priyankasharma1538@gmail.com

R. C. Purohit
Department of Soil and Water Conservation Engineering, College of Technology and Engineering, Maharana Pratap University of Agriculture and Technology, Udaipur 313001, Rajasthan, India. E-mail: purohitrc.yahoo.co.in

Rajavel Manickam
Meteorological Centre, India Meteorological Department, Raipur, Chhattisgarh, India

S. B. Gadge
Department of Irrigation and Drainage Engineering, Dr. Annasaheb Shinde College of Agricultural Engineering (Dr. ASCAE), Mahatma Phule Agricultural University (MPKV), Rahuri 413722, India

S. Moulick
Department of Civil Engineering, Kalinga Institute of Industrial Technology (KIIT) University, Bhubaneswar, Odisha, India. E-mail: sanjib_moulick72@yahoo.co.uk

S. S. Burark
Department of Agricultural Economics and Management, RCA, Maharana Pratap University of Agriculture and Technology, Udaipur 313001, Rajasthan, India

S. Senthilvel
Department of Land and Water Management, Agricultural Engineering College & Research Institute (AEC & RI), Pallapuram Post, Kumulur, Trichy 621712, Tamil Nadu, India

S. V. Kottiswaran
Department of Soil and Water Conservation Engineering, Agricultural Engineering College and Research Institute (AEC & RI), Tamil Nadu Agricultural University, Coimbatore 641003, Tamil Nadu, India. E-mail: kotti.1958@gmail.com

S. Vanitha
Department of Land and Water Management, Agricultural Engineering College & Research Institute (AEC & RI), Pallapuram Post, Kumulur, Trichy 621712, Tamil Nadu, India

Saranya Jeyalakshmi
Center for Engineering Innovation, University of Windsor, Windsor, Ontario, N9B 3P4, Canada

Sunil D. Gorantiwar
Department of Irrigation and Drainage Engineering, Dr. Annasaheb Shinde College of Agricultural Engineering (Dr. ASCAE), Mahatma Phule Agricultural University (MPKV), Rahuri 413722, India. E-mail: sdgorantiwar@gmail.com

Surendran Udayar Pillai
Centre for Water Resources Development and Management, Calicut, Kerala, India. E-mail: u.surendran@gmail.com

T. Arthi
Department of Land and Water Management, Agricultural Engineering College & Research Institute (AEC & RI), Pallapuram Post, Kumulur, Trichy 621712, Tamil Nadu, India. E-mail: arthi.elakia@gmail.com

Truptimayee Suna
Department of Soil and Water Conservation Engineering, Agricultural Engineering College and Research Institute (AEC & RI), Tamil Nadu Agricultural University, Coimbatore 641003, Tamil Nadu, India

Umapati Satishkumar
Department of Soil and Water Engineering, College of Agriculture Engineering, University of Agricultural Sciences, Raichur 584104, Karnataka, India

V. D. Paradkar
Department of Irrigation and Drainage Engineering, Dr. Annasaheb Shinde College of Agricultural Engineering (Dr. ASCAE), Mahatma Phule Agricultural University (MPKV), Rahuri 413722, India

Vasantgouda Roti
Department of SWCE, College of Technology, G. B. Pant University of Agriculture & Technology (GBPUA&T), Pantnagar 263145, Uttarakhand, India

LIST OF ABBREVIATIONS

AICRP	All India Coordinated Research Project
APAR	absorbed photosynthetically active radiations
ASM	available soil moisture
ASMD	available soil moisture depletion
AWHC	available water holding capacity
BCR	benefit–cost ratio
BPM	black plastic mulch
CCS	commercial cane sugar
CD	critical difference
CEC	cation exchange capacity
CIFA	Central Institute of Freshwater Aquaculture
CPE	cumulative pan evaporation
CR	capillary rise
CU	coefficient of uniformity
CWR	crop water requirement
DAP	days after planting
DAT	days after transplanting
DP	deep percolation
DSP	double-side planting
DU	distribution of uniformity
DWR	daily water requirement
EC	electrical conductivity
EDT	total distribution efficiency
ET	evapotranspiration
ETc	crop evapotranspiration
ETm	maximum evapotranspiration
FAO	Food and Agricultural Organization
FRBD	factorial randomized block design
FUE	fertilizer use efficiency
HDP	high-density planting
IAI	integrated aquaculture-cum-irrigation
IARI	Indian Agricultural Research Institute
IBCR	internal benefit–cost ratio

ICAR	Indian Council of Agricultural Research
ICID	International Commission on Irrigation and Drainage
IIT	Indian Institute of Technology
IMC	Indian major carp
INCID	Indian National Committee on Irrigation and Drainage
IW	irrigation water
IWUE	irrigation water use
LDPE	low-density polyethylene
MAP	monoammonium phosphate
MIS	micro irrigation systems
MKP	monopotassium phosphate
MSL	mean sea level
NF	normal fertilizer
NUE	nutrient use efficiency
PE	pan evaporation
PFDC	Precision Farming Development Centre
PM	Penman–Monteith
PM	plastic mulching
PVC	polyvinyl chloride
RAM	readily available moisture
RDF	recommended dose of fertilizer
RH	relative humidity
RMSE	root mean square error
SAR	sodium absorption ratio
SAU	State Agricultural Universities
SD	stocking density
SDF	surface drip fertigation
SED	standard error of deviation
SF	subsurface flow
SOP	sulphate of potash
SPD	strip plot design
SR	survival rate
SSD	subsurface drip irrigation
SSDF	subsurface drip fertigation
SSP	single-side planting
TAM	total available soil moisture
TAN	total ammonia nitrogen
TAN	total available nitrogen

TDM	total dry matter
TNAU	Tamil Nadu Agricultural University
TNPFP	Tamil Nadu Precision Farming Project
TSS	total soluble salts
TSS	total suspended solid
UCC	uniformity coefficient of Christiansen
UCMQ	uniform coefficient of the minor quarter
UDHP	ultrahigh-density planting
UP	urea phosphate
WALMI	Water and Land Management Institute
WMO	World Meteorological Organization
WP	mean wilting point
WR	water requirement
WSF	water-soluble fertilizer
WUE	water use efficiency
WWC	World Water Council

PREFACE 1

Inadequate design is a serious chronic cause of failure of micro irrigation systems.
However if the irrigator uses the help of a professional engineer for designing the system,
He can live a full productive and joyful life.
Giving back is very important to me, as it defines who I am.
I am an ordinary irrigation expert, as I still live like the reader.
I just can't see you, but I can enjoy that you have read my books on micro irrigation.
God bless you as you browse through my books that have been prepared for you only.
I can assure you that drip irrigation can potentially provide a high coefficient of uniformity and distribution efficiency, only if properly designed.

—Megh R. Goyal, Drip Man

According to https://en.wikipedia.org/wiki/Vertical_farming, "vertical farming (VF) is the practice of producing food in vertically stacked layers, vertically inclined surfaces, and/or integrated in other structures. The modern idea of vertical farming facilitates automation of water application and fertigation." Several potential advantages of VF are: reliable harvest throughout the year, minimum overhead despite high initial cost, low energy use, low labor cost, low water use, reduced incidence of pests and diseases, reduced use of pesticides, low processing cost, high crop yield, better quality produce, crop diversity, eco-friendly, lowering of temperature inside the building, mental therapy for homeowners, increase in property value, among others.

It is my opinion that VF requires a vertical irrigation system; and a vertical drip irrigation system (VDIS) will be water and energy efficient compared to other systems. The design of a VDIS is similar to a land drip irrigation system (LDIS). Basic components of VDIS are: pump, filtration system, fertigation unit, controls and accessories, automation unit, main with submains, drip laterals, and emitter types. Designers and irrigators must ensure uniform water application at the upstream and downstream ends. Acceptable values of coefficient of uniformity (CU) and distribution efficiency (DE) for VDIS will be same as for LDIS. Perhaps one can use pressure-compensating emitters or micro-tubes of different lengths and diameters. I invite future contributions related to VDIS to my book series. My vision for micro irrigation technology has been expanding everyday and globally.

After my first textbook, *Drip/Trickle or Micro Irrigation Management* by Apple Academic Press, Inc., and response from international readers, Apple Academic Press, Inc. has published for the world community a 10-volume series on "Research Advances in Sustainable Micro Irriga-tion," edited by me. The website appleacademicpress.com gives details on these 10 book volumes.

This current book volume is one of the future volumes under book series "Innovations and Challenges in Micro Irrigation." Both book series are a must for those interested in irrigation planning and management, namely, researchers, scientists, educators, and students.

The mission of these two book series is to serve as reference manuals for graduate and undergraduate students of agricultural, biological, and civil engineering; horticulture, soil science, crop science, and agronomy. I hope that they will also be a valuable reference for professionals who work with micro irrigation and water management, for professional training institutes, technical agricultural centers, irrigation centers, agricultural extension services, and other agencies that work with micro irrigation programs.

The current book volume, *Engineering Interventions in Sustainable Trickle Irrigation: Water Requirements, Uniformity, Fertigation, and Crop Performance* discusses crop water requirements, fertigation tech-nology, and performance of agricultural crops under best management practices. The book presents research studies on drip-irrigated tomato, chilies, cucumber, eggplant, cabbage, garlic, sugarcane, maize, cashew nut, sapota, banana, mango, and blueberries.

The contributions by the cooperating authors to this book have been most valuable in the compilation of this volume. Their names are mentioned in each chapter and in the list of contributors. This book would not have been written without the valuable cooperation of budding engineer Basamma K. Aladakatti (coeditor of this book) and these investigators, many of whom are renowned scientists who have worked in the field of micro irrigation throughout their professional careers.

My colleague Aladakatti completed her Master of Technology in Soil & Water Engineering from AEC & RI at Tamil Nadu Agricultural University (TNAU). She is now a PhD Research Scholar in the Department of Soil and Water Engineering, College of Technology and Engineering at Maharana Pratap University of Agriculture and Technology, Udaipur, India.

I would like to thank Apple Academic Press, Inc. (AAP) for making every effort to publish the book when the diminishing water resources are a major issue worldwide.

We request that reader offer us your constructive suggestions that may help to improve the next edition.

I express my deep admiration to my wife Subhadra Devi Goyal for understanding and collaboration during the preparation of this book. While I am writing this preface, I was invited to be guest speaker by Dr. B. J. Pandian, Director of Water Technology Centre at TNAU at the National Congress on New Challenges and Advances in Sustainable Micro Irrigation, held March 1–3, 2017. It was my honor to be the guest speaker at this congress.

As an educator, there is a piece of advice to one and all in the world: "Permit that our almighty God, our Creator, excellent Teacher and Micro Irrigation Designer, irrigate our life with His Grace of rain trickle by trickle, because our life must continue trickling on …"

—Megh R. Goyal, PhD., PE
Senior Editor-in-Chief

PREFACE 2

I am happy to have an opportunity to contribute my piece of work and am privileged to be represented along side Dr. Megh R. Goyal as a coeditor for this book. It is very satisfying to me to have shared my research papers with the readers of this book.

The book is split into four major parts, which is a good way to put different concepts bifurcated and to see them explained comprehensively. Although this book is a work of compilation of different research studies contributed by different research scientists, yet content of each chapter has been thoroughly screened for its relevancy and detail.

Improving agricultural water use efficiency (WUE) is vitally important in many parts of the world due to the decreasing availability of water resources and the increasing competition for water between different users that have limited water resources.

Micro irrigation is an effective tool for conserving water resources, and studies have revealed significant water savings ranging from 40% to 70% under drip irrigation compared with surface irrigation. This book covers valuable research studies that evaluate the crop water and fertigation requirements, optimum irrigation and fertigation scheduling, effect of plastic mulch thickness on the yield, and growth parameters for various drip-irrigated crops.

I would like to express my sincere thanks to Dr. Megh R. Goyal (Father of Irrigation Engineering in Puerto Rico, eminent engineer, and chief editor of this book series) for inviting me to join him as a coeditor for this valuable book volume. His zeal is the guiding spirit in all the work he does. I would like to express my sincere gratitude to all the scientists and research scholars who have contributed immensely during the preparation of this book. Sincere thanks to Apple Academic Press, Inc., one of the distinguished publishes available to agricultural engineering fraternity for accepting to publish our work and supporting us at all stages.

I wish to thank my MTech advisor, Dr. K. Shanmugasundaram, for his kind support and constant encouragement. I am thankful to my PhD advisor, Dr. R. C. Purohit, for his candid suggestions and assistance given during the course of this endeavor.

I wish to express my glowing gratitude to my affectionate and loving family members for their constant encouragement, prayers, love, care, and affection showered upon me in the pursuit of my educational ambition for which I am eternally grateful. I express my deep sense of thanks to my senior and junior friends for their moral support and encouragement given to me to complete this work successfully.

I hope this book will reach the community of all agricultural engineers who have been innovating different techniques in spite of challenges in adapting micro irrigation.

—Basamma K. Aladakatti, PhD, Coeditor

WARNING/DISCLAIMER

PLEASE READ CAREFULLY

The goal of this volume, *Engineering Interventions in Sustainable Trickle Irrigation: Water Requirements, Uniformity, Fertigation, and Crop Performance,* is to guide the world community on how to efficiently design for economical crop production.

The editors and the publisher have made every effort to make this book as accurate as possible. However, there still may be grammatical errors or mistakes in the content or typography. Therefore, the contents in this book should be considered as a general guide and not a complete solution to address any specific situation.

The editors and publisher shall have neither liability nor responsibility to any person, organization, or entity with respect to any loss or damage caused, or alleged to have caused, directly or indirectly, by information or advice contained in this book. Therefore, the purchaser/reader must assume full responsibility for the use of the book or the information therein.

The mention of commercial brands and trade names are only for technical purposes. It does not mean that a particular product is endorsed over another product or equipment not mentioned. The editors and publisher do not have any preference for a particular product.

All weblinks that are mentioned in this book were active at the time of publication. The editors and the publisher shall have neither liability nor responsibility if any of the weblinks are inactive at the time of reading of this book.

ABOUT THE SENIOR EDITOR-IN-CHIEF

 Megh R. Goyal, PhD, PE, is, at present, a Retired Professor in Agricultural and Biomedical Engineering from the General Engineering Department in the College of Engineering at the University of Puerto Rico–Mayaguez Campus; and Senior Acquisitions Editor and Senior Technical Editor-in-Chief in Agricultural and Biomedical Engineering for Apple Academic Press, Inc.

He received his BSc degree in Engineering in 1971 from Punjab Agricultural University, Ludhiana, India; his MSc degree in 1977 and PhD degree in 1979 from the Ohio State University, Columbus; his Master of Divinity degree in 2001 from Puerto Rico Evangelical Seminary, Hato Rey, Puerto Rico, USA.

Since 1971, he has worked as Soil Conservation Inspector (1971); Research Assistant at Haryana Agricultural University (1972–1975) and the Ohio State University (1975–1979); Research Agricultural Engineer/ Professor at Department of Agricultural Engineering of UPRM (1979–1997); and Professor in Agricultural and Biomedical Engineering at General Engineering Department of UPRM (1997–2012). He spent one-year sabbatical leave in 2002–2003 at Biomedical Engineering Department, Florida International University, Miami, USA.

He was first agricultural engineer to receive the professional license in Agricultural Engineering in 1986 from College of Engineers and Surveyors of Puerto Rico. On September 16, 2005, he was proclaimed the "Father of Irrigation Engineering in Puerto Rico for the Twentieth Century" by the American Society of Agricultural and Biological Engineers, Puerto Rico Section, for his pioneering work on micro irrigation, evapotranspiration, agroclimatology, and soil and water engineering. During his professional career of 45 years, he has received awards such as: Scientist of the Year, Blue Ribbon Extension Award, Research Paper Award, Nolan Mitchell Young Extension Worker Award, Agricultural Engineer of the Year, Citations by

Mayors of Juana Diaz and Ponce, Membership Grand Prize for ASAE Campaign, Felix Castro Rodriguez Academic Excellence, RashtryaRatan Award and Bharat Excellence Award and Gold Medal, Domingo Marrero Navarro Prize, Adopted son of Moca, Irrigation Protagonist of UPRM, Man of Drip Irrigation by Mayor of Municipalities of Mayaguez/Caguas/Ponce and Senate/Secretary of Agriculture of ELA, Puerto Rico.

The Water Technology Centre of Tamil Nadu Agricultural University in Coimbatore, India recognized Dr. Goyal as one of the experts "who rendered meritorious service for the development of the micro irrigation sector in India" by bestowing *"Award of Outstanding Contribution in Micro Irrigation."* This award was presented to Dr. Goyal during the inaugural session of the National Congress on "New Challenges and Advances in Sustainable Micro Irrigation on March 1, 2017, held at Tamil Nadu Agricultural University.

He has authored more than 200 journal articles and textbooks and edited over 50 books including: "Elements of Agroclimatology (Spanish) by UNISARC, Colombia"; two "Bibliographies on Drip Irrigation."

Apple Academic Press Inc. (AAP) has published his books, namely: "Management of Drip/Trickle or Micro Irrigation," and "Evapotranspiration: Principles and Applications for Water Management," 10-volume set on *"Research Advances in Sustainable Micro Irrigation."* Readers may contact him at: goyalmegh@gmail.com.

ABOUT THE EDITOR
BASAMMA K ALADAKATTI

 Basamma K. Aladakatti is a Senior Research Fellow (Sujala-III Project) at Department of Soil Science and Agril. Chemistry, College of Agriculture, UAS, Dharwad- 580005, India; and she finished her PhD in the Department of Soil and Water Engineering from the College of Technology and Engineering, Maharana Pratap University of Agriculture and Technology (MPUAT), Udaipur, India. She completed her BTech and MTech degrees from the University of Agricultural Sciences, Raichur, and Tamil Nadu Agricultural University, Coimbatore, India, respectively.

She is an awardee of the INSPIRE fellowship from the Ministry of Science & Technology, Government of India, for her PhD and received a Junior Research Fellowship award from the Indian Council of Agricultural Research (ICAR), New Delhi. She has published several full-length research papers in national and international journals as well as more than 20 abstracts and four full-length papers in various symposiums and conferences on various significant topics of agriculture engineering. She has received two best paper awards. Her career pursuits include working on research projects pertaining to refinement of micro irrigation and drought analysis and forecasting.

OTHER BOOKS ON MICRO IRRIGATION TECHNOLOGY FROM APPLE ACADEMIC PRESS, INC.

Management of Drip/Trickle or Micro Irrigation

Evapotranspiration: Principles and Applications for Water Management

Book Series: Research Advances in Sustainable Micro Irrigation
Senior Editor-in-Chief: Megh R. Goyal, PhD, PE
  Volume 1: Sustainable Micro Irrigation: Principles and Practices
  Volume 2: Sustainable Practices in Surface and Subsurface Micro
            Irrigation
  Volume 3: Sustainable Micro Irrigation Management for Trees and Vines
  Volume 4: Management, Performance, and Applications of Micro
            Irrigation Systems
  Volume 5: Applications of Furrow and Micro Irrigation in Arid and
            Semi-Arid Regions
  Volume 6: Best Management Practices for Drip Irrigated Crops
  Volume 7: Closed Circuit Micro Irrigation Design: Theory and
            Applications
  Volume 8: Wastewater Management for Irrigation: Principles and
            Practices
  Volume 9: Water and Fertigation Management in Micro Irrigation
Volume 10: Innovation in Micro Irrigation Technology

Book Series: Innovations and Challenges in Micro Irrigation
Senior Editor-in-Chief: Megh R. Goyal, PhD, PE
Volume 1: Principles and Management of Clogging in Micro Irrigation
Volume 2: Sustainable Micro Irrigation Design Systems for Agricultural
          Crops: Methods and Practices

Volume 3: Performance Evaluation of Micro Irrigation Management: Principles and Practices

Volume 4: Potential Use of Solar Energy and Emerging Technologies in Micro Irrigation

Volume 5: Micro Irrigation Management: Technological Advances and Their Applications

Volume 6: Micro Irrigation Engineering for Horticultural Crops: Policy Options, Scheduling, and Design

Volume 7: Micro Irrigation Scheduling and Practices

Volume 8: Engineering Interventions in Sustainable Trickle Irrigation

Book Series: Innovations in Agricultural and Biological Engineering
Senior Editor-in-Chief: Megh R. Goyal, PhD, PE

- Dairy Engineering: Advanced Technologies and Their Applications
- Developing Technologies in Food Science: Status, Applications, and Challenges
- Engineering Interventions in Agricultural Processing
- Engineering Practices for Agricultural Production and Water Conservation: An Interdisciplinary Approach
- Emerging Technologies in Agricultural Engineering
- Flood Assessment: Modeling and Parameterization
- Food Engineering: Emerging Issues, Modeling, and Applications
- Food Process Engineering: Emerging Trends in Research and Their Applications
- Food Technology: Applied Research and Production Techniques
- Modeling Methods and Practices in Soil and Water Engineering
- Processing Technologies for Milk and Dairy Products: Methods Application and Energy Usage
- Soil and Water Engineering: Principles and Applications of Modeling
- Soil Salinity Management in Agriculture: Technological Advances and Applications
- Technological Interventions in the Processing of Fruits and Vegetables
- Technological Interventions in Management of Irrigated Agriculture
- Engineering Interventions in Foods and Plants

- Technological Interventions in Dairy Science: Innovative Approaches in Processing, Preservation, and Analysis of Milk Products
- Novel Dairy Processing Technologies: Techniques, Management, and Energy Conservation
- Sustainable Biological Systems for Agriculture: Emerging Issues in Nanotechnology, Biofertilizers, Wastewater, and Farm Machines
- State-of-the-Art Technologies in Food Science: Human Health, Emerging Issues and Specialty Topics
- Scientific and Technical Terms in Bioengineering and Biological Engineering

PART I
Crop Irrigation Requirements

CHAPTER 1

WATER REQUIREMENTS FOR HORTICULTURAL CROPS UNDER MICRO IRRIGATION

JAYAKUMAR MANICKAM[1,*], SURENDRAN UDAYAR PILLAI[2], and RAJAVEL MANICKAM[3]

[1]Regional Coffee Research Station, Coffee Board, Chundale, Kerala, India

[2]Water Management (Agriculture) Division, Centre for Water Resources Development and Management, Calicut, Kerala, India

[3]Meteorological Centre, India Meteorological Department, Raipur, Chhattisgarh, India

*Corresponding author. E-mail: agrokumar2013@gmail.com

CONTENTS

ABSTRACT

Assessment of water requirement of horticulture crops is prerequisite for efficient use of water in limited water availability and increased crop productivity. Different methods available for estimation of crop water requirement from historical to present are discussed in this chapter. Selection of a method to determine water requirement of crop depends on availability of weather data. Location specific studies are conducted to evaluate different methods of estimation of water requirement to find out best method with good accuracy. Water requirement of horticultural crops are reviewed in this chapter. This information can be used to compute water requirement of crops under micro irrigation.

1.1   INTRODUCTION

1.1.1   WATER REQUIREMENT OF CROPS

It is defined as the quantity of water, regardless of its source, required by a crop or diversified pattern of crops in a given period for its normal growth and development under field conditions at a given place. In other words, it is the total quantity of water required to mature an adequately irrigated crop. It is expressed in depth per unit time. Water requirement, if considered as a demand, includes the quantity of water needed to meet the losses due to evapotranspiration (ET), plus the losses during the application of irrigation water (unavoidable losses) and the additional quantity of water required for special operations, such as land preparation, transplanting, leaching of salts below the crop root zone, frost control, etc.

1.1.2   EVAPOTRANSPIRATION

$$\text{Evapotranspiration (ET)} = \text{Evaporation} + \text{Transpiration} \qquad (1.1)$$

Evaporation is a diffusive process by which water from natural surfaces, such as free water surface, bare soil, from live or dead vegetation foliage (intercepted water, dewfall, guttation, etc.) is lost in the form of vapor to the atmosphere. Likewise, transpiration is a process by which water is lost in the form of vapor through plant surfaces, particularly leaves. In this process, water is essentially absorbed by the plant roots due to water potential gradients and it moves upward through the stem and is ultimately lost into the atmosphere through numerous minute stomata in the plant leaves. It is basically an evaporation process. However, unlike evaporation from a water or soil surface, plant structure and stomatal behavior operating in conjunction with the physical principles governing evaporation modify transpiration.

Thus, ET is a combined loss of water from the soil (evaporation) and plant (transpiration) surfaces to the atmosphere through vaporization of liquid water, and is expressed in depth per unit time (e.g., mm/day). ET is the largest and one of the most basic components of the hydrologic cycle. It plays an important role in the water and energy balance on the earth surface and has a major role in agricultural production processes. Crops require water for ET to meet atmospheric evaporative demands, photosynthetic activity, and other metabolic activities. Amount of water actually used for photosynthetic activity is less than 1% of total water requirement of the crop. Therefore, crop water requirement is simply water required for evaporation from soil surface and transpiration from the crop canopy.

Quantification of ET is required in the context of many issues:

- Management of water resources in agriculture.
- Designing of irrigation projects on sound economic basis.
- Fixing cropping patterns and working out the irrigation requirements of crops.
- Scheduling of irritations.
- Classifying regions climatologically for agriculture.

ET depends on crop and climate factors. Micro irrigation systems ideally require estimation of ET on a daily basis to plan for the volume of water discharge. Common methods to estimate crop ET include lysimeter, estimation of reference crop evapotranspiration (ET_o), pan evaporation, and soil water balance method.

This chapter reviews the different methods of estimation of ET and water requirement of horticultural crops.

1.2   FACTORS AFFECTING ET

Water losses to the atmosphere are primarily determined by both environmental and plant factors, besides to a certain extent by management factors. The environmental effect on ET is called atmospheric demand or evaporative demand of the atmosphere. Weather parameters, soil, crop characteristics, management, and environment are factors affecting ET.

1.2.1   WEATHER PARAMETERS

Radiation, air temperature, humidity, and wind speed are weather parameters affecting ET. Several procedures have been developed to assess the evaporation rate from these parameters. The evaporation power of the atmosphere is expressed by ET_o. ET_o represents the ET from a standardized vegetated surface.

Evapotranspiration process is determined by the amount of energy available to vaporize water. Solar radiation is the largest energy source and is able to change large quantities of liquid water into water vapor. The solar radiation absorbed by the atmosphere and the heat emitted by the earth increase the air temperature. The sensible heat of the surrounding air transfers energy to the crop and exerts as such a controlling influence on the rate of ET. In sunny, warm weather the loss of water by ET is greater than that in cloudy and cool weather. Well-watered fields in hot dry arid regions consume large amounts of water due to the abundance of energy and the desiccating power of the atmosphere. In humid tropical regions, notwithstanding the high energy input, the high humidity of the air will reduce the ET demand. In such an environment, the air is already close to saturation, so that less additional water can be stored and hence the ET rate is lower than in arid regions. The process of vapor removal depends to a large extent on the wind and air turbulence which transfers large quantities of air over the evaporating surface. When vaporizing water, the air above the evaporating surface becomes gradually saturated with water vapor. If this air is not continuously replaced with drier air, the driving force for water vapor removal and the ET rate decreases.

1.2.2   SOIL EVAPORATION

Water evaporates from the soil surface at rates comparable to evaporation from free water surface as long as the surface is wet and the soil is not shaded by plants or mulches. Evaporation from a wet soil surface occurs during what has been called "first-stage drying." "Second-stage drying" begins when the soil surface becomes visibly dry. This generally occurs 1–5 days after irrigation or precipitation. During the initial portion of second-stage drying, soil evaporation rates are controlled by hydraulic properties, which determine the rate at which water will move through the soil and to the soil surface. During the latter portion of this stage, most of the water evaporated at the soil surface moves through the soil in the form of vapor. During second-stage drying evaporation decreases approximately as the square root of the time elapsed.

1.2.3   SOIL WATER FOR PLANTS

The contribution of soil evaporation to the total ET decreases as the plant cover increases. Thus with increasing plant cover, the ET depends mainly on the soil water status and on the ability of plant roots to extract available water. Certainly, as soil water becomes less available to plants, transpiration decreases. There are other cases, generally near midday under a high radiation and heat load, when with adequate water, plants cannot extract water from the soil at a rate sufficient to meet the evaporative demand. When such a situation occurs, stomatal diffusion resistance will increase and the transpiration rate will decrease.

1.2.4   CROP FACTORS

Crop type, variety, and development stage influence the ET. Differences in resistance to transpiration, crop height, crop roughness, reflection, ground cover, and crop rooting characteristics also affect the ET of different crops under identical environmental conditions. Crop ET under standard conditions (ET_c) refers to the evaporating demand from crops that are grown in large fields under optimum soil water, excellent management and environmental conditions, and achieve full production under the given climatic conditions.

1.2.5   MANAGEMENT AND ENVIRONMENTAL CONDITIONS

Soil salinity, poor land fertility, limited application of fertilizers, the presence of hard or impenetrable soil horizons, the absence of control of diseases and pests and poor soil management may limit the crop development of crop and reduce the ET. Other factors affecting ET are ground cover, plant density, and the soil water content. The effect of soil water content on ET is conditioned primarily by the magnitude of the water deficit and the type of soil. On the other hand, too much water will result in water logging which might damage the root and limit root water uptake by inhibiting respiration.

Regimes of management practices can also influence the climatic and crop factors affecting ET. Cultivation practices and the type of irrigation method can alter the microclimate, affect the crop characteristics or affect the wetting of the soil and crop surface. A windbreak reduces wind velocities and decreases the ET rate of the field directly beyond the barrier. The effect can be significant especially in windy, warm, and dry conditions although ET from the trees themselves may offset any reduction in the field. Soil evaporation in a young orchard, where trees are widely spaced, can be reduced by using a well-designed drip or trickle irrigation system. The drippers apply water directly to the soil near trees, thereby leaving the major part of the soil surface dry, and limiting the evaporation losses. The use of mulches, especially when the crop is small, is another way of substantially reducing soil evaporation. Antitranspirants, such as stomata-closing, film-forming or reflecting material, reduce the water losses from the crop and hence the transpiration rate.

1.3   ET TERMINOLOGIES

1.3.1   REFERENCE CROP EVAPOTRANSPIRATION (ET$_O$)

ET rate from a reference surface, not short of water, is called the reference crop evapotranspiration or reference evapotranspiration and is denoted as ET$_0$. The reference surface is a hypothetical grass reference crop with specific characteristics. The only factors affecting ET$_0$ are climatic parameters. Consequently, ET$_0$ is a climatic parameter and can be computed from weather data. ET$_0$ expresses the evaporating power of the atmosphere at a specific location and time of the year and does not consider the crop characteristics and soil factors.

1.3.2 CROP EVAPOTRANSPIRATION UNDER STANDARD CONDITIONS (ET$_c$)

ET$_c$ is the evapotranspiration from disease-free, well-fertilized crops, grown in large fields, under optimum soil water conditions, and achieving full production under the given climatic conditions.

1.4 METHODS TO ESTIMATE CROP EVAPOTRANSPIRATION

In 1960, Jensen and Haise suggested the following equation to estimate daily ET of full crop cover[27]

$$ET = (0.014\ T - 0.37)\ R_s \qquad (1.2)$$

where T is in °F and R_s is solar radiation in mm/day or in/day.

1.4.1 BLANEY–CRIDDLE METHOD

The most widely known empirical ET-estimating method used in the USA in the 1950s and 1960s was the Blaney–Criddle (B–C) method. The procedure was first proposed by Blaney and Morin in 1942.[8] It was modified later by Blaney and Criddle.[9–11] The equation is:

$$U = KF = \sum kf \qquad (1.3)$$

where U = estimated CU (or ET), in, F = the sum of monthly CU factors, f, for the period ($f = t \cdot p/100$ where t = mean monthly air temperature, °F, and p = mean monthly percent of annual daytime hours (daytime is defined as the period between sunrise and sunset); K = empirical CU coefficient (irrigation season or growing period); and k = monthly CU coefficient.

1.4.2 TRANSITION METHODS IN THE UNITED STATES

Estimating methods in the United States began to change in the 1960s from methods based primarily on mean air temperature to methods considering both temperature and solar radiation as listed below.

$$\text{ET} = (0.014\ T_f - 0.37)\ R_s \qquad \text{(Alfalfa reference ET)}$$
$$\text{ET} = (0.0082\ T_f - 0.19)\ R_s \qquad \text{(Grass reference ET, Florida)}$$
$$\text{ET} = (0.0023\ TD_c^{0.5}\ (T_c + 17.8)\ R_s \quad \text{(Grass reference ET)} \qquad (1.4)$$

where T_c is mean air temperature in °C, R_s is solar radiation in mm/day or in/day, and TD is the difference between maximum and minimum daily air temperature.

1.4.3   METHODS BASED ON VAPOR PRESSURE

Bowen ratio is the ratio of temperature to vapor pressure gradients. The $\Delta t/\Delta e^{12}$ and energy balance concepts were not incorporated at an early date into methods for estimating ET as they were for estimating evaporation from water surfaces. In contrast to the development of largely empirical methods in the United States, Penman[34] in the United Kingdom took a basic approach and related ET to energy balance and rates of sensible heat and water vapor transfer. Penman equation was based on the physics of the processes, and it laid the foundation for current ET-estimating methodology using standard weather measurements of solar radiation, air temperature, humidity, and wind speed. The Penman equation[35–37] stands out as the most commonly applied *physics-based* equation. Later, a surface resistance term was added.[30,38] The modern combination equation applied to standardized surfaces is currently referred to as the Penman–Monteith equation (PM). It represents the state of the art in estimating hourly and daily ET.

Other methods of estimating and measuring ET range from eddy covariance and energy balance using Bowen ratio or sensible heat flux based on surface temperature, radiosonde measurements of complete boundary layer profiles of temperature and humidity, and energy balance estimates based on satellite imagery.

1.4.4   JENSEN METHOD

Jensen described the process of using the rate of ET from a well-watered crop with an aerodynamically rough surface like alfalfa with 30–50 cm of growth as a measure of potential ET, or E_o.[28] ET for a given crop could be related to E_o using a coefficient, now known as a crop coefficient:

$$E_t = K_c \times E_o \qquad (1.5)$$

where K_c is a dimensionless coefficient similar to that proposed by Van Wijk and de Vries[44] representing the combined effects of resistance to water movement from the soil to the evaporating surfaces, resistance to diffusion of water vapor from the evaporating surfaces through the laminar boundary layer, resistance to turbulent transfer to the free atmosphere, and relative amount of radiant energy available as compared to the reference crop.

At that time, methods other than those based on air temperature were not well known. In order to facilitate the understanding of the $E_o \times K_c$ process, illustrating the change in the K_c as crop cover enabled users to visualize how the coefficient changed from a value near 0.15 for the bare soil to 1.0 at full cover.

1.4.5   LYSIMETER

In lysimeters, the crop grows in isolated tanks filled with either disturbed or undisturbed soil. In precision weighing lysimeters, where the water loss is directly measured by the change of mass, evapotranspiration can be obtained with an accuracy of a few hundredths of a millimeter, and small time periods such as an hour can be considered. In nonweighing lysimeters, the evapotranspiration for a given time period is determined by deducting the drainage water, collected at the bottom of the lysimeters, from the total water input. As lysimeters are difficult and expensive to construct and as their operation and maintenance require special care, their use is limited to specific research purposes.

1.4.6   ET ESTIMATED FROM PAN EVAPORATION

Evaporation from an open water surface provides an index of the integrated effect of radiation, air temperature, air humidity, and wind on evapotranspiration. The pan has proved its practical value and has been used successfully to estimate reference evapotranspiration by observing the evaporation loss from a water surface and applying empirical coefficients to relate pan evaporation to ET_o. ET_o is computed by the following formula.

$$ET_o = K_p \times E_p \tag{1.6}$$

where E_p is the open pan evaporation (mm/day) and K_p is the pan coefficient.

1.4.7 ENERGY BALANCE METHOD

Evaporation of water requires relatively large amounts of energy, either in the form of sensible heat or radiant energy. Therefore, the evapotranspiration process is governed by energy exchange at the vegetation surface and is limited by the amount of energy available. Because of this limitation, it is possible to predict the evapotranspiration rate by applying the principle of energy conservation. The energy arriving at the surface must be equal to the energy leaving the surface for the same time period.

All fluxes of energy should be considered when deriving an energy balance equation. The equation for an evaporating surface can be written as:

$$R_n - G - LET - H = 0 \tag{1.7}$$

where R_n is the net radiation, H is the sensible heat, G is the soil heat flux, and LET is the latent heat flux.

The various terms can be either positive or negative. Positive R_n supplies energy to the surface and positive G, LET and H remove energy from the surface. The equation is restricted to the four components: R_n, LET, H, and G. Other energy terms, such as heat stored or released in the plant, or the energy used in metabolic activities, are not considered because these terms account for only a small fraction of the daily net radiation and can be considered negligible when compared with the other four components.

The latent heat flux (LET) representing the evapotranspiration fraction can be derived from the energy balance equation if all other components are known. Net radiation (R_n) and soil heat fluxes (G) can be measured or estimated from climatic parameters. Measurements of the sensible heat (H) are, however, complex and cannot be easily obtained. H requires accurate measurement of temperature gradients above the surface.

1.4.8   MASS TRANSFER METHOD

This approach considers the vertical movement of small parcels of air (eddies) above a large homogeneous surface. The eddies transport material (water vapor) and energy (heat, momentum) from and toward the evaporating surface. By assuming steady state conditions and that the eddy transfer coefficients for water vapor are proportional to those for heat and momentum, the evapotranspiration rate can be computed from the vertical gradients of air temperature and water vapor via Bowen ratio. Other direct measurement methods use gradients of wind speed and water vapor. These methods and other methods such as eddy covariance require accurate measurements of vapor pressure, and air temperature or wind speed at different levels above the surface. Therefore, their application is restricted to primarily research situations.

1.4.9   SOIL WATER BALANCE

Evapotranspiration can also be determined by measuring the various components of the soil water balance. The method consists of assessing the incoming and outgoing water flux into the crop root zone over some time period. Irrigation (I) and rainfall (P) add water to the root zone. Portions of I and P might be lost by surface runoff (RO) and by deep percolation (DP) that will eventually recharge the water table. Water might also be transported upward by a capillary rise (CR) from a shallow water table toward the root zone or even transferred horizontally by subsurface flow in (SF_{in}) or out of (SF_{out}) the root zone. In many situations, however, except under conditions with large slopes, SF_{in} and SF_{out} are minor and can be ignored. Soil evaporation and crop transpiration deplete water from the root zone. If all fluxes other than evapotranspiration (ET) can be assessed, the evapotranspiration can be deduced from the change in soil water content (DSW) over the time period:

$$ET = I + P - RO - DP + CR \pm DSF \pm DSW \qquad (1.8)$$

Some fluxes such as subsurface flow, deep percolation, and CR from a water table are difficult to assess and short time periods cannot be considered. The soil water balance method can usually only give us ET estimates over long time periods of the order of week-long or 10-day periods.

1.4.10   ET COMPUTED FROM METEOROLOGICAL DATA

Owing to the difficulty of obtaining accurate field measurements, ET is commonly computed from weather data. Several empirical or semiempirical equations have been developed for assessing crop or ET_0 from meteorological data. Some of the methods are only valid under specific climatic and agronomic conditions and cannot be applied under conditions different from those under which they were originally developed. FAO Penman–Monteith method is now recommended as the standard method for the computation of the reference evapotranspiration, ET_0.[3] The ET from crop surfaces under standard conditions is determined by K_c that relate ET_c to ET_0. The ET from crop surfaces under nonstandard conditions is adjusted by a water stress coefficient (K_s) and/or by modifying the K_c.

1.4.10.1   PENMAN–MONTEITH EQUATION

In 1948, Penman combined the energy balance with the mass transfer method and derived an equation to compute the evaporation from an open water surface from standard climatological records of sunshine, temperature, humidity, and wind speed. This so-called combination method was further developed by many researchers and extended to cropped surfaces by introducing resistance factors. The final form of Penman–Monteith method by FAO is given below to estimate ET_0:

$$ET_o = \frac{0.408\Delta(R_n - G) + \gamma \dfrac{900}{T+273} u_2 (e_s - e_a)}{\Delta + \gamma(1 + 0.34\, u_2)} \tag{1.9}$$

where ET_0 is reference evapotranspiration (mm/day), R_n is net radiation at the crop surface (MJ/m²/day), G is soil heat flux density (MJ/m²/day), T is mean daily air temperature at 2 m height (°C), u_2 is wind speed at 2 m height (m/s), e_s is saturation vapor pressure (kPa), e_a is actual vapor pressure (kPa), $(e_s - e_a)$ is saturation vapor pressure deficit (kPa), Δ is slope vapor pressure curve (kPa/°C), and γ is the psychrometric constant (kPa/°C).

FAO Penman–Monteith equation is a close, simple representation of the physical and physiological factors governing the evapotranspiration process. By using the FAO Penman–Monteith definition for ET_0, one may calculate K_c at research sites by relating the measured crop evapotranspiration (ET_c) with the calculated ET_0 (i.e., $K_c = ET_c/ET_0$). In the K_c approach, differences in the crop canopy and aerodynamic resistance relative to the hypothetical reference crop are accounted for the K_c. The K_c factor serves as an aggregation of the physical and physiological differences between crops and the reference definition.

1.4.11 CALCULATION OF CROP EVAPOTRANSPIRATION (ETC) UNDER STANDARD CONDITIONS

No limitations are placed on crop growth or evapotranspiration from soil water and salinity stress, crop density, pests and diseases, weed infestation, or low fertility. ET_c is determined by the K_c approach whereby the effect of the various weather conditions are incorporated into ET_0 and the crop characteristics into the K_c:

$$ET_c = K_c \, ET_0 \qquad (1.10)$$

The effect of both crop transpiration and soil evaporation are integrated into a single K_c. The K_c incorporates crop characteristics and averaged effects of evaporation from the soil. For normal irrigation planning and management purposes, for the development of basic irrigation schedules, and for most hydrologic water balance studies, average K_c are relevant and more convenient than the K_c computed on a daily time step using a separate crop and soil coefficient. When values for K_c are needed on a daily basis for specific fields of crops and for specific years, one must use transpiration and evaporation coefficient (K_{cb} and K_e). The calculation procedure for crop evapotranspiration, ET_c, consists of:

- Identifying the crop growth stages, determining their lengths, and selecting the corresponding K_c coefficients (Table 1.1);
- Adjusting the selected K_c coefficients for frequency of wetting or climatic conditions during the stage;[3]
- Constructing the K_c curve (allowing one to determine K_c values for any period during the growing period); and
- Calculating ET_c as the product of ET_0 and K_c.

For in depth study, the author is encouraged to consult "Goyal, M. R. *Management of Drip/Trickle or Micro Irrigation* (Chapters 2 and 3); Apple Academic Press Inc.: Oakville, ON, 2014."

TABLE 1.1   Crop Coefficient for Major Crops.

Crop	K_c (initial stage)	K_c (midstage)	K_c (end stage)
Cereals	0.15	1.10	0.25
Legumes (Leguminosae)	0.15	1.10	0.50
Oil crops	0.15	1.10	0.25
Roots and tubers	0.15	1.00	0.85
Small vegetables	0.15	0.95	0.85
Sugarcane	0.15	1.20	0.70
Vegetables—cucumber family (Cucurbitaceae)	0.15	0.95	0.70
Vegetables—solanum family (Solanaceae)	0.15	1.10	0.70

Adapted from: Goyal, M. R. *Management of Drip/Trickle or Micro Irrigation* (Chapter 1); Apple Academic Press Inc.: Oakville, ON, 2014; p 50.

1.4.12   SOFTWARE PACKAGES TO COMPUTE CROP WATER REQUIREMENT

1.4.12.1   CROPWAT

CROPWAT is a decision support system developed by FAO for planning and management of irrigation. It is a practical tool to carry out standard calculations for ET_o, crop water requirements and crop irrigation requirements, and more specifically the design and management of irrigation schemes. The development of irrigation schedules and evaluation of rainfed and irrigation practices are based on daily water balance using various options for water supply and irrigation management conditions. CROPWAT uses Penman–Montieth equation for estimating ET_o. Estimation of crop water requirement through CROPWAT follows steps below:

- ET_o values measured or calculated using the FAO Penman–Montieth equation based on decade/monthly climatic data: minimum and maximum air temperature, relative humidity, sunshine duration, and wind speed.
- Rainfall data (daily/decade/monthly data); monthly rainfall is divided into a number of rainstorms each month.

- A cropping pattern consisting of the planting date, K_c data files (including K_c values, stage days, root depth, depletion fraction) and the area planted (0–100% of the total area), a set of K_c files are provided in the program.
- In addition, for irrigation scheduling the model requires information on: soil type, total available soil moisture, maximum rooting depth, initial soil moisture depletion (% of total available soil moisture)
- Scheduling criteria: Several options can be selected regarding the calculation of application timing and application depth (e.g., 80 mm every 14 days or irrigate to return the soil back to field capacity when all the easily available moisture has been used).

When all the data are entered CROPWAT automatically calculates the results and output generated in the form of tables or plotted in graphs, the time step of the results can be of a convenient time step: daily, weekly, decade, or monthly. The output parameters for each crop in the cropping pattern are:

- Reference crop evapotranspiration—ET_o (mm/period).
- Crop K_c—average values of crop coefficient for each time step.
- Effective rainfall (mm/period)—the amount of water that enters the soil.
- Crop water requirement—CWR or ET_m (mm/period).
- Irrigation requirements—IWR (mm/period).
- Total available soil moisture—TAM (mm).
- Readily available moisture—RAM (mm).
- Actual evapotranspiration—ET_c (mm).
- Ratio of actual crop evapotranspiration to the maximum crop evapotranspiration—ET_c/ET_m (in %).
- Daily soil moisture deficit (mm).
- Irrigation interval (days) and irrigation depth applied (mm).
- Lost irrigation (mm)—irrigation water that is not stored in soil (i.e., either surface runoff or percolation).
- Estimated yield reduction due to crop stress (when ET_c/ET_m falls below 100%).

Based on the review of literature, several ET_o estimation methods were chosen based on combination theory, solar radiation, temperature, and pan evaporation (FAO-56 Penman–Monteith, FAO-24 Penman,

Kimberly–Penman, FAO-24 Radiation, Priestley–Taylor, Hargreaves, FAO-24 Blaney–Criddle, and FAO-24 Pan Evaporation) and were compared for the calculation of ET_0 for the long-term meteorological data of Kozhikode (Fig. 1.1). The results showed that Blaney–Criddle predicted the higher ET_0 values and others are almost in similar lines and CROPWAT has been considered as the standard method for calculation of ET. Since the publication of the FAO Paper 56,[3] the Penman–Monteith method has been recognized as an accurate formula from various studies and has been recommended by the International Commission on Irrigation and Drainage (ICID) and World Meteorological Organization (WMO).[40,41]

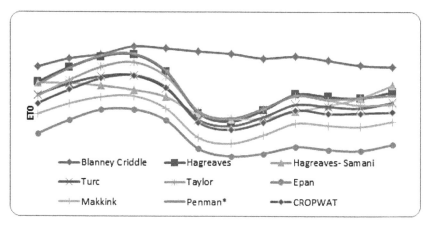

FIGURE 1.1   Variability of ET_0 by different methods for Kozhikode in different months.

1.4.12.2 DAILY ET

Daily ET is a simple calculator for estimating daily ET_0 using data collected from a conventional weather station. The minimum input data required are maximum and minimum air temperature, relative humidity, sunshine duration, and wind speed. The outputs are daily or monthly ET_0 in mm. Daily ET program was developed by Cranfield University. While it does not use any crop related data, it is still a practical tool, which can be used alongside other packages on occasional or day to day basis. However, it is not meant to replace any full featured software, which might have been developed to assist in the prediction of crop water requirements or in performing advanced irrigation scheduling. The program runs under any version of Windows and can accept data in different forms: wet and dry bulb temperature in place of relative humidity, sunshine duration in place of solar radiation. ET_o can be calculated using Penman method or Penman–Montieth or FAO modified Penman method.

1.4.12.3 ET_O CALCULATOR

This software was developed by FAO and the calculation of ET_o is based on Penman–Montieth equation. Version 3.1 was developed in 2009. There is a menu for data base management, selection of climate station and ET_o calculation. The user can select a data file or create a new file in which meteorological data from climate stations are stored. The user can import climatic data from a file. In selected climate station section, the name of the selected file and the corresponding name of the climatic station and country are displayed. The user can update the station characteristics, expand or shorten the data range, alter the thresholds of the data limits and examine the available meteorological data. ET_o calculation section contains the ET_o calculator. In the corresponding data and ET_o menu, the climatic parameters to calculate ET_o can be selected, meteorological data can be updated, specified, and plotted and results can be exported.

1.5   CROP WATER REQUIREMENTS

1.5.1   VEGETABLE CROPS

The length of the total growing season and each growth stage of the crop are important when estimating crop water needs. The growth of an annual crop can be divided into four stages:[3]

- Initial (establishment): from sowing to 10% ground cover.
- Crop development: from 10% to 70% ground cover.
- Midseason (fruit formation): including flowering and fruit set or yield formation.
- Late-season: including ripening and harvest.

Table 1.2 shows the range of typical root depths for various vegetable crops. Rooting depth and depletion fraction (p), which is the fraction of

TABLE 1.2 Root Depth and Depletion Fraction of Some Selected Vegetable Crops.

Crop	Tomatoes	Onions		Watermelon	Carrot	Lettuce	Broccoli	Cabbage
		Green	Dry					
Root depth (m)	0.5–1.5	0.3–0.6	0.3–0.6	0.8–1.5	0.5–1.0	0.3–0.5	0.4–0.6	0.5–0.8
Depletion fraction (p)[a]	0.4–0.5	0.3–0.4	0.3–0.4	0.4–0.5	0.4–0.5	0.3–0.4	0.4–0.5	0.4–0.5

[a]The values for depletion (p) apply when $ET_c \approx 5$ mm/day, otherwise $p = p + 0.04 (5 - ET_c)$.

Adapted from: Goyal, M. R. *Management of Drip/Trickle or Micro Irrigation* (Chapter 1); Apple Academic Press Inc.: Oakville, ON, 2014; p 59.

total available water that can be depleted from the root zone before moisture stress occurs, are vital factors in determining how much water should be applied and when.

K_c is a factor for estimating crop water requirements based on ET_o. K_c varies between crops and growth stages, which reflect the changing characteristics of a plant over the growing season. Crop type and growth stages are the major factors influencing the K_c. As the crop grows, the ground cover, crop height, and leaf area change. Differences in the crop evapotranspiration rate over the various growth stages will change the K_c as shown in Table 1.3.

Soil moisture-based water budget method to compute water requirement of vegetable crops (example in Table 1.4) was developed at California University.[24] Critical growth conditions are defined in Table 1.5. Modified Penmann method and K_c were used to estimate water requirement of vegetable crops in Raichur, India,[29] as shown in Table 1.6.

TABLE 1.3 Crop Coefficient (K_c) for Various Growth Stages of Selected Vegetable Crops.[16]

Crop	Growth stage				
	Initial	Development	Midseason	Late	At harvest
Cabbage	0.4[a]–0.5[b]	0.7–0.8	0.95–1.1	0.9–1.0	0.8–0.95
Carrots	0.4–0.6	0.6–0.75	1.0–1.15	0.8–0.9	0.7–0.80
Cucumber	0.4–0.5	0.7–0.8	0.95–1.05	0.8–0.9	0.65–0.75
Lettuce	0.3–0.5	0.6–0.7	0.95–1.1	0.9–1.0	0.8–0.95
Onion dry	0.4–0.6	0.7–0.8	0.95–1.1	0.85–0.9	0.75–0.85
Onion green	0.4–0.6	0.6–0.75	0.95–1.05	0.95–1.05	0.95–1.05
Pepper	0.3–0.4	0.6–0.75	0.95–1.1	0.85–1.0	0.8–0.9
Tomato	0.4–0.5	0.7–0.8	1.05–1.25	0.8–0.95	0.6–0.65
Watermelon	0.4–0.5	0.7–0.8	0.95–1.05	0.8–0.9	0.65–0.75

[a]The first crop reading is for high humidity and low wind conditions.
[b]The second reading is for low humidity and strong wind conditions.

Source[16]: de Carvalho, D. F.; da Silva, D. G.; da Rocha, H. S.; de Almeida, W. S.; da Sousa, E. S. Evapotranspiration and Crop Coefficient for Potato in Organic Farming. *Engenharia Agrícola* **2013**, *33*(1), 201–211. https://dx.doi.org/10.1590/S0100-69162013000100020; https://creativecommons.org/licenses/by-nc/4.0/deed.en

TABLE 1.4 Moisture Balance Sheet for Scheduling Irrigation in a Tomato Crop.

Crop: tomatoes	Effective root depth (Drz) = 0.55 m, p = 0.4, TAW = 180 mm/m,
Soil type: clay	RAW = 0.4 × 180 = 72 mm Net irrigation
Month: January	depth = Drz × RAW = 0.55 × 72 = 39.6 (rounded to 40 mm)

Day	A	B	$C = A \times B$	D	$E = D - 5$ mm	F	$H = (E + F) - C$
	ET$_0$ (mm/ day)	Crop coeffi- cient (K_c)	Crop water use (ET$_c$) (mm/day)	Rainfall (mm)	Effective rain (mm)^a	Irrigation application d_{net} (mm)	Cumulative soil water deficit (mm)
1	7.6	0.85	6.5	0	0	0	−6.5
2	8.6	0.85	7.3	3.8	0	0	−13.8
3	8.6	0.85	7.3	0.4	0	0	−21.1
4	8.8	0.85	7.5	0	0	0	−28.6
5	7.1	0.85	6.0	0	0	0	−34.6
6	9.1	0.85	7.7	0	0	40	Irrigation
7	6.4	0.85	5.4	0	0	0	0.00
8	3.4	0.85	2.9	0	0	0	−2.9
9	6.2	0.85	5.3	6	1	0	−8.2
10	6.3	0.85	5.4	3.2	0	0	−13.6
11	4.3	0.85	3.7	4.6	0	0	−17.3
12	7.7	0.85	6.5	1.4	0	0	−23.8
13	8.7	0.85	7.4	17.8	12.8	0	−11.0
14	7.2	0.85	6.1	0	0	0	−17.1
15	7.0	0.85	6.0	0	0	0	−23.1
16	8.4	0.85	7.1	0	0	0	−30.2

^{a}To calculate effective rainfall, during spring, summer, and autumn periods, subtract 5 mm from each of the daily rainfall totals. Assume rainfalls of 5 mm or less to be nonsignificant (zero). In winter, all the rainfall is assumed to be effective.

Total available water—TAW, readily available water—RAW, depletion fraction—p.

TABLE 1.5 Critical Growth Stage of Crops, and Crop Total Water Use, for Determining Irrigation Water Needs.[16]

Crop	Water requirement (cm/ha)	Critical need stage
Asparagus	63–115	Establishment and fern development
Bean, green	63–95	Bloom and pod set
Bean, pinto	95–125	Bloom and pod set
Beet, table	63–95	Establishment and early growth
Broccoli	125–160	Establishment and heading
Cabbage	125–190	Uniform throughout growth

TABLE 1.5 *(Continued)*

Crop	Water requirement (cm/ha)	Critical need stage
Cantaloupe	83–125	Establishment vining to first net
Carrot	63–95	Emergence through establishment
Cauliflower	125–190	Establishment and 6–7 leaf stage
Celery	190–223	Uniform, last month of growth
Collards/kale	75–90	Uniform throughout growth
Corn, sweet	125–223	Establishment, tassel elongation, ear development
Cowpea	63–95	Bloom, fruit set, pod development
Cucumber, pickle	95–125	Establishment, vining, fruit set
Cucumber, slicer	125–160	Establishment, vining, fruit set
Eggplant	125–223	Bloom through fruit set
Garlic	95–125	Rapid growth to maturity
Lettuce	50–75	Establishment
Mustard green	63–95	Uniform throughout growth
Okra	95–125	Uniform throughout growth
Onion	160–190	Establishment, bulbing to maturity
Pepper, bell	160–223	Establishment, bloom set
Pepper, jalapeno	160–190	Uniform throughout growth
Potato	125–255	Vining, bloom, tuber initiation
Potato	125–255	Vining, bloom, tuber initiation
Pumpkin	160–190	2–4 weeks after emergence, bloom, fruit set, development
Pumpkin	160–190	2–4 weeks after emergence, bloom, fruit set, development
Radish, red globe	33–63	Rapid growth and development
Spinach	63–95	Uniform throughout growth, after each cut, if needed
Squash	45–63	Uniform throughout growth
Sweet potato	63–125	Uniform until 2–3 weeks prior to anticipated harvest
Tomato	125–160	Bloom through harvest

Source[16]: de Carvalho, D. F.; da Silva, D. G.; da Rocha, H. S.; de Almeida, W. S.; da Sousa, E. S. Evapotranspiration and Crop Coefficient for Potato in Organic Farming. *Engenharia Agrícola* **2013,** *33*(1), 201–211. https://dx.doi.org/10.1590/S0100-69162013000100020; https://creativecommons.org/licenses/by-nc/4.0/deed.en

TABLE 1.6   Water Requirement of Vegetable Crops in Raichur.[29]

Crop	Season	Water requirement (mm)
Brinjal	Kharif	360
Chilli	Kharif	530
Cluster bean	Kharif	396
Cowpea	Kharif	443
Cucumber	Kharif	374
Cucumber	Rabi	362
French bean	Kharif	287
French bean	Rabi	263
Garlic	Kharif	556
Ladies finger	Kharif	326
Onion	Kharif	536
Radish	Kharif	168
Radish	Rabi	165
Ridge gourd	Kharif	383
Snake gourd	Kharif	420
Tomato	Kharif	531
Winged bean	Kharif	487

Adapted from: Manohar, N.; Nagaraj, K.; Patil, M. G. Studies on Evapotranspiration and Consumptive Use of Different Crops, Part II: Vegetable Crops. *Karnataka J. Agric. Sci.* **2000,** *13*, 384–388. Open access article. http://14.139.155.167/test5/index.php/kjas/article/viewFile/4234/4468

1.5.2   WATER REQUIREMENT OF VEGETABLE CROPS IN GREENHOUSE

Canopy development and management of some greenhouse horticultural crops is quite different than that of crops under outdoor conditions. Differences in plant spacing, crop height (use of vertical supports and pruning practices) and in aerodynamic properties may affect the K_c values. Moreover, the proportion of diffuse radiation in a greenhouse is higher than outdoor.[5] Thus, it is questionable whether the standard K_c values, determined experimentally outdoors, can be used directly to determine the ET of the greenhouse crops. Lysimeter was used to measure seasonal ET values for melon, green bean, sweet pepper, and watermelon for 3 years (1992, 1993, and 1998) in Cajamar, Almeria on the coastal area of southeast Spain.[33] In melon, ET varied from 177 to 298 mm, while pepper had ET values of 353–371 mm. ET of green bean is 174, sweet pepper is 353–371 and watermelon is 170 mm. In Mediterranean areas, the seasonal ET of greenhouse horticultural crops is quite low when compared to that of irrigated melon, green bean, watermelon and sweet pepper crops grown outdoors.[6,7,19,20,25] This is due, firstly, to a lower evaporative demand inside a plastic greenhouse, which is 30–40% lower than outdoors throughout the entire greenhouse cropping season.[22] Secondly, greenhouse cultivation in the Mediterranean areas is mostly concentrated in periods of low evaporative demand (autumn, winter, and spring), whereas irrigated crops outdoors are often grown during high evaporative demand periods. Moreover, the whitening of the external plastic cover as temperature increases in spring is common in most Mediterranean greenhouses to reduce the air temperature inside. This practice also reduces the greenhouse transmission coefficient for solar radiation, and, therefore, it decreases further the evaporative demand indoors.

1.5.2.1 TOMATO

Water requirements related to reference evapotranspiration (ET_o) in mm/period are given by the K_c for different crop development stages, or: during the initial stage 0.4–0.5 (10–15 days), the development stage 0.7–0.8 (20–30 days), the midseason stage 1.05–1.25 (30–40 days), the late-season stage 0.8–0.9 (30–40 days) and at harvest 0.6–0.65.[21] Total water requirements (ET_m) after transplanting, of a tomato crop grown in the field for 90–120 days, are 400–600 mm, depending on the climate. With pan coefficient method, ET_m for tomato under net house conditions using drip irrigation were 52.720, 61.451, 69.607, and 79.524 L/plant for 60, 80, 100, and 120% ET levels, respectively in Raichur, India during summer.[17] Typical daily ET_c values for the southwest Florida area during the fall and spring tomato production periods are 900, 2200, 4000, 4600, and 4000 gallons per acre per day during 0–20, 20–40, 40–60, 60–80, and 80–100 days after transplanting respectively.[39] Early season ET with pan evaporation method was about 4–5 mm/day and was about 6 mm/day in late season in Florida under drip irrigation.[23] Tomato in southwest Florida requires 153 mm during fall and 200 mm during spring season as estimated with lysimeter.[14]

1.5.2.2 POTATO

For high yields, the crop water requirements (ET_m) for a 120–150 day crop are 500–700 mm, depending on climate. The relationship between maximum evapotranspiration (ET_m) and reference evapotranspiration (ET_o) is given by the K_c which is: during the initial stage 0.4–0.5 (20–30 days), the development stage 0.7–0.8 (30–40 days), the midseason stage 1.05–1.2 (30–60 days), the late-season stage 0.85–0.95 (20–35 days), and at maturity 0.7–0.75.[21] Potato grown under optimal conditions (well-fertilized, well-irrigated, well-drained soils, pest-free stand, and uniform and optimum canopy) requires about 400–550 mm of water per growing season in southern Alberta. Potato roots grow to an effective water extraction depth of 60 cm and obtain 70% of the plants' seasonal water from the upper 30-cm depth. Water use rates for potato begin at about 0.4 mm per day when the crop sprouts (emerges) and increase to as high as 7 mm per day when the potato canopy completely shades the ground and tubers are bulking.

Potato water demand decreases as the crop achieves full tuber bulking and maturation.[18] Drip irrigated potato in Abu-Ghraib-Baghdad, Iraq requires 415–447 mm in 10 irrigations.[4] Calculated seasonal water requirement for the potato crop using Blaney–Criddle method was 491 mm in the first and 451 mm in the second season. While with Penman method, it was 524 mm in 1998–99 and 475 mm in 1999–2000 season in North Khartoum State, Sudan.[1] In Brazil, organically grown potato requires 109.6 mm of water as estimated through soil water balance, while the ET_c accumulated from FAO's K_c method was 142.2 and 138 mm, respectively, considering the classical values and the values adjusted to the local climatic conditions.[15]

1.5.2.3 ONION AND GARLIC

Total water consumption depends on local conditions and varies from 400–1000 mm per season. After sowing, irrigation of 30–40 mm is required for germination, followed by light daily irrigation until seedling emergence. It is critical to keep the soil wet during this period.[31] Daily crop water requirement of onion is 5–7.25 mm/day and seasonal crop water requirement may vary from 400–775 mm under range of environments.[32] Garlic water requirements in a 2-year study period were 546.5 mm and 519.2 mm/season, respectively during 2008 and 2009, in Hamedan, Iran under lysimeter study.[2]

1.5.2.4   BROCCOLI

Daily ET rate through lysimeter was below 0.1 mm/h during 00:00 to 08:00 h and thereafter increased and reached up to a maximum of 0.75 mm/h on 14:00 h and decreased in the following hours to reach nearly 0.2 mm/h on 19:00 h in San Joaquin Valley of California. ET was below 0.2 mm/h after 19:00 PM.[13]

1.5.2.5   BELL PEPPER

Daily ET rate was in the range of 2–6 during sowing to 18 days after sowing to 4–7 mm/day during 19–65 days after sowing and the trend was increasing during 42–65 days after sowing. ET was 4–9 mm/day during 65–120 days after sowing.[43]

1.5.3   FRUIT CROPS

Drip irrigation offers scope for enormous savings in water usage and it is the most useful system to boost horticultural production in areas with limited water resources. Research work carried out at Tamil Nadu Agricultural University, India[42] indicates the water saving ranged from nearly 40% to 68% with an additional yield benefit of 14–98% over the conventional irrigation methods. Drip irrigation has potential in rainfed areas with meager water resources during the periods other than in the rainy seasons. Most of the fruit crops require drip irrigation during the period of flowering to fruit development in order to increase the fruit set and improve the fruit size reflecting on final yield. However, the drip irrigation has to be dispensed 10–15 days before the expected harvesting period in order to improve the sweetness of the fruits. Banana is a water-loving plant. After banana plantation the soil should not be allowed to dry completely. Drip irrigation is ideal for banana cultivation. Irrigation requirement of banana under drip irrigation varied from 5 to 18.6 L/day/plant at early and peak growth stages of the plant, respectively. The frequency of drip irrigation can be daily or on alternate days. Daily water requirement of various fruit crops is given in Table 1.7.

TABLE 1.7   Daily Water Requirements of Various Fruit Crops.[42]

Crops	Crops water requirement (L/day/plant)
Amla	15–25
Banana	20–25
Citrus	22–30
Grapes	15–25
Guava	22–30
Mango	30–50
Papaya	15–25
Sapota	20–30

Adapted from[42]: Unpublished TNAU Annual Report, Tamil Nadu Agricultural University, Coimbatore, 2013, p 232; Public domain.

For the young trees which are in their pre-bearing stage, one-third of the recommended dose of water may be given and slowly increased to reach the level indicated in Table 1.7 during full bearing. Water requirement of horticultural crops in India under micro irrigation system based on the data on research work carried by SAUs, IARI, WALMIs, and AICRP on Water Requirement (WR) is presented in Table 1.8.[26]

TABLE 1.8   Water Requirements of Horticultural Crops in India Under Micro irrigation System.[26]

Crop	Irrigation depth (cm)	Crop	Irrigation depth (cm)
Banana	97.0	Okra	8.6
Beet	18.0	Onion	26.0
Bitter gourd	33.0	Papaya	73.0
Broccoli	60.0	Pomegranate	16.0
Cauliflower	18.0	Potato	27.5
Chili	41.7	Radish	11.0
Cucumber	24.0	Sweet potato	25.0
Eggplant	64.0	Tomato	10.7
Grape	28.0	Watermelon	25.0

Adapted from: *Indian National Committee on Irrigation and Drainage*, INCID, New Delhi, India, 2004; www.cwc.gov.in/main/INCID/welcome.html and http://www.icid.org/icid_data.html; Public domain.

1.6   SUMMARY

This chapter comprehensively reviewed concepts of crop water require-ment, ET and available methods of estimation of ET rate for horticul-tural crops. Calculation of crop water requirement on daily basis is an essential requirement in micro irrigation system as such systems reduce water losses and improve water use efficiency. If micro irrigation system is a component of precision agriculture, precise estimation of daily water requirement should be the priority. Selection of the methods to estimate crop water requirement depends on the availability of weather data. FAO Penman–Monteith method is recommended for determining reference ET_o. The international scientific community has accepted this equation as the most precise because of its good results when compared with other equations in various regions and climates of the world and these values can be utilized for computation of crop water requirement on a daily basis. Water requirement of horticultural crops in India under micro irrigation system based on the data on research work carried out by SAUs, IARI, WALMIs, and AICRP on water requirement (WR) is also presented.

KEYWORDS

- Blaney–Criddle method
- crop evapotranspiration
- crop water requirement
- evaporation
- evapotranspiration
- FAO Penman–Monteith method
- micro irrigation
- reference crop evapotranspiration
- water use efficiency

REFERENCES

1. Abubaker, B. M. A.; Shuang-En, Y.; Guang-Cheng, S.; Alhadi, M.; ElSiddig, A. Effect of Irrigation Levels on the Growth, Yield and Quality of Potato. *Bulg. J. Agric. Sci.* **2014**, *20*, 303–309.

2. Abyaneh, H. Z.; Varkeshi, M. B.; Ghasemi, A.; Marofi, S.; Chayjan, R. A. Determination of Water Requirement, Single and Dual Crop Coefficient of Garlic (*Allium sativum*) in the Cold Semi-arid Climate. *Aust. J. Crop Sci.* **2011**, *5*, 1050–1054.

3. Allen, R. G.; Pereira, L. S.; Raes, D.; Smith, M. *Crop Evapotranspiration—Guidelines for Computing Crop Water Requirements;* FAO Irrigation and Drainage Paper 56, Rome, 1998; p 213.

4. Ati, A. S.; Iyada, A. D.; Najim, S. M. Water Use Efficiency of Potato (*Solanum tuberosum* L.) Under Different Irrigation Methods and Potassium Fertilizer Dates. *Ann. Agric. Sci.* **2012**, *57*, 99–103.

5. Baille, A. Energy Cycle in Greenhouses. In *Greenhouse Ecosystems, Ecosystems of the World 20*; Stanhil, G., Enoch, H. Z., Eds.; Elsevier: Amsterdam, 1999; pp 214–245.

6. Barros, L. C. G.; Hanks, R. J. Evapotranspiration and Yield of Beans Affected by Mulch and Irrigation. *Agron. J.* **1993**, *85*, 692–697.

7. Beese, F.; Horton, R.; Wierenga, P. J. Growth and Yield Response of Chili Pepper to Trickle Irrigation. *Agron. J.* **1982**, *34*, 30–34.

8. Blaney, H. F.; Morin, K. V. Evaporation and Consumptive Use of Water Formulas. *Am. Geophys. Union Trans.* **1942**, *1*, 76–82.

9. Blaney, H. F.; Criddle, W. D. *Determining Water Requirements in Irrigated Areas from Climatological Data*; Univ. of California Fact Sheet, 1945, 17 p.

10. Blaney, H. F.; Criddle, W. D. *Determining Water Requirements in Irrigated Areas from Climatological and Irrigation Data*; USDA-SCS Technical Publication TP-96, 1950, 50 p.

11. Blaney, H. F.; Rich, L. R.; Criddle, W. D. Consumptive Use of Water. *Trans. Am. Soc. Civ. Eng.* **1952**, *117*, 948–1023.

12. Bowen, I. S. The Ratio of Heat Losses by Conduction and by Evaporation from Any Water Surface. *Phys. Rev.* **1926**, *27*, 779–787.

13. Bryla, R. D.; Trout, T. J.; Ayars, J. E. Weighing Lysimeters for Developing Crop Coefficients and Efficient Irrigation Practices for Vegetable Crops. *Hortic. Sci.* **2010**, *45*, 1597–1604.

14. Clark, G. A.; Stanley, C. D.; Csiziinsky, A. A.; Smajstrla, A. G.; Zazueta, F. S. *Water Requirements and Crop Coefficients for Tomato Production in Southwest Florida*; A Cooperative Research Project, Final Report Submitted to Southwest Florida Water Management District 2379 Broad Street, Bmksville, FL 34609 by University of Florida, Gulf Coast Research and Education Center 5007, 60th Street East Bradenton, FL 34203, 1993; p 15.

15. de Carvalho D. F.; da Silva, D. G.; da Rocha, H. S.; de Almeida, W. S.; Sousa, E. S. Evapotranspiration and Crop Coefficient for Potato in Organic Farming. *Engl. Agric. Jaboticabal* **2013**, *33*, 201–211.

16. Doorenbos, J.; Kassam A. H. *Yield Response to Water;* FAO Irrigation and Drainage Paper No. 33, FAO, Rome, Italy, 1979; p 25.

17. Dunage, V. S.; Balakrishnan, P.; Patil, M. G. Water Use Efficiency and Economics of Tomato Using Drip Irrigation Under Net House Conditions. *Karnataka J. Agric. Sci.* **2009,** *22,* 133–136.

18. Efetha, A. *Irrigation Scheduling for Potato in Southern Alberta*; Alberta Agriculture and Rural Development, 2011. http://www1.agric.gov.ab.ca/$Department/deptdocs. nsf/all/agdex13571 (accessed on Dec 25, 2015).

19. Erdem, Y.; Yuksel, A. N. Yield Response of Watermelon to Irrigation Shortage. *Hortic. Sci.* **2003,** *98,* 365–383.

20. Fabeiro, C.; Martin de Santa Olalla, F.; De Juan, J. A. Production of Muskmelon (*Cucumis melo L.)* Under Controlled Deficit Irrigation in a Semi-arid Climate. *Agric. Water Manag.* **2002,** *54,* 93–105.

21. FAO. *Crop Water Information: Tomato and Potato*; Water and Land Management Division, 2015. www.fao.org (accessed on Dec 25, 2015).

22. Fernández, M. D. Water Needs and Irrigation Programming for Horticultural Crops in Greenhouse (Spanish). Doctoral Thesis, Universidad de Almería, España, 2000; p 152.

23. Fischer, J. R. *Water and Nutrient Requirements for Drip-irrigated Vegetables in Humid Regions*; Southern Cooperative Series Bulletin 363, University of Florida, 1992; pp 1–17.

24. Hartz, T. K. *Water Management in Drip Irrigated Vegetable Production*; Department of Vegetable Crops, University of California, Davis, CA 95616, 1999. http://vric. ucdavis.edu (accessed on Dec 25, 2015).

25. Hegde, D. M.; Srinivas, K. Plant Water Relations and Nutrient Uptake in French Bean. *Irrig. Sci.* **1990,** *11,* 51–56.

26. INCID, ICID. Past Efforts for Promoting Micro-irrigation (Chapter 5). In *Report of Task Force on Micro Irrigation by Indian National Committee on Irrigation & Drainage (NCID)*; 2004; p 122. http://www.icid.org/icid_data.html and www.cwc. gov.in/main/INCID/welcome.html (accessed on Dec 25, 2015).

27. Jensen, M. E.; Haise, H. R. *Estimating Evapotranspiration from Solar Radiation*, Proceedings of Journal of Irrigation and Drainage, Division of American Society of Civil Engineers, 1963, 89(IR4); pp 15–41 (Closure, 91(IR1), pp 203–205).

28. Jensen, M. E. Water Consumption by Agricultural Plants. In *Water Deficits and Plant Growth*; Kozlowski, T. T., Ed.; Academic Press: New York, 1968; Vol. II, pp 1–22.

29. Manohar, N.; Nagaraj, K.; Patil, M. G. Studies on Evapotranspiration and Consumptive Use of Different Crops, Part II: Vegetable Crops. *Karnataka J. Agric. Sci.* **2000,** *13,* 384–388.

30. Monteith, J. L. *Evaporation and the Environment in the State and Movement of Water in Living Organisms,* 19th Symposium, Society for Experimental Biology, Swansea, Cambridge University Press, 1965; pp 205–234.

31. Naandan, J. *Onion*; NaanDan Jain Irrigation Ltd., 2014. www.naandanjain.com (accessed on Dec 25, 2015).

32. Netafim. Onion—Best Practices, 2015, https://www.netafim.com/crop/onion/best-practice (accessed on Dec 25, 2015).

33. Orgaza, F.; Fernandez, M. D.; Bonachelac, S.; Gallardoc, M.; Fereresa, E. Evapotranspiration of Horticultural Crops in an Unheated Plastic Greenhouse. *Agric. Water Manag.* **2005,** *72,* 81–96.

34. Penman, H. L. Natural Evaporation from Open Water, Bare Soil and Grass. *Proc. R. Soc. Lond. Ser. A* **1948**, *193*, 120–146.
35. Penman, H. L. Evaporation: An Introductory Survey. *Neth. J. Agric. Res.* **1956**, *4*, 9–29.
36. Penman, H. L. *Vegetation and Hydrology*; Tech. Common. 53, Commonwealth Bureau of Soils: Harpenden, England, 1963; p 120.
37. Penman, H. L.; Long, I. F. Weather in Wheat. *Q. J. R. Meteorol. Soc.* **1960**, *18*, 16–50.
38. Rijtema, P. E. *Analysis of Actual Evapotranspiration*; Agricultural Research Report No. 69, Centre for Agricultural Publication and Documents, Wageningen, The Netherlands, 1965; p 111.
39. Stanley, C. D.; Clark, G. A. *Water Requirements for Drip-irrigated Tomato Production in Southwest Florida*; Florida Cooperative Extension Service, Institute of Food and Agricultural Sciences, University of Florida, 2003. http://edis.ifas.ufl.edu (accessed on Dec 25, 2015).
40. Surendran, U.; Sushanth, C. M.; Mammen, G.; Joseph, E. J. Modeling the Impacts of Increase in Temperature on Irrigation Water Requirements in Palakkad District—A Case Study in Humid Tropical Kerala. *J. Water Clim. Chang.* **2014**, *5*(3), 471–487.
41. Surendran, U.; Sushanth, C. M.; Mammen, G.; Joseph, E. J. Modeling the Crop Water Requirement Using FAO-CROPWAT and Assessment of Water Resources for Sustainable Water Resource Management: A Case Study in Palakkad District of Humid Tropical Kerala, India. *Aquat. Procedia* 2015, *4*, 1215–1219.
42. *TNAU Annual Report;* Tamil Nadu Agricultural University: Coimbatore, 2013; p 232.
43. Trout, T.; Gartung, G. Use of Crop Canopy Size to Estimate Crop Coefficients for Vegetable Crops. *ASCE Conf. Proc.* **2006**, *200*, 297–303.
44. Van-Wijk, W. R.; de Vries, D. A. Evapotranspiration. *Neth. J. Agric. Sci.* **1954**, *2*, 105–119.

APPENDIX
Glossaray of Technical Terms

Blaney–Criddle equation is a relatively simplistic method for calculating evapotranspiration. When sufficient meteorological data are available the Penman–Monteith equation is usually preferred. However, the Blaney–Criddle equation is ideal when only air temperature datasets are available for a site.

Bowen ratio: In meteorology and hydrology, the Bowen ratio is used to describe the type of heat transfer in a water body. Heat transfer can either occur as sensible heat (differences in temperature without evapotranspiration) or latent heat (the energy required during a change of state, without a change in temperature).

Crop coefficients are properties of plants used in predicting evapotranspiration (ET). The most basic crop coefficient, Kc, is simply the ratio of ET observed for the crop studied over that observed for the well-calibrated reference crop under the same conditions.

Evaporation is a type of vaporization of a liquid that occurs from the surface of a liquid into a gaseous phase that is not saturated with the evaporating substance.

Evapotranspiration (ET) is the sum of evaporation and plant transpiration from the Earth's land and ocean surface to the atmosphere.

Lysimeter (from Greek λύσις (loosening) and the suffix *meter*) is a measuring device which can be used to measure the amount of actual evapotranspiration which is released by plants (usually crops or trees).

Penman equation describes evaporation (E) from an open water surface, and was developed by Howard Penman in 1948. Penman's equation requires daily mean temperature, wind speed, air pressure, and solar radiation to predict E. Simpler hydrometeorological equations continue to be used where obtaining such data is impractical, to give comparable results within specific contexts, for example, humid vs arid climates.

Penman–Monteith equation: Like the Penman equation, the Penman–Monteith equation (after Howard Penman and John Monteith) approximates net evapotranspiration (ET), requiring as input daily mean temperature, wind speed, relative humidity, and solar radiation.

Reference evapotranspiration sometimes incorrectly referred to as potential ET, is a representation of the environmental demand for evapotranspiration and represents the evapotranspiration rate of a short green crop (grass), completely shading the ground, of uniform height and with adequate water status in the soil profile.

Transpiration is the process of water movement through a plant and its evaporation from aerial parts, such as leaves, stems and flowers.

Surface runoff (also known as overland flow) is the flow of water that occurs when excess storm water, meltwater, or other sources flow over the Earth's surface.

Vapor pressure or equilibrium vapor pressure is defined as the pressure exerted by a vapor in thermodynamic equilibrium with its condensed phases (solid or liquid) at a given temperature in a closed system.

CHAPTER 2

WATER REQUIREMENT OF DRIP-IRRIGATED TOMATO INSIDE A SHADE NET HOUSE

PRIYANKA SHARMA*, MAHESH KOTHARI, and PRADIP DALAVI

Department of Irrigation and Drainage Engineering, College of Technology & Engineering, Maharana Pratap University of Agriculture and Technology, Udaipur 313001, India

Corresponding author. E-mail: priyankasharma1538@gmail.com

CONTENTS

ABSTRACT

A field experiment was conducted at the plasticulture farm of Department of Irrigation Water Management Engineering, College of Technology of Agriculture Engineering, Udaipur, Rajasthan, to evaluate the water requirement of tomato crop inside the shade net house. The experiment was set in randomized block design with four deficit irrigation treatments [drip irrigation with 100%, 80%, 60% and 40% of crop evapotranspiration (ET_c)] and five replications under a shade net house. Tomato (*Lycopersicon esculentum* Mill, badshah variety) plants were grown under the shade net house of 50% shade. The crop water requirement of tomato crop inside the shade net house was determined by gravimetric method. The study reveals that the actual water requirement for drip irrigated tomato crop in semiarid region inside the shade net house could be recommended between 1.62 and 4.58 mm day^{-1}.

2.1 INTRODUCTION

Tomato (*Lycopersicon esculentum* Mill) is an important vegetable crop grown worldwide for both fresh and processing markets.[10] In terms of acreage, it is the largest vegetable crop grown worldwide. The fruit is cultivated where climatic conditions are favorable and the seasonal water requirement is 300–600 mm.[12] Crops grown in open fields of a semidry climate were subjected to direct sunlight, high temperatures, and wind resulting in high crop evapotranspiration (ET_c), therefore, demanding large amounts of water. In contrast, shade houses favor plant growth; since plants are less stressful, direct sunlight was avoided, temperature is lower, humidity is higher, wind speed reduced, and ET_c was low.[5] Plastic nets are extensively used for shading plants in hot and sunny regions to reduce the solar radiation levels and to improve the environment for plant growth.[1]

Protected cultivation techniques, including net house technology, provides an optimum environmental medium for better crop growth in order to get maximum yield and high-quality products.[3] Although tomato is extensively grown under open conditions, yet it can be grown under shade net house during the off-season to produce better quality fruits with increased productivity.[9] Irrigation water requirement of 23–31% pan evaporation has been used for plants grown under 70% light reduction. In addition, water use efficiency was increased under shady conditions.[7]

Deficit irrigation is a strategy that allows the crop to sustain some degree of water deficit in order to reduce costs and potentially increase income.[11] Deficit irrigation consists of irrigating the root zone with less water than required for evapotranspiration. However, deficit irrigation for most vegetables such as tomato has been extensively studied, but with contrasting results.[2] For example, Zegbe-Dominguez et al.[18] revealed that tomato dry mass yield did not decrease under deficit irrigation compared to full irrigation, besides making a 50% saving in water and approximately 200% increase in irrigation water use efficiency and relevant fruit quality attributes were improved.

The use of drip irrigation and fertigation saves water and fertilizer and gives better plant yield and quality.[4] Drip irrigation is the slow and regular application of water, directly to the root zone, through a network of economically designed plastic pipes with low-discharge emitters.[8] The advantage of using a drip irrigation system is that it can significantly reduce soil evaporation and increase water use efficiency by creating a low, wet area in the root zone. Due to water shortages in many parts of the world today, drip irrigation is becoming quite popular.[13,15]

For tomato crops cultivated under Indian greenhouse, it is recommended that daily amount of required water for different growing system varies from 0.89 to 2.31 L plant^{-1} day^{-1}.[16,17] They also noted that the irrigation water should be given on every alternate day.

The present investigation is aimed at the evaluation of water requirement of tomato crop under the shade net house using the gravimetric method.

2.2 MATERIAL AND METHODS

A field experiment was conducted at the Plasticulture Farm of College of Technology and Engineering, MPUAT Udaipur, during the period of February 2013 to July 2013. The location corresponds to 24°35′ N latitude and 73°44′ E longitude and at an elevation of 582.17 m above the mean sea level (MSL). The soil type at the site is sandy loam. The shade net house of dimensions was 28 m × 36 m, that is, 1008 m^2 with a level of 50% shading (shade factor = 0.50) of the green colored net.

2.2.1 CROP WATER REQUIREMENT

Crop water requirement inside the shade net house was found out by the gravimetric method. The available soil moisture (ASM) content at which irrigation should be applied is a good criterion as it indicates moisture status of the soil and its availability to plants. Researchers[6,14] studied these criteria, and they suggested applying irrigation when 40% ASM is depleted (i.e., when ASM is 60%).

ASM content = Field Capacity – Permanent Wilting Point   (2.1)

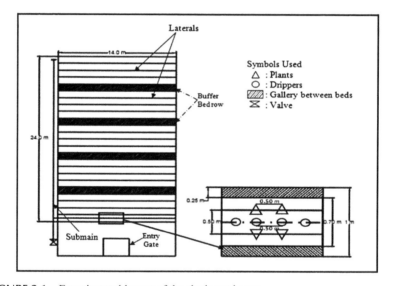

FIGURE 2.1   Experimental layout of the shade net house.

2.2.2 NET DEPTH OF IRRIGATION

The quantity of water to be applied to the field was calculated by the following formula:

$$d = \sum_{i=1}^{n} \frac{M_{1i} - M_{2i}}{100} \times A_i \times D_i,$$ (2.2)

where, d = net depth of water applied during an irrigation, (cm); n = number of soil layers sampled in the root zone depth D; M_{1i} = moisture content in the ith layer of soil at field capacity, %; M_{2i} = moisture content in the ith layer of soil before irrigation, %; A_i = apparent specific gravity of the ith layer of the soil, gm/cm^3; and D_i = depth of ith layer of the soil (cm).

2.2.3 IRRIGATION TREATMENTS

The irrigation treatments were based on ET_c for a crop that was calculated by using climatological parameters inside and outside the shade net house. The irrigation treatments were:

T$_1$: Drip irrigation with 100% of ET_c
T$_2$: Drip irrigation with 80% of ET_c
T$_3$: Drip irrigation with 60% of ET_c
T$_4$: Drip irrigation with 40% of ET_c

All treatments were arranged randomly with five replication R$_1$, R$_2$, R$_3$, R$_4$, and R$_5$ for each treatment as a block (Fig. 2.1).

2.3 RESULTS AND DISCUSSION

The net water applied during the growth period of tomato crop for 100%, 80%, 60%, and 40% of ET_c under the shade net house is shown in Table 2.1. This table shows that the ET_c under 100%, 80%, 60%, and 40% ET_c were 495.00, 396.00, 297.00, and 198.00 mm, respectively, for the entire crop period, that is, 154 days. The minimum value of ET_c was 40% and the maximum was observed under 100% of ET_c.

TABLE 2.1 Measured Crop Evapotranspiration Under the Shade Net House.

Week	Crop evapotranspiration under the shade net house (ET_c), mm day^{-1}			
	100%	80%	60%	40%
1	1.614	1.320	1.076	0.907
2	1.729	1.375	1.201	1.004
3	1.863	1.776	1.118	1.083
4	2.426	2.084	1.456	1.203
5	2.726	2.340	1.636	1.373
6	3.264	2.464	1.958	1.389
7	3.231	1.896	1.939	1.517

TABLE 2.1 *(Continued)*

Week	Crop evapotranspiration under the shade net house (ET_c), mm day^{-1}			
	100%	80%	60%	40%
8	3.397	2.718	2.038	1.583
9	3.528	2.822	2.246	1.851
10	3.834	2.924	2.401	2.489
11	3.916	3.769	2.466	2.340
12	4.037	4.219	2.547	1.891
13	4.246	4.307	3.359	1.970
14	4.367	3.919	2.817	2.247
15	4.582	3.666	2.914	2.401
16	4.209	3.467	3.479	2.480
17	3.853	3.224	2.561	1.823
18	3.557	2.845	2.134	1.707
19	3.241	2.593	2.483	1.655
20	2.731	2.197	2.346	1.641
21	2.631	2.105	2.119	1.052
22	1.734	1.387	1.040	0.694
$ET_{c\ in}$ (mm)	495.00	416.00	297.00	198.00

Variation in the weekly crop water requirement inside the shade net house over the growth period of tomato is shown in Figure 2.2. Figure 2.2 reveals that the net water requirement for drip irrigated tomato crop varied from 1.614 to 4.582 mm per day. The higher water requirement in the later growth period of tomato may be due to a higher temperature, sunshine hours, and wind speed during the latter growth period of the tomato crop.

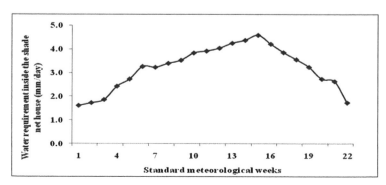

FIGURE 2.2 Variation in crop water requirement under the shade net house.

2.4 SUMMARY

To achieve satisfactory water requirement, irrigations are required for the entire growing season in processing tomato cultivated under the shade net house. Based on the results obtained from the present investigation in this chapter, it has been concluded that the crop water requirement inside the shade net house of drip irrigated tomato crop was 495.0 mm. The study reveals that the actual water requirement for drip irrigated tomato crop in semiarid region inside the shade net house could be recommended between 1.62 and 4.58 mm day^{-1}.

KEYWORDS

- drip irrigation
- irrigation regimes
- shade net house
- tomato
- water requirement

REFERENCES

1. Abdel-Ghany, A. M.; Al-Helal, I. M. Characterization of Solar Radiation Transmission Through Plastic Shading Nets. *Sol. Energy Mater. Sol. Cells* **2010,** *94*(8), 1371–1378.
2. Dorji, K.; Behboudian, M. H.; Zegbe-Dominguez, J. A. Water Relations, Growth, Yield, and Fruit Quality of Hot Pepper Under Deficit Irrigation and Partial Rootzone Drying. *Sci. Hortic.* **2005,** *104*, 137–149.
3. Dunage, V. S.; Balakrishnan, P.; Patil, M. G. Water Use Efficiency and Economics of Tomato Using Drip Irrigation Under Net House Conditions. *Karnataka J. Agric. Sci.* **3009,** *22*(1), 133–136.
4. Harmanto, S. V.; Babel, M.; Tantau, H. Water Requirement of Drip Irrigated Tomatoes Grown in Greenhouse in Tropical Environment. *Agric. Water Manag.* **2005,** *71*(3), 225–242.
5. Hashem, F. A.; Medany, M. A.; El-Moniem, E. A.; Abdallah, M. M. F. Influence of Green-house Cover on Potential Evapotranspiration and Cucumber Water Requirements. *Ann. Agric. Sci.* **2011,** *56*(1), 49–55.

6. Islam, T.; Sarker, H.; Alam, J.; Harun-ur-Rashid, M. Water Use and Yield Relationships of Irrigated Potato. *Agric. Water Manag.* **1990,** *18,* 173–179.

7. Jifon, L. J.; Syvertsen, J. P. Moderate Shade Can Increase Net Gas Exchange and Reduce Photoinhibition in Citrus Leaves. *Tree Physiol.* **2003,** *23,* 119–127.

8. Liu, H.; Huang, G. Laboratory Experiment on Drip Emitter Clogging with Fresh Water and Treated Sewage Effluent. *Agric. Water Manag.* **2009,** *96*(5), 745–756.

9. Mantur, S. M.; Patil, S. R. Influence of Spacing and Pruning on Yield of Tomato Grown Under Shade House. *Karnataka J. Agric. Sci.* **2008,** *2*(1), 97–98.

10. Opiyo, A. A.; Ying, T. J. The Effect of 1-methylcycopropene Treatment on the Shelf-life and Quality of Cherry Tomato (*Lycopersicon esculentum* var. cerasiforme) Fruit. *Int. J. Food Sci. Technol.* **2005,** *40,* 665–673.

11. Owusu-Sekyere, J.; Asante, P.; Osei-Bonsu, P. Water Requirement, Deficit Irrigation and Crop Coefficient of Hot Pepper (*Capsicum frutescens*) Using Irrigation Interval of Four (4) Days. *J. Agric. Biol. Sci.* **2010,** *5*(5), 72–78.

12. Owusu-Sekyere, J.; Sam-Amoah, L.; Teye, E.; Osei, B. Crop Coefficient (kc), Water Requirement and the Effect of Deficit Irrigation on Tomato in the Coastal Savannah Zone of Ghana. *Int. J. Sci. Nat.* **2012,** *3*(1), 83–87.

13. Powell, N. L.; Wright, F. S. Subsurface Mircroirrigated Corn and Peanut: Effect on Soil pH *Agric. Water Manag.* **1998,** *36,* 169–180.

14. Sagoo, G. A.; Khan, A. E.; Himyatullah, H.; Khan, A. M.; Ahmad, K. H. Morphological Response of Autumn Planted Sugarcane to Available Soil Moisture Depletion and Planting Geometry on Different Soils Under Arid Conditions. *Sarhad J. Agric.* **2010,** *26*(2), 187–194.

15. Sahin, U.; Anapal, O.; Donmez, M. F.; Sahin, F. Biological Treatment of Clogged Emitters in a Drip Irrigation System. *J. Environ. Manag.* **2005,** *76*(4), 338–341.

16. Tiwari, G. N. *Greenhouse Technology for Controlled Environment*; Narosa Publishing House: New Delhi, 2003; pp 67–77.

17. Tiwari, K. N.; Singh, A.; Mal, P. K. Economic Feasibility of Raising Seedlings and Vegetables Production Under Low Cost Plastic Tunnel. In *Platiculture On-line Publication*; International Committee of Plastics in Agriculture (CIPA): Paris, 2000.

18. Zegbe-Dominguez, J. A.; Behboudian, M. H.; Lang, A.; Clothier, B. E. Deficit Irrigation and Partial Root Zone Drying Maintain Fruit Dry Mass and Enhance Fruit Quality in Petopride Processing Tomato (*Lycopersicon esculentum* Mill.). *Sci. Hortic.* **2003,** *98,* 505–510.

CHAPTER 3

OPTIMIZED IRRIGATION SCHEDULING FOR DRIP-IRRIGATED TOMATO USING THE FAO AQUACROP MODEL

SARANYA JEYALAKSHMI*

Center for Engineering Innovation, University of Windsor, Windsor, Ontario N9B 3P4, Canada

Corresponding author. E-mail: saranyaj55@gmail.com

CONTENTS

Edited and adopted partially from "Saranya Jeyalakshmi, Effect of plastic mulch on yield of tomato under drip irrigation. Unpublished M. Tech. Thesis for Department of Irrigation and Drainage Engineering, Dr. PDKV, Akola, Maharashtra, India."
Note: 1 US$ = 60.00 Rs. (Indian Rupees); One q (quintal) = 100.00 kg.

ABSTRACT

Simulation models that quantify the effects of water on yield at the farm level are valuable tools in water and irrigation management. FAO has developed a yield-response to water model, named AquaCrop, which simulates attainable yields of the major field and vegetable crops culti-vated worldwide. The model pays attention to the fundamental processes involved in crop productivity and in the responses to water, from a physi-ological and agronomic background perspective. The objectives of this study were to develop optimal cropping pattern and irrigation schedules under water deficit condition for the study area using a generic crop growth model. A set of sensitive parameters for tomato were calibrated using the experimental data from the research farm of Irrigation and Drainage Engi-neering, Dr. PDKV, Akola, Maharashtra for the period of October 2013 to February 2014 and validated using next set of data from the same field for the period from August to December 2014. The experiment considered of five treatments with four replications. The treatments include four levels of drip irrigation viz.namely, 40%, 60%, 80%, and 100% evapotranspiration (ET) with BPM and one control treatment of 100% ET evapotranspiration without BPM. In case of growth and yield attributes, treatment with 80% ET evapotranspiration along with BPM was significantly superior over all other treatments with maximum fruit yield of 33 t/ha. The verification test shows that the model slightly overestimates canopy cover, and biomass. Water productivity values of 31.6 g m^{-2} wasere considered to evaluate the model performance. While linear function between observed tuber yields and estimated by Aqua Crop always had always a correlation coefficient greater than 0.85 ($p < 0.001$),. Tthe AquaCrop model was able to accu-rately simulate soil water content of root zone, crop biomass, and grain yield, with normalized root mean square error (RMSE) less than 10%.

3.1   INTRODUCTION

Water is a precious natural resource, a basic human need, and prime national asset. Globally, 3790 km^3 of fresh water is annually used whereas 69% of this total fresh water is used for agriculture. As the world becomes increasingly dependent on the production of irrigated lands, irrigated agri-culture faces serious challenges that threaten its suitability. It is prudent to make efficient use of water and bring more area under irrigation through available water resources. This can be achieved by introducing advanced and sophisticated methods of irrigation and improved water management practices.

Among the management practices for increasing water use efficiency (WUE), mulching (organic and inorganic) is an appropriate approach to enhance efficiency level of irrigation besides improving crop yield.[1,2,4,9,11] It also makes possible the application of fertilizers and other chemicals along with irrigation water to match the plant requirements at various growth stages. India being an agriculture country should have a good respect towards water conservation strategies. Already India is suffering from great stress of water scarcity. Each and every drop of water is important for but unfortunately, because of carelessness, we often waste a huge amount of water. One of such practices is over and excessive irrigation. Mulching is a soil and water conserving practice, in which any suitable material is used to spread over the ground between rows of crop or around the tree trunk. This practice helps to retain soil moisture, reduces weed growth and enhances soil structure.[8,10]

India is the second largest producer of vegetable crops in the world. Tomato ranks third in priority after potato and onion in India but ranks second after potato in the world. The total land area under cultivation of this crop was 5312 ha during 2005 and the area increased to 0.87 million ha in 2012. Productivity level of tomato in the country is 20.2 t/ha, which is 10% less than that of the global average.

This chapter discusses effects of irrigation scheduling and mulching on the performance of tomato crop. Economic analysis and benefit–cost ratio (BCR) for tomato production are also presented.

3.2 MATERIALS AND METHODS

Simulation models that quantify the effects of water on yield at the farm level are valuable tools in water and irrigation management. FAO has developed a yield response to water model, named AquaCrop, which simulates attainable yields of the major field and vegetable crops cultivated worldwide.[3,12] The model focuses on the fundamental processes involved in crop productivity and in the responses to water, from a physiological and agronomic background perspective.

The objectives of this study were to develop optimal cropping pattern and irrigation scheduling under water deficit conditions for the study area using a generic crop growth model. A set of sensitive parameters for tomato were calibrated using the experimental data from the research farm of Irrigation and Drainage Engineering, Dr. PDKV, Akola, Maharashtra for the period of October 2013 to February 2014 and validated using next set of data from the same field for the period from August to December 2014.

The experimental site was fairly uniform and leveled. Akola is situated in Western Vidarbha region of Maharashtra state and comes under subtropical zone. It is situated at an altitude of 307.4 m above mean sea level (MSL) at the intersection of 20°40' N latitude and 77°02' E longitude. Average annual precipitation is 760 mm, out of which approximately 86% is received during June through September. The climate of the area is semiarid, characterized by three distinct seasons: mainly summer being hot and dry from March to May, the warm and rainy monsoon from June to October and winter with mild cold from November to February. The mean annual maximum and minimum temperatures are 48.23°C and 22.05°C in summer and 32.88°C and 14.35°C in winter, respectively. The soil at the site is clay with a pH of 7.9. Tomato of variety Phule Raja (seedlings of 30 days old) was transplanted at a spacing of 90 cm × 50 cm on 23rd October 2013. The experiment was laid out in randomized block design having five treatments and four replications, with a plot size of 3.5 m × 5.4 m each. The treatments were as follows:

T$_1$ Irrigation scheduling at 40% evapotranspiration (ET) with silver polyethylene mulch under drip irrigation.

T$_2$ Irrigation scheduling at 60% ET with silver polyethylene mulch under drip irrigation.

T$_3$ Irrigation scheduling at 80% ET with silver polyethylene mulch under drip irrigation.

T$_4$ Irrigation scheduling at 100% ET with silver polyethylene mulch under drip irrigation.

T$_5$ Irrigation scheduling at 100% ET without silver polyethylene mulch under drip irrigation.

The LLDPE film of 50 μ thickness silver-black color was used for mulching. In this experiment, silver color was used on the upper side and black on lower side. The lateral lines of 16 mm diameter were laid along the crop rows and for each row of the crop there was a single lateral. The laterals were provided with inline dripper of 4 lph discharge capacity. The main line was directly connected to a 3 hp pump to lift water from a tank. The manifold unit included a sand filter, screen filter, a pressure gauge, and control valve. The duration of delivery of water to each treatment was controlled with the help of control valves provided at the inlet end of each lateral.

3.2.1 IRRIGATION SCHEDULING

Before transplanting, common irrigation was applied on October 22, 2013 to the experimental plots, to bring the soil to field capacity. The first irrigation was given on October 23, 2013 and the irrigation after this day (after transplanting) was scheduled on alternate days. The quantity of water to be applied per treatment was calculated as follows (De Krupp's formula):[3,12]

$$Q = A \times B \times C \times D \tag{3.1}$$

where Q = water requirement per plant (L/plant); A = ET_o = $E_{pan} \times K_p$; $B = K_c$; C = canopy factor; D = area allotted per plant (m²). The value of pan coefficient was taken as 0.7. Water requirement of tomato crop was estimated on alternate day basis.

3.2.2 ECONOMIC ANALYSIS

BCR for each treatment was worked out to compare the net returns. The seasonal system cost of drip irrigation system includes depreciation, prevailing bank interest rate, and repair and maintenance cost of the system. The fixed cost of drip irrigation system was determined. For this purpose, the life period of polyvinyl chloride items was considered as 10 years; and for galvanized iron (GI) items and motor, pump was 25 years. Standard market rates were considered for each item. Fixed cost, operating cost, net return, and BCR for each treatment were worked out. The cost of cultivation included expenses incurred in land preparation, intercultural operations, fertilization, crop protection measures, irrigation water, and harvesting. Therefore, total seasonal cost included depreciation, interest, repairs and maintenance cost, cost of cultivation, and cost of mulching. The income from produce was calculated using the prevailing average market price of tomato at Rs. 1000/100 kg.

3.3   RESULTS AND DISCUSSION

3.3.1   PERFORMANCE OF TOMATO CROP

By utilizing water equivalent to drip irrigation at 100% evapotranspiration (ET) with and without plastic mulching (PM) by adopting drip irrigation at 40%, 60%, and 80% ET with PM, the percentage increase in irrigated area over 100% was 116.31%, 55.9%, and 21.45%, respectively.

TABLE 3.1   Comparison of Water Utilization in Drip-irrigated Tomatoes.

Treatment	Water applied (ha cm)	Water saving (%)	Area would be irrigated by applying water equivalent to 100%	Percentage increase in area over 100% ET
T1	13.37	53.77	2.16	116.31
T2	18.55	35.85	1.55	55.9
T3	23.73	17.95	1.21	21.45
T4	28.92	0	0	0
T5	28.92	0	0	0

Considering the area of plantation and its plant population if one cannot irrigate the field with the requirement of 100% crop ET, then one can take the privilege of other treatments by changing the water requirement to a suitable ET level. The total water requirement of tomato crop was 130.13 L/plant (Table 3.1). However, one can reduce the water requirement up to 60.2 L/plant with 40% ET in water scarce situations and can bring some additional area under cultivation. This shows the major advantage of water saving in drip irrigation with lower ET levels and also the introduction of plastic mulch in *Rabi* season.

Table 3.2 shows that all growth characteristics were influenced by different irrigation levels and PM. The plant height, number of branches, leaf area index (LAI) and crop yield were found to be higher for treatment T_3. The treatment T_3 with 80% ET was found to be significantly superior to all other treatments. The weed count was highest for control treatment, that is, for treatment without plastic mulch as compared to treatments with plastic mulch. Practically no weeds were observed under the PM. The treatment T_5 (100% ET without PM) was significantly superior to all other treatments in terms of weed count. The weed count was observed maximum in control treatment with no mulch. However, in the treatments with mulch, the weed count was minimum, and it was increased with increase in the ET level. These results are also in agreement with those reported by other investigators.[5,7–9]

TABLE 3.2 Performance of Tomato Crop Under Different Irrigation Levels and Plastic Mulching.

Treatment	Average at 30, 60, 90, and 120 DAT			Weed count (per m²)	Yield (q/ha)
	Plant height (cm)	Number of branches	Leaf area index		
T_1	86.8	10.25	0.77	12	253.04
T_2	100.94	11.5	1.01	19	261.24
T_3	117.81	16	1.27	22	355.82
T_4	77.25	11.5	1.07	25	277.78
T_5	63.21	8	0.74	39	210.58
Mean	**89.22**	**11.45**	**1.58**	**23**	271.69
F test	Sig.	Sig.	Sig.	Sig	Sig.
SE (m) ±	6.24	0.72	0.028	0.87	3.56
CD at 5%	19.24	2.25	6.49	2.66	10.99
CV (%)	12.59	12.47	2.762	9.65	10.50

DAT, days after transplanting. 1 quintal (q) = 100 kg.

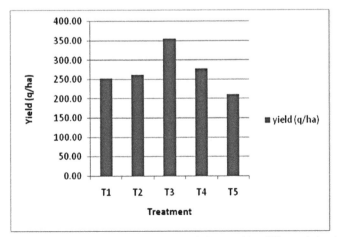

FIGURE 3.1 Tomato yield in each treatment as influenced by different irrigation levels and plastic mulching.

The percentage increase in yield due to mulching in 100% ET level was 31.92% and the treatment T_5 (80% ET with PM) recorded 68.97% higher yield over treatment T_5 (control). Hence, from the results in Table 3.2 and Figure 3.1, it can be concluded that the treatment T_3 (80% ET with PM) was observed superior among all the treatments, due to adequate moisture availability and proper air water ratio in the root zone during the crop period. Similar findings were also reported by Gupta et al.[6]

3.3.2 ECONOMIC ANALYSIS

The BCR varied from 2.32 to 3.42 (Table 3.3). It was maximum for treatment T_3 (80% ET with PM), while it was minimum for treatment T_5. In mulch treatments, B:C ratio was found the maximum for treatment T_3 followed by T_2 and T_1. It is observed that treatments with plastic mulch resulted in higher BCR compared to treatment without mulch. In all cases, BCR is more than 1.

TABLE 3.3   Cost Analysis of Tomato Production.

Treatment	Yield of tomato (100 kg/ha)	Gross return (Rs./ha)	Total cost (Rs./ha)	Net return (Rs./ha)	BCR (2)/(3)
	(1)	(2)	(3)	(4)	
T1	253.04	253040	103895	149145	2.44
T2	261.24	261240	103937	157303	2.51
T3	355.82	355820	103978	251842	3.42
T4	277.78	277780	104020	173760	2.67
T5	210.58	210580	90622	119958	2.32

The market rate of tomato was Rs. 1000/100 kg.

3.3.3 AQUACROP MODEL

The verification test showed that the AquaCrop model slightly overestimated canopy cover and biomass. Water productivity (WP) values of 31.6 g m^{-2} were considered to evaluate the model performance. A linear function between observed yield and estimated by AquaCrop model had always a correlation coefficient greater than 0.85 ($p < 0.001$). The AquaCrop model was able to accurately simulate soil water content in the root zone, crop biomass, and grain yield, with normalized root mean square error (RMSE) less than 10%.

3.4 SUMMARY

The field experiment consisted of randomized block design having five treatments with four replications at research farm of Irrigation and Drainage Engineering, Dr. PDKV, Akola for its response to different irrigation levels and plastic mulch treatments on water saving and yield of tomato during October 2013 to February 2014. The treatment with irrigation at 100% ET without PM resulted in less yield (210.58 (100 kg)/ha) compared to treatment with irrigation equal to 40% ET with PM (253.04 (100 kg)/ha). Treatment T1 (40% ET with PM) recorded highest WUE (100 kg/ha-cm) (21.87) followed by treatment T3 (15.38), T2 (15.06 q/ha cm), T4 (9.61), and T5 (7.28). The B:C ratio was found the maximum for treatment T3 (80% ET with PM: 3.42) while it was minimum for control treatment T5 (2.32). B:C ratio for mulch treatments was higher than that for nonmulched treatment. Under silver mulch, the B:C ratio for tomato production is higher compared to that of control treatment in 100% ET level without PM. On the basis of the results in this chapter, it is concluded that if water is not the limitation, tomato should be grown under PM irrigated with drip irrigation at 80% replenishment of crop ET. If water is the constraint, then tomato should be grown under plastic mulch with drip irrigation having irrigation scheduling at 40% of ET. These results are only valid for a site in Akola or for similar climatic conditions.

KEYWORDS

- **AquaCrop model**
- **benefit–cost ratio**
- **drip irrigation**
- **evapotranspiration**

- **irrigation scheduling**
- **water saving**
- **water use efficiency**

REFERENCES

1. Baye, B. Effect of Mulching and Amount of Water on the Yield of Tomato Under Drip Irrigation. *J. Hortic. For.* **2006,** *3*(7), 200–206.
2. Diaz-Perez, J. C. In *Drip Irrigation Levels Affect Plant Growth and Fruit Yield of Bell Pepper,* Georgia Water Resources Conference, USA, 2009; pp 27–29.
3. Doorenbos, J.; Kassam, A. H. Yield Response to Water. *Irrig. Drain. Eng.* **1979,** *33,* 257.
4. Favati, F.; Stella, L.; Galgan, F.; Miccolis, M. Processing Tomato Quality as Affected by Irrigation Scheduling. *Sci. Hortic.* **2009,** *122,* 562–571.
5. Gudugi, I. A. S.; Odofin, A. J.; Adeboye, M. K.; Oladiran, J. A. Agronomic Characteristics of Tomato as Influenced by Irrigation and Mulching. *Adv. Appl. Sci. Res.* **2012,** *3*(5), 2539–2543.
6. Gupta, A. J.; Chattoo, M. A. Techno-economic Evaluation of Drip Irrigation and Fertigation Practices in Capsicum Under Kashmir Conditions. *Veg. Sci.* **2009,** *36*(3), 309–314.
7. Hatami, S.; Nourjou, A.; Henareh, M.; Pourakbar, L. Comparison Effects of Different Methods of Black Plastic Mulching and Planting Patterns on Weed Control, WUE and Yield in Tomato Crops. *Int. J. AgriSci.* 2007, *2*(10), 928–934.
8. James, A.; Griffin, R. W.; John, A. Evaluation of the Effects of Plastic Mulches (Red, Black, Olive) and Control (Bare Ground) on the Growth and Yield of Tomato. *Agric. Int. J.* **2013,** *1*(2), 38–46.
9. Jamil, M.; Munir, M.; Gasim, J. B.; Rehman, K. Effect of Different Types of Mulches and Their Duration on the Growth and Yield of Garlic. *Int. J. Agric. Biol.* **2003,** *4,* 588–591.
10. Jawale, P. V.; Dixit, M. S. Economic Feasibility of Drip Irrigation System in Akola District. Unpublished B. Tech. Thesis, Dr. PDKV, Akola, 2007; pp 21–22.
11. Kanwar, D. P. S.; Dikshit, S. N.; Sharma, G. L.; Patel, K. L.; Sarnaik, D. A. Studies on Effect of Growth and Yield: Attributing Characters of Sweet Pepper Under Black Polyethylene Mulch. *J. Soil Crops* **2013,** *23*(1), 73–77.
12. Michael, A. M. *Irrigation: Theory and Practice*; Vikas Publishing House Pvt. Ltd., New Delhi,; 2009; p 487.

PART II
Irrigation Application Uniformity

CHAPTER 4

EMITTER WATER DISTRIBUTION UNIFORMITY: COMPENSATED HYDRAULIC GRADIENT

S. VANITHA* and S. SENTHILVEL

Department of Land and Water Management, Agricultural Engineering College & Research Institute (AEC & RI), Pallapuram Post, Kumulur, Trichy 621712, Tamil Nadu, India

Corresponding author. E-mail: vanitha.subramani636@gmail.com

CONTENTS

ABSTRACT

Drip irrigation has proved to be a success in terms of water and increased yield in a wide range of crops. Its ability of small and frequent irrigation applications has created interest because of decreased water requirements, possible increased production, and better quality produce. Drip irrigation can potentially provide high application efficiency and achieve high application uniformity. The hydraulic gradient compensation maintains uniform operating pressure head and discharges with high order of water distribution uniformity realizing a maximum yield of crop. In drip irrigation systems, the ultimate water distribution uniformity depends on the spatial distribution of the operating pressure heads at the emission points and the corresponding emitter discharges. In 2 lph of emitter arrangement with compensated hydraulic gradient, the water distribution uniformity was 97.8% and irrigation usage efficiency ranged from 17.98 to 20.69 kg/ha/mm.

4.1 INTRODUCTION

Water remains as the indispensable natural resource anchoring and fortifying all forms of life in the world. Agriculture maintains its cult status as the primary consumer of water in India. However, the exponentially growing population and the latest developments by way of industrialization and urbanization have created miscellaneous water needs to keep up the environmental equilibrium between the biotic and abiotic components. According to UNO, water crisis is the major threat to mankind in 21st century according to Meghanatha et al.[1]

India is perhaps the second largest producer of vegetables next only to China. Okra or ladies' finger, which is also known as "*Bhindi*," is one of the important vegetables of India. It is grown throughout the tropical and subtropical regions and also in the warmer parts of the temperate regions. Okra has a good potential as an export crop and accounts for 60% of the export of fresh vegetables. It is cultivated on 0.349 million ha with the production of 3.344 million t and productivity of 9.6 metric t/ha. The major okra-producing states are Uttar Pradesh, Bihar, Orissa, West Bengal, Andhra Pradesh, and Karnataka. The crop is also used in paper industry as well as for the extraction of fiber.

The properties of emitters that play a vital role in designing a drip irrigation system are: discharge variation due to manufacturing tolerance, closeness of discharge–pressure relationship to design specifications, emitter discharge exponent, operating pressure range, pressure loss in laterals due to insertions of emitters, and stability of the discharge–pressure relationship over a long period of time.

Hence, this research study formulates to find the impact of compensating hydraulic gradient along laterals on water distribution uniformity under drip irrigation. Its values are found to be greater for pressure compensating emitters than for non-compensating emitters according to Özekici and Sneed.[2]

4.2 MATERIALS AND METHODS

The experiment was conducted in PFDC farm (Eastern Block-NA4) of Tamil Nadu Agricultural University, Coimbatore. The farm is located at 11°N latitude and 77°E longitude with an altitude of 427 m above MSL. From the meteorological data available in the Department of Meteorology at TNAU, the 60 year average weather data were collected. The relative humidity ranged from 61% (14.22 h) to 90% (07.22 h). The maximum and minimum temperatures recorded during the study period were 37.7°C and 15.6°C, respectively. The maximum and minimum sunshine hours were 10.8 and 0.2 h per day, respectively. The mean bright sunshine hour per day was 7.4 h with mean solar radiation of about 429 cal cm^{-2} day^{-1}. The lowest and the highest evaporation rate were 1.6 and 7.4 mm per day and the corresponding solar radiation values were recorded as 257.3 and 348.4 cal cm^{-2} day^{-1}, respectively. The climatic conditions were favorable for raising an okra crop. The soil in the experimental field is sandy clay loam in texture having a pH of 8.25 and electrical conductivity of 0.17 dS m^{-1}. The irrigation water had a pH of 7.25 and electrical conductivity of 5.53 dS m^{-1} and carbonate is absent.

4.2.1 IRRIGATION SCHEDULING: DEPTH, DURATION, AND FREQUENCY OF IRRIGATION

Hydraulic design features on the layout of drip irrigation for reference crop okra (*Abelmoschus esculentus*). The spacing between lateral and plant was 45 and 40 cm, respectively. The depth of irrigation for okra within the effective root zone of 60 cm is based on the available water holding capacity of the effective root zone and the allowable soil moisture depletion of 50%.

$$d = \text{AWHC} \times \text{ASMD}\%$$
$$= \frac{(FC - WP) \times D \times \text{ASMD}\%}{100}, \tag{4.1}$$

where d = depth equivalent of irrigation in cm of water, AWHC = available water holding capacity of the effluent root zone, ASMD = available soil moisture depletion, and FC = mean field capacity, WP = mean wilting point, D = effective root zone depth.

$$d = \frac{(32.8 - 16.3) \times 60 \times \left(\dfrac{50}{100}\right)}{100} = 4.95 \text{ cm} \approx 5 \text{ cm}$$

Therefore, depth of irrigation was 5 cm.

Volume of irrigation water required in liters = wetting area in meter square × peak evapotranspiration in mm = 0.45 × 0.45 × 12 × 0.7 × 0.8 = 1.3608 ≈ 1.4 L

For the given duration of 100 days for okra, 20 irrigations were scheduled at a rate of 3.45 cm of water per irrigation (7 L = 7000 cm³).

Wetting area = 45 × 45 = 2025 cm²

Therefore, depth of irrigation = 3.45 cm

However, if the entire effective root zone of 60 cm is to be wetted, then the irrigation depth is calculated as 5 cm based on AWHC and ASMD. If the irrigation scheduling is by using drip layout, the percentage of water saving is: (5 − 3.5)/5 or (1.5/5) × 100 = 30%.

The depth of irrigation scheduling equals to 3.45 cm of water per irrigation with an irrigation frequency of 5 days. For the seasonal length of 100 days, 20 irrigations were given amounting to 69 cm of water.

Total volume of water required per liter = 2 (50 × 0.7) = 70 L/twin drip lateral

If emitter discharge = 2 lph, then duration of irrigation = 7/2 = 3.5 h = 210 min

4.3 RESULTS AND DISCUSSION

4.3.1 UNIFORMITY COEFFICIENT FOR 2 LPH DESIGNATED LATERAL LAYOUT (FIG. 4.1 AND TABLE 4.1)

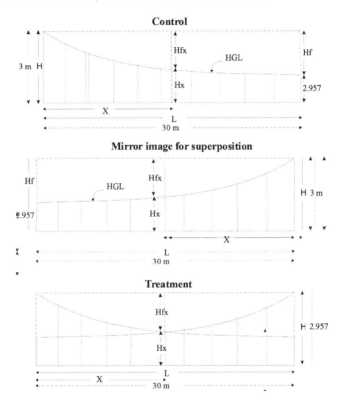

FIGURE 4.1 Hydraulic gradient compensation by superposition of mirror image (original length 30 m/2 lph).

4.3.2 SIMULATION AND PROJECTION OF UNIFORMITY COEFFICIENT (FOR UNIT LATERAL LENGTH OF 100 M)

Since the uniformity coefficient values are exaggerated in a small and compact experimental layout (Table 4.1), the reliable estimates can be made by way of simulating the same over a longer unit length for projection (Table 4.2).

TABLE 4.1 Discharge Versus Operating Pressure Head at Emission Points.

Primary lateral/2 lph											
Primary emission point in X	0	3	6	9	12	15	18	21	24	27	30
H in m	3	2.989	2.981	2.974	2.969	2.965	2.962	2.960	2.959	2.958	2.957
q in lph (q_1)	2	1.999	1.997	1.993	1.990	1.988	1.986	1.985	1.984	1.983	1.982

Secondary lateral/2 lph											
Secondary emission point in X	30	27	24	21	18	15	12	9	6	3	0
H in m	2.957	2.958	2.959	2.960	2.962	2.965	2.969	2.974	2.981	2.989	3
q in lph (q_2)	1.982	1.983	1.984	1.985	1.986	1.988	1.990	1.993	1.997	1.999	2
$q_1 + q_2$	3.982	3.982	3.981	3.978	3.976	3.976	3.976	3.978	3.981	3.982	3.982

TABLE 4.2 Discharge Versus Operating Pressure Head at Emission Points (Simulated for 100 m length).

Primary lateral/2 lph																
X	0	6	12	18	24	30	36	42	48	54	60	66	72	78	84	90	96
H_x	3	2.8	2.6	2.5	2.4	2.3	2.2	2.1	2.0	2.0	1.9	1.9	1.9	1.9	1.8	1.8	1.8
q	1.9	1.9	1.8	1.8	1.7	1.7	1.7	1.6	1.6	1.6	1.6	1.6	1.5	1.5	1.5	1.5	1.5
Secondary lateral/2 lph																
X	96	90	84	78	72	66	60	54	48	42	36	30	24	18	12	6	0
H_x	1.8	1.8	1.8	1.9	1.9	1.9	1.9	2.0	2.0	2.1	2.2	2.3	2.4	2.5	2.6	2.8	3
q	1.5	1.5	1.5	1.5	1.5	1.6	1.6	1.6	1.6	1.6	1.7	1.7	1.7	1.8	1.8	1.9	1.9
Q	3.5	3.5	3.4	3.4	3.3	3.3	3.3	3.3	3.3	3.3	3.3	3.3	3.3	3.4	3.4	3.5	3.5

4.3.3 UNIFORMITY COEFFICIENT FOR 2 LPH DESIGNATED LATERAL LAYOUT (100 M): FIGURE 4.2

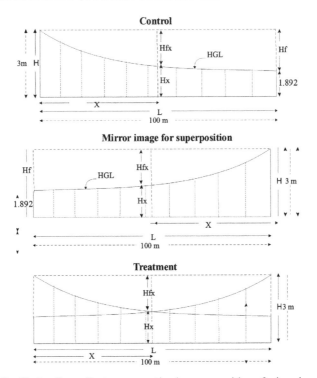

FIGURE 4.2 Hydraulic gradient compensation by superposition of mirror image (original length 100 m/2 lph).

4.3.4 IRRIGATION USAGE EFFICIENCY

4.3.4.1 IUE FOR 2 LPH LATERAL WITHOUT HYDRAULIC GRADIENT COMPENSATION

In the table, the yield of okra realized 7407.40 kg/ha against a depth of irrigation 412.02 mm of water. The irrigation water usage efficiency is projected 17.98 kg/ha/mm of irrigation.

In the plot without hydraulic gradient compensation, the head reaches earlier giving good crop stand and yield; while tail reaches where slight lagging begins, possibly due to the gradual reduction of operating pressure head from head to a tail end with a proportional decrease in the emitter discharge.

4.3.4.2 IUE FOR 2 LPH LATERAL WITH HYDRAULIC GRADIENT COMPENSATION

Table 4.3 indicates the okra yield of 8518.51 kg/ha against a depth of irrigation 411.736 mm of water. The irrigation water usage efficiency is 20.69 kg/ha/mm of water. In this plot of hydraulic gradient compensation, the crop stand was good and uniform right to head to tail end possibly due to compensated discharges variations along the lateral line.

TABLE 4.3 Okra Yield and Irrigation Water Usage Efficiency.

Particular	Yield (kg/ ha)	Irrigation depth (mm)	IUE (kg/ha/mm of water)
2 lph without hydraulic gradient compensation: control	7407.40	412.02	17.98
2 lph with hydraulic gradient compensation	8518.51	411.736	20.69

4.4 SUMMARY

In case of drip irrigation systems, the ultimate water distribution uniformity depends on the spatial distribution of the operating pressure heads at the emission points and the corresponding emitter discharges. The hydraulic gradient compensated drip lateral layout registered high order of water distribution uniformity. In case of 2 lph arrangement with hydraulic gradient compensated, the water distribution uniformity was 97.8% and irrigation usage efficiency ranged from 17.98 to 20.69 kg/ha/mm.

KEYWORDS

- **compensation**
- **distribution uniformity**
- **drip lateral**
- **emitter discharge**
- **okra**
- **spatial distribution**
- **uniformity**

REFERENCES

1. Meghanatha, R. A.; Shankhdha, D.; Shankhdhar, S. C. Physiological Characterization of Rice Genotypes Under Periodic Water Stress. *Indian J. Plant Physiol.* **2007,** *12*(2), 189–193.
2. Özekici, B.; Sneed, R. E. Manufacturing Variations for Various Trickle Irrigation Online Emitters. *Appl. Eng. Agric.* **1995,** *11*(2), 235–240.

CHAPTER 5

ASSESSMENT OF COEFFICIENT OF UNIFORMITY AND DISTRIBUTION UNIFORMITY OF MINI SPRINKLERS

AMOL B. KHOMNE[1,*], R. C. PUROHIT[1], P. K. SINGH[1], and S. S. BURARK[2]

1Department of Soil and Water Conservation Engineering, College of Technology and Engineering, Maharana Pratap University of Agriculture and Technology, Udaipur 313001, Rajasthan, India

2Department of Agricultural Economics and Management, RCA, Maharana Pratap University of Agriculture and Technology, Udaipur 313001, Rajasthan, India

*Corresponding author. E-mail: khomneamol@gmail.com

CONTENTS

ABSTRACT

The basic aim of the mini sprinkler irrigation method, as in other irrigation methods, is to apply irrigation water as uniformly as possible to the root zone of the crop. The study was aimed to evaluate the uniformity of different mini sprinklers in the Department of Soil and Water Engineering at College of Technology and Engineering, Udaipur. Three commercially available mini sprinklers of different companies (C-1, C-2, and C-3) with nozzle size 2.5 × 1.8 mm, 2.4 × 1.8 mm, and 2.3 × 1.8 mm, respectively, were tested for their uniformity performance in terms of uniformity coefficient and distribution uniformity. The experiment was conducted at four different operating pressures ranging from 2 to 3.5 kg/cm^2 with increment of 0.5 kg/cm^2 and at two different riser heights of 50 and 100 cm. All the mini sprinklers were operated for spacing of 10 m × 10 m. The uniformity coefficient and distribution uniformity of mini sprinklers were studied for single-head mini sprinkler pattern method. The best-fit relation between operating pressure and riser height was established and polynomial relationship was found to be best among other relations on the basis of value of their regression coefficient (r^2). From the study, it was also observed that the values of uniformity coefficient and distribution uniformity were more for 100 cm riser height as compare to 50 cm.

5.1   INTRODUCTION

In recent years, the more efficient pressurized irrigation systems (e.g., drip, trickle, and sprinkler irrigation) have become more usable instead of open channel irrigation system to increase the water use efficiency.[2] The irrigation uniformity is an important indicator for the evaluation of performance of sprinkler irrigation system. Therefore, it must be considered during the design and installation of the system.[8] The most effective factor affecting the irrigation uniformity is operating pressure. Too high pressure causes small droplets that lead to higher water distribution near sprinkler, and too low pressure produces large-size droplets that fall further away from sprinkler.[5] The degree of uniformity of water distribution depends on operating pressure, nozzle diameter, and riser height. The performance of sprinkler nozzle determines the productivity and efficiency of the whole system.[15]

It is necessary to have information on the uniformity of application at different riser heights in order to achieve an optimum height of riser for an acceptable uniformity of application. This will help in determining the criteria for use of mini sprinkler in relation to crop height and to make a technically sound design. By determining highest uniformity of water distribution for optimum riser height, operating pressure, and proper sprinkler spacing, only required optimum number of sprinklers per unit area can be used, and per hectare cost of the irrigation systems can be reduced.

Keeping all above points in view, there is strong need to generate the basic information regarding pressure–discharge relationship, uniformity coefficient, uniformity of distribution, wetting diameter for optimum design, and operation of sprinkler irrigation systems. Considering these aspects, the study in this chapter was carried out to assess the coefficient of uniformity (CU) and distribution of uniformity (DU) of mini sprinklers.

5.2 MATERIALS AND METHODS

5.2.1 EXPERIMENTAL SITE

The experiment was conducted at the Plasticulture Farm of College of Technology and Engineering, Udaipur. The site is situated at the southern part of Rajasthan and lies in Aravalli ranges. The area is characterized by subtropical continental semihumid monsoon-type climate. The location of study area is situated at 24°35′N latitude and 73°44′E longitude and at an altitude of 582 m above mean sea level. The location of study area is shown in Figure 5.1.

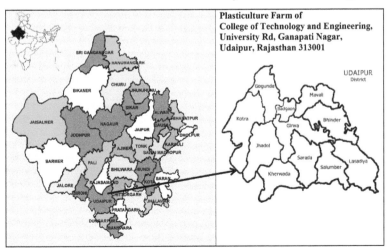

FIGURE 5.1   Location map of study area.

5.2.2　DESCRIPTION OF EXPERIMENTAL SETUP

The experimental setup consisted of water source (open well), pump (5 HP), filtration unit (i.e., sand filter of capacity 40 m³/h and screen filter of capacity 40 m³/h), main line (PVC pipe of Class II 90 mm diameter), submain pipe (PVC pipe of Class II 63 mm diameter), bypass assembly, laterals (JISL—Class II of 32 mm diameter), mini sprinklers with risers, and other fittings and accessories.

5.2.3　SPECIFICATIONS OF MINI SPRINKLER UNDER STUDY

Mini sprinkler is an important component of the sprinkler irrigation system. In this experiment, mini sprinklers from three different companies (such as Company-1, Company-2, and Company-3: C-1, C-2, C-3) were selected to study their performance under different operating pressures ranging from 2 to 3.5 kg/cm². The CU and DU of mini sprinklers were studied for 50 and 100 cm riser heights. The precipitation depth of mini sprinklers was collected in catch cans, placed at grid of 2 m × 2 m spacing.[13] Table 5.1 shows specifications of mini sprinklers.

TABLE 5.1　Specifications of Mini Sprinklers.

Company name	Nozzle color	Nozzle size (mm)	Operating pressure (kg/cm²)	Discharge (L/h)	Radius (m)
C-1	Purple and green	2.5 × 1.8	2	480	9.5
			2.5	535	10.5
			3	590	10.5
			3.5	635	10.5
C-2	Yellow and green	2.4 × 1.8	2	490	9
			2.5	530	9.1
			3	590	9.4
			3.5	630	9.5
C-3	Yellow and green	2.3 × 1.8	2	470	9.25
			2.5	510	9.5
			3	570	9.5
			3.5	–	–

5.2.4 COEFFICIENT OF UNIFORMITY

The CU of all three mini sprinklers of different companies (C-1, C-2, C-3) was determined at operating pressures of 2–3.5 kg/cm² with an increment of 0.5 kg/cm² at both the riser heights of 50 and 100 cm. The observations were recorded in three replications. CU was determined by using Christiansen's formula[11] as follows:

$$CU = \left[1 - \frac{\Sigma x}{mn}\right] \times 100, \qquad (5.1)$$

where CU = coefficient of uniformity (%), m = average value of all observations (mm), n = total number of observation points, and x = numerical deviation of individual observations from the average application rate (mm).

5.2.5 DISTRIBUTION UNIFORMITY

DU indicates the degree to which the water is applied uniformly over the area. If less water is applied in one part of the field and more water is applied in other part, then it is considered as poor DU. The DU of three mini sprinklers was determined at operating pressures of 2–3.5 kg/cm² at both riser heights of 50 and 100 cm. DU was determined using equation by Keller and Merriam.[7]

$$DU = \frac{\text{Average low quarter depth of water caught}}{\text{Average depth of water caught}} \times 100, \qquad (5.2)$$

where DU = distribution uniformity of mini sprinkler (%).

5.2.6 PRESSURE, RISER HEIGHT: RELATIONSHIPS OF CU AND DU

The mathematical relationships (linear, logarithmic, power, polynomial, and exponential) between pressure versus CU and pressure versus DU were developed for 50 and 100 cm riser heights using the observation data. The best-fit equation was deduced on the basis of regression coefficient (R^2).

5.3 RESULTS AND DISCUSSION

5.3.1 COEFFICIENT OF UNIFORMITY

The CU of single mini sprinkler head of all three mini sprinklers was calculated for 50 and 100 cm riser height, under different operating pressures ranging from 2 to 3.5 kg/cm^2.

5.3.1.1 CU OF MINI SPRINKLERS AT 50 CM RISER HEIGHT

Mini sprinkler was placed at a height of 50 cm on the riser stand. The observed data were used to determine the CU of a single mini sprinkler head pattern under different operating pressures (Table 5.2 and Fig. 5.2). Data in Table 5.2 show that the CU was increased initially, when the operating pressure was increased from 2.0 to 3.0 kg/cm^2, and further increase in operating pressure from 3.0 to 3.5 kg/cm^2 caused reduction in the CU. The graphical presentation of pressure to discharge relationships is depicted in Figure 5.1. The similar findings have been observed by researchers Bishaw and Olumana[1] and Kara et al.[6]

TABLE 5.2 Coefficient of Uniformity of a Single Mini Sprinkler at Different Operating Pressures at a Riser Height of 50 cm.

Company name	Pressure (kg/cm²)	Coefficient of uniformity (%)			
		R-I	R-II	R-III	Mean
C-1	2	30.15	30.17	30.13	30.15
	2.5	46.10	46.10	46.26	46.16
	3	48.54	54.65	47.23	50.14
	3.5	44.18	49.28	42.98	45.48
C-2	2	32.19	36.15	31.42	33.25
	2.5	43.62	41.35	39.07	41.35
	3	43.48	48.78	42.49	44.92
	3.5	39.98	44.75	39.18	41.30
C-3	2	32.50	32.50	32.53	32.51
	2.5	40.85	40.80	40.90	40.85
	3	47.85	47.85	47.88	47.86
	3.5	45.21	43.11	47.32	45.21

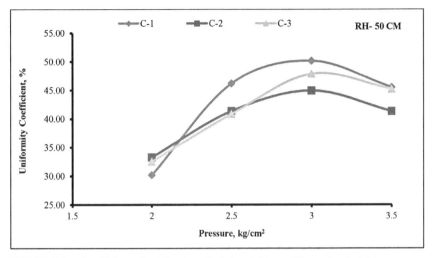

FIGURE 5.2   Coefficient of uniformity of mini sprinkler at 50 cm riser height.

It was revealed that CU increased with increase in operating pressure up to the certain limit, and then it started to decrease due to excessive breakup of the jet as studied by Sahoo et al.[14]

Among the mini sprinklers tested, the mini sprinkler C-1 produced highest average uniformity of 50.14%. Data also showed that as the operating pressure was increased from 2.0 to 3.0 kg/cm², the CU was increased from 30.15% to 50.14%, 33.25% to 44.92%, and 32.51% to 47.86% for C-1, C-2, and C-3 mini sprinklers, respectively.

Further increase in operating pressure from 3.0 to 3.5 kg/cm² caused decrease in CU from 50.14% to 45.48%, 44.92% to 41.30%, and 47.86% to 45.21% for the mini sprinklers C-1, C-2, and C-3, respectively. This may be due to the fact that the increase in operating pressure from 2 to 3.0 kg/cm² resulted in the increase of radius of throw and the average depth of water slightly from center to a certain distance and thereafter, it was decreased gradually. The wind velocity factor during field tests could affect the uniformity of distribution pattern according to Yacoubi et al.[16]

As the mini sprinklers of all the companies had different nozzle sizes, therefore it was found that as nozzle size of mini sprinkler increases, the CU also increases from C-3 to C-1. Similar results have been observed by Dehkordi et al.[3]

The relationships between operating pressure (*P*) and CU for single mini sprinkler head and for 50 cm riser were developed by fitting polynomial, exponential, linear, logarithmic, and power relations (Table 5.3). The polynomial relationship was the best-fit equation between pressure and CU among all other relationships. The maximum value of regression coefficient for the polynomial relationship was found for C-1 mini sprinkler and minimum for C-3.

TABLE 5.3 Relationship Between the Pressure and Coefficient of Uniformity of Single Mini Sprinkler Head for 50 cm Riser Height.

Company name	Type of relationship	Equation	Regression coefficient (R^2)
C-1	Polynomial	$U = -20.66\ P^2 + 123.6\ P - 134.3$	0.997
	Exponential	$U = 20.46\ e^{0.263P}$	0.553
	Linear	$U = 9.994\ P + 15.49$	0.537
	Logarithmic	$U = 28.72\ \ln(P) + 14.54$	0.619
	Power	$U = 19.99\ P^{0.754}$	0.634
C-2	Polynomial	$U = -11.70\ P^2 + 69.93\ P - 59.91$	0.995
	Exponential	$U = 26.70\ e^{0.146P}$	0.541
	Linear	$U = 5.544\ P + 24.95$	0.526
	Logarithmic	$U = 15.94\ \ln(P) + 24.42$	0.606
	Power	$U = 26.34\ P^{0.421}$	0.622
C-3	Polynomial	$U = -10.98\ P^2 + 69.43\ P - 62.83$	0.974
	Exponential	$U = 21.90\ e^{0.229P}$	0.754
	Linear	$U = 9.022\ P + 16.79$	0.751
	Logarithmic	$U = 25.12\ \ln(P) + 16.72$	0.813
	Power	$U = 21.84\ P^{0.64}$	0.819

Legend: P = operating pressure, U = coefficient of uniformity, C = company.

5.3.1.2   CU OF MINI SPRINKLER AT 100 CM RISER HEIGHT

For the determination of CU of a single mini sprinkler head, the mini sprinklers of C-1, C-2, and C-3 were placed at height of 100 cm on riser rod. The graphical representation of data from Table 5.4 is shown in Figure 5.3.

Figure 5.3 shows that CU was increased initially, when the operating pressure increased from 2.0 to 3.0 kg/cm²; however, with further increase in operating pressure from 3.0 to 3.5 kg/cm^2, the CU values were decreased. The similar results have been observed by Mandave et al.[10]

The highest CU of 55.09%, 47.71%, and 49.51% were obtained in C-1, C-2, and C-3 mini sprinklers respectively, at an operating pressure of 3.0 kg/cm^2, which were significantly superior than the other operating pressures.

Among the mini sprinklers tested, the mini sprinkler C-1 produced highest average uniformity for different pressure range of 2–3.5 kg/cm^2 as compared to other mini sprinklers, C-2 and C-3. Data also indicated that as the operating pressure was increased from 2.0 to 3.0 kg/cm^2, the CU was increased from 39.77% to 55.09%, 35.12% to 47.71%, and 34.01% to 49.51% for C-1, C-2, and C-3 mini sprinklers, respectively.

With further increase in operating pressure from 3.0 to 3.5 kg/cm^2, the CU was decreased from 55.09% to 49.44%, 47.71% to 37.58%, and 49.51% to 43.12% for all the mini sprinklers. The wind velocity factor during field tests could affect the CU of mini sprinklers.

The polynomial relationship was the best fit among all other functions for the mini sprinklers of three different companies at 100 cm riser height (Table 5.5). The maximum value of coefficient of determination for polynomial function was found for C-1 and minimum for C-3 mini sprinkler.

TABLE 5.4   Coefficient of Uniformity of a Single Mini Sprinkler at Different Operating Pressures at a Riser Height of 100 cm.

Company ID	Pressure (kg/cm^2)	Coefficient of uniformity (%)			
		R-I	R-II	R-III	Mean
C-1	2	39.75	39.75	39.82	39.77
	2.5	49.65	48.79	49.66	49.37
	3	55.00	55.00	55.27	55.09
	3.5	49.12	50.32	48.87	49.44
C-2	2	34.83	34.88	35.64	35.12
	2.5	39.31	44.54	46.2	43.35
	3	47.45	46.74	48.93	47.71
	3.5	39.68	36.18	36.87	37.58
C-3	2	34.58	33.53	33.91	34.01
	2.5	43.85	38.66	41.33	41.28
	3	49.09	50.06	49.39	49.51
	3.5	37.90	44.42	47.04	43.12

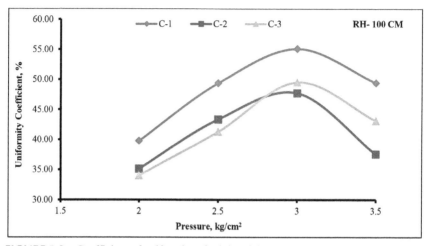

FIGURE 5.3   Coefficient of uniformity of mini sprinkler at 100 cm riser height.

TABLE 5.5   Relationships Between the Operating Pressure (P) and Coefficient of Uniformity (U) of a Single Mini Sprinkler at a Riser Height of 100 cm.

Company ID	Type of relationship	Equation	Regression coefficient (R^2)
C-1	Polynomial	$U = -10.05\ P^2 + 58.96\ P - 62.58$	0.977
	Exponential	$U = 31.61\ e^{0.152P}$	0.519
	Linear	$U = 6.946\ P + 29.31$	0.497
	Logarithmic	$U = 20.01\ \ln(P) + 28.59$	0.576
	Power	$U = 31.14\ P^{0.438}$	0.599
C-2	Polynomial	$U = -7.39\ P^2 + 44.20\ P - 45.50$	0.994
	Exponential	$U = 34.48\ e^{0.059P}$	0.077
	Linear	$U = 2.347\ P + 34.48$	0.071
	Logarithmic	$U = 8.040\ \ln(P) + 32.97$	0.116
	Power	$U = 33.23\ P^{0.203}$	0.125
C-3	Polynomial	$U = -5.8\ P^2 + 36.07\ P - 34.60$	0.921
	Exponential	$U = 25.44\ e^{0.178P}$	0.553
	Linear	$U = 2.347\ P + 34.48$	0.071
	Logarithmic	$U = 8.040\ \ln(P) + 32.97$	0.116
	Power	$U = 25.13\ P^{0.508}$	0.624

Tables 5.2 and 5.4 indicate that with increase in riser height from 50 to 100 cm, the highest CU was 4.95%, 2.79%, and 1.65% for C-1, C-2, and C-3 mini sprinklers at 3.00 kg/cm² operating pressure. Therefore, it was observed that CU was increased with increase in riser height. The similar results have been reported by El-Shafei et al.[4] and Montero et al.[12]

5.3.2 DISTRIBUTION UNIFORMITY

The DU of single mini sprinkler head of all the mini sprinklers was calculated for riser height of 50 and 100 cm under different operating pressures ranging from 2 to 3.5 kg/cm².

5.3.2.1 DU OF SINGLE MINI SPRINKLER AT 50 CM RISER HEIGHT

The values for DU were determined for single mini sprinkler (Table 5.6). The graphical representation of results obtained from this table is shown in Figure 5.4.

TABLE 5.6 Distribution Uniformity of Mini Sprinkler at Different Operating Pressures and at 50 cm Riser Height.

Company ID	Pressure (kg/cm²)	Distribution uniformity (%)			
		R-I	R-II	R-III	Mean
C-1	2	14.90	14.91	14.89	14.90
	2.5	22.62	22.62	22.70	22.64
	3	22.41	25.24	21.81	23.15
	3.5	20.25	22.58	19.70	20.84
C-2	2	12.83	14.41	12.53	13.26
	2.5	20.14	19.09	18.04	19.09
	3	19.68	22.08	19.23	20.33
	3.5	18.18	20.34	17.81	18.78
C-3	2	13.92	14.09	16.11	14.71
	2.5	18.26	17.49	19.03	18.26
	3	22.30	22.56	22.65	22.50
	3.5	20.25	20.25	20.29	20.26

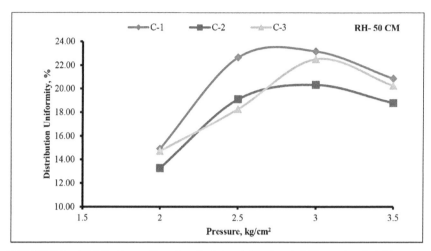

FIGURE 5.4   Distribution uniformity of mini sprinkler head at 50 cm riser height.

The results in Table 5.6 show that at 50 cm riser height, the 3.0 kg/ cm^2 operating pressure gave higher DU compared to all other operating pressures. The high DU at 3.0 kg/cm^2 operating pressure was 23.15% for C-1, 20.33% for C-2, and 22.50% for C-3 mini sprinklers compared to their corresponding DU at 2.0, 2.5, and 3.5 kg/cm^2 operating pressures. The effect of wind velocity is also observed on all distribution patterns.

It was observed that the DU was increased with increase in operating pressure from 2.0 to 3.0 kg/cm^2; however with further increase in operating pressure from 3.0 to 3.5 kg/cm^2, the DU values were found to be decreased. As the operating pressure increases from 2.0 to 3.0 kg/cm^2, the DU was increased from 14.90% to 23.15%, 13.26% to 20.33%, and 14.71% to 22.50% for C-1, C-2, and C-3 mini sprinklers, respectively. The results followed the findings of El-Shafei et al.[4] and Mandave et al.[9]

The linear, logarithmic, power, polynomial, and exponential relationships were fitted to the data. The polynomial relationship was found to be best to correlate operating pressures and DU in all mini sprinklers. The maximum value of regression coefficient was found for C-3, while the minimum value was found for C-2 as shown in Table 5.7.

TABLE 5.7 Relationships Between Operating Pressure (P) and Distribution Uniformity (DU) of a Mini Sprinkler for 50 cm Riser Height.

Company ID	Type of relationship	Equation	Regression coefficient (R^2)
C-1	Polynomial	$DU = -10.053\ P^2 + 58.96\ P - 62.58$	0.977
	Exponential	$DU = 11.40\ e^{0.205P}$	0.423
	Linear	$DU = 3.668\ P + 10.29$	0.390
	Logarithmic	$DU = 10.79\ \ln(P) + 9.694$	0.472
	Power	$DU = 11.06\ P^{0.602}$	0.505
C-2	Polynomial	$DU = -5.8\ P^2 + 36.07\ P - 34.60$	0.921
	Exponential	$DU = 9.590\ e^{0.221P}$	0.547
	Linear	$DU = 3.559\ P + 8.076$	0.534
	Logarithmic	$DU = 10.23\ \ln(P) + 7.730$	0.616
	Power	$DU = 9.398\ P^{0.635}$	0.628
C-3	Polynomial	$DU = -7.39\ P^2 + 44.20\ P - 45.50$	0.994
	Exponential	$DU = 9.833\ e^{0.233P}$	0.691
	Linear	$DU = 4.177\ P + 7.444$	0.665
	Logarithmic	$DU = 11.69\ \ln(P) + 7.353$	0.727
	Power	$DU = 9.782\ P^{0.654}$	0.755

5.3.2.2   DU OF MINI SPRINKLER HEAD AT 100 CM RISER HEIGHT

For the determination of DU of single mini sprinkler, the mini sprinkler was placed at a height of 100 cm on riser rod and the precipitation depth was collected in catch cans placed in a grid for C-1, C-2, and C-3 mini sprinklers. The values were worked out for single mini sprinkler head pattern under different operating pressures (Table 5.8), and the results are shown in Figure 5.5.

Table 5.8 and Figure 5.5 show that the DU was increased with increase in operating pressure from 2.0 to 3.0 kg/cm²; however with further increase in operating pressure from 3.0 to 3.5 kg/cm², the DU was found to be decreased. It was revealed that DU increases with increase in operating pressure up to the certain limit, and then it start to decrease due to excessive breakup of the jet as shown by Sahoo et al.[14]

It was found that as the operating pressure increases from 2.0 to 3.0 kg/cm², the DU increases from 16.61% to 28.87%, 15.23% to 24.57%, and

16.10% to 24.06% for C-1, C-2, and C-3 mini sprinklers, respectively. The results are in agreement with the findings of Mandave et al.[10]

TABLE 5.8 Distribution Uniformity of Mini Sprinkler at Different Operating Pressures for 100 cm Riser Height.

Company ID	Pressure (kg/cm²)	Distribution uniformity (%)			
		R-I	R-II	R-III	Mean
C-1	2	16.61	16.61	16.60	16.61
	2.5	23.97	23.97	24.06	24.00
	3	27.95	31.47	27.20	28.87
	3.5	21.80	21.80	21.90	21.83
C-2	2	14.75	16.00	14.96	15.23
	2.5	21.14	20.99	20.83	20.99
	3	24.79	24.69	24.25	24.57
	3.5	17.88	19.37	17.22	18.16
C-3	2	16.00	16.00	16.30	16.10
	2.5	20.10	19.26	20.94	20.10
	3	22.78	23.05	26.36	24.06
	3.5	18.67	17.81	19.54	18.67

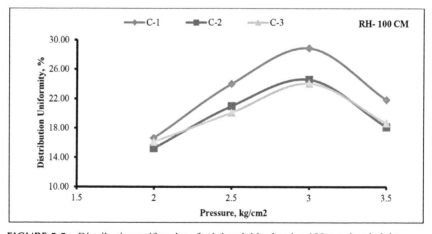

FIGURE 5.5 Distribution uniformity of mini sprinkler head at 100 cm riser height.

TABLE 5.9 Relationships Between Operating Pressure (P) and Distribution Uniformity (DU) of Mini Sprinkler for 100 cm Riser Height.

Company ID	Type of relationship	Equation	Regression coefficient (R^2)
C-1	Polynomial	$DU = -14.44\ P^2 + 83.52\ P - 93.15$	0.943
	Exponential	$DU = 12.88\ e^{0.201P}$	0.316
	Linear	$DU = 4.106\ P + 11.53$	0.271
	Logarithmic	$DU = 12.35\ \ln(P) + 10.59$	0.342
	Power	$DU = 12.36\ P^{0.599}$	0.392
C-2	Polynomial	$DU = -12.17\ P^2 + 69.40\ P - 75.29$	0.935
	Exponential	$DU = 13.33\ e^{0.136P}$	0.187
	Linear	$DU = 2.471\ P + 12.94$	0.159
	Logarithmic	$DU = 7.768\ \ln(P) + 12.04$	0.220
	Power	$DU = 12.74\ P^{0.425}$	0.253
C-3	Polynomial	$DU = -9.39\ P^2 + 53.98\ P - 54.76$	0.869
	Exponential	$DU = 13.84\ e^{0.125P}$	0.233
	Linear	$DU = 2.336\ P + 13.30$	0.205
	Logarithmic	$DU = 7.129\ \ln(P) + 12.67$	0.266
	Power	$DU = 13.41\ P^{0.379}$	0.299

Among the mini sprinklers tested, the mini sprinkler of C-1 gave highest values of DU under all operating pressures compared to C-2 and C-3. As the mini sprinklers of all the companies had different nozzle sizes, therefore it was found that as nozzle size of mini sprinkler increases, the uniformity also increases from C-3 to C-1. Similar results have been observed by Dehkordi et al.[3]

The linear, logarithmic, power, polynomial, and exponential relationships were developed for operating pressure versus DU (Table 5.9). The polynomial relationship was found to be best to correlate operating pressures and DU. The maximum value of regression coefficient was found for C-1, while the minimum value was found for C-3 mini sprinkler.

Tables 5.6 and 5.8 indicate that as riser height increases from 50 to 100 cm, the highest value of DU was also increased by 5.72%, 4.24%, and 1.56% for C-1, C-2, and C-3 mini sprinklers at 3 kg/cm^2 operating pressure. Therefore, it was observed that DU was increased with increase in riser height. These results were confirmed by El-Shafei et al.[4] and Montero et al.[12]

5.4 SUMMARY

The purpose of mini sprinkler irrigation method, as in other irrigation methods, is to apply irrigation water as uniformly as possible to the root zone of the crop. The study in this chapter evaluated the uniformity of different mini sprinklers. Three commercially available mini sprinklers of different companies (C-1, C-2, and C-3 with nozzle size 2.5 × 1.8, 2.4 × 1.8, 2.3 × 1.8 mm, respectively) were tested for CU and DU. The experiment was conducted at four different operating pressures ranging from 2 to 3.5 kg/cm² with increment of 0.5 kg/cm² and at two different riser heights of 50 and 100 cm. All the mini sprinklers were operated for spacing of 10 m ×10 m. The polynomial relationships were found to be best among all other equations. From the study, it was also observed that the values of CU and DU were higher for 100 cm riser height compared to 50 cm riser height.

KEYWORDS

- **distribution uniformity**
- **drip irrigation**
- **mini sprinkler**
- **regression coefficient**
- **riser height**
- **sprinkler nozzle**
- **uniformity coefficient**

REFERENCES

1. Bishaw, D.; Olumana, M. Evaluating the Effect of Operating Pressure and Riser Height on Irrigation Water Application Under Different Wind Conditions in Ethiopia. *Asia Pac. J. Energy Environ.* **2015,** *2,* 41–48.
2. Cobo, M. T. C.; Poyato, E. C.; Montesinos, P.; Diaz, J. A. R. New Model for Sustainable Management of Pressurized Irrigation Networks. Application to Bembezar MD Irrigation District (Spain). *Sci. Total Environ.* **2014,** *8,* 473–474.

3. Dehkordi, D. K.; Mohsenifar, K.; Fardipour, S.; Khodabakhshi, H. Evaluation of Coefficient of Uniformity of Four Useful Sprinklers in Khuzestan Province Under Different Conditions. *Int. J. Agric. Biosci.* **2016,** *5,* 8–14.

4. El-Shafei, A.; Allam, K. A.; El-Abedin, T. Heterogeneity Analysis of Sprinkler Irrigation in Peanut Fields. *Misr J. Agric. Eng.* **2008,** *25,* 58–86.

5. Haman, D. Z.; Smajstrla, A. G.; Pitts, D. J. *Uniformity of Sprinkler and Micro-irrigation Systems for Nurseries*; The Institute of Food and Agriculture Science, University of Florida: Florida, 2003; p 5.

6. Kara, T.; Ekmekci, E.; Apan, M. Determining the Coefficient of Uniformity and Water Distribution Characteristics of Some Sprinklers. *Pak. J. Biol. Sci.* **2008,** *11,* 214–219.

7. Keller, J.; Merriam, J. L. *Farm Irrigation System Evaluation: A Guide for Management*; Agricultural and Irrigation Engineering Department, Utah State University: Logan, Utah, 1978; p 271.

8. Han, W and Wu, P Evaluation Model Development for Sprinkler Irrigation Uniformity Based on Catch-can Data. *Afr. J. Biotechnol.* **2011,** *10,* 14796–14802.

9. Mandave, V. R.; Jadhav, S. B. Performance Evaluation of Portable Mini-sprinkler Irrigation System. *Int. J. Innov. Res. Sci. Eng. Technol.* **2014,** *3,* 177–184.

10. Mandave, V. R.; Patil, M.; Pawar, P. Hydraulic Performance Evaluation. In *Investigation of Mini-sprinkler Irrigation System*; Lambert Academic Publishing: Germany, 2014; p 110.

11. Michael, A. M. *Irrigation: Theory and Practice*; Vikas Publishing House Pvt. Ltd.: New Delhi, 1978; p 64.

12. Montero, J.; Tarjuelo, J. M.; Ortega, J. F. Heterogeneity Analysis of the Irrigation in Fields with Medium Size Sprinklers. *Agric. Eng. Int. CIGR J.* **2000,** *2,* 1682–1130.

13. Suliman, M. M.; Mohamed, A. E.; Makki, E. K. Effect of Different Solid-set Sprinkler Patterns on Water Distribution and Losses Under Shambat Conditions, Sudan. *Univ. Khartoum J. Agric. Sci.* **2010,** *18,* 166–184.

14. Sahoo, N.; Pradhan, P. L.; Anumala, N. K.; Ghosal, M. K. Uniform Water Distribution from Low Pressure Rotating Sprinklers. *Agric. Eng. Int. CIGR J.* **2008,** *10,* 1–10.

15. Wilson, T. P.; Zoldoske, D. F. Evaluating Sprinkler Irrigation Uniformity, 1997. http://cwi.csufresno.edu/wateright/evalsprink.asp (accessed Feb 2, 2016).

16. Yacoubi, S.; Zayani, K.; Slatni, A.; Playan, E. Assessing Sprinkler Irrigation Performance Using Field Evaluations at the Medjerda Lower Valley of Tunisia. *Engineering* **2012,** *4,* 682–691.

CHAPTER 6

WATER FOOTPRINT IN DRIP-IRRIGATED BLUEBERRIES: ARGENTINA

A. PANNUNZIO[1,*], E. A. HOLZAPFEL[2], P. TEXEIRA SORIA[1], and F. BOLOGNA[3]

Department of Soil and Water Engineering, University of Buenos Aires, Av. San Martin 4500, Buenos Aires, Argentina
Department of Soil and Water Engineering, University of Concepción, Casilla 537, Chillán, Chile
Engineering Section, Berries del Sol S.A., Ruta 28 km 5, Colonia Ayuí, Entre Ríos, Argentina
Corresponding author. E-mail: pannunzio@agro.uba.ar

CONTENTS

ABSTRACT

This chapter discusses water footprint research in blueberries under drip irrigation in Concordia, Argentine. Water footprint is an important aspect that is related to the basin, well ruled and legal monitoring of rights and obligations of all social actors.

6.1   INTRODUCTION

Water is vital, multifunctional, and scare resource and these characteristics generate a strong competition among water users. A thorough planning is required following the scope of economic and sustainable development. During the seventeenth century, the deficient management of water resources became relevant and the paradigm of integrated management of hydric resources[22] was suggested.

The increase of human population generates the need to improve the efficiency of food production. At this point, irrigation is a basic tool, however water availability is restricted and it obliges farmers to progress in water productivity. [26–29] Irrigation uses about 70% of total fresh water available and the water application efficiency is around 40%.[18]

The use of indicators is basic to reach sustainability[1] and to improve irrigation water management.[7] The water footprint is a good indicator of efficiency of water used in different processes, and it includes direct and indirect use of water to produce goods or services during a certain period.

Water footprint concept was proposed in the 21st century by Hoekstra.[19] This concept divides water use considering its source and contamination associated with productive process. In 1993, Allan introduced virtual water concept, measuring the water contained in each product and the water used during the process.[2] Both concepts answer the requirements of quantification of this water use, however water footprint implies volume of water used and virtual water implies the flow of water as a net balance of water of a country. Chapagain and Hoekstra estimated water footprint for several countries for the period between 1997 and 2001.[9] Currently, water footprint is classified as: (a) blue water that is related with the water use from superficial and groundwater sources,[39] (b) green water that is related with the rainwater used by the crop, and (c) gray water that is the contaminated water used during the process.[13] Considering the difficulties of measuring gray water, this water is not thoroughly studied as blue and green waters.

The differentiation of three types of water is crucial due to environmental implications, thus requiring several management policies.[33]

Water footprint in agriculture has been considered by several authors. Mekkonen and Hoekstra[23,24] quantified blue, green, and gray water for global production of 126 crops. Ridoutt et al.[36] calculated water footprint for mango. Deuret et al.[10] calculated water footprint for kiwi crop. Herath et al.[17] studied gray water of potato crop in New Zeeland. In 2013, the Instituto de Investigaciones Agropecuarias de Chile edited the book "Determinación de la Huella del Agua y estrategias de manejo de los Recursos Hídricos," for several crops including blueberry crop.

The general values for blueberries water footprint are around 341, 334, and 170 L/kg for green, blue, and gray footprints, resulting in a total of 845 L/kg.[25] These values are consistent with those obtained in Chile between 400 and 800 L/kg.[8]

As an example of the importance of water footprint as an indicator, the ISO 14.046 Standard quantifies water environmental impacts and the possibilities of reducing its effects. In 2014, the *Autoridad del Agua de la Provincia de Buenos Aires* incorporated water footprint to establish water cost for Argentine farmers. In this country, water footprints for agricultural production have been studied by various authors[31] with studies of water footprint in rice production in *Entre Ríos*.[6,11]

The knowledge of water footprint in blueberry (*Vaccinium corymbosum* L.) crop is an important issue to plan efficient water use, improving productivity, sustainability, and competitiveness of irrigated crops.[22,31] Blueberry in Argentine is destined to the northern hemisphere markets, taking advantage of commercial window during the period of September–December. Early growing varieties are required for the production of blueberries in September. These varieties need antifrost protection system during winter because some frost occurs in Concordia area of Argentine[14], while plant sensitivity to low temperature is high. Sprinkler, micro-sprinkler, and mini-sprinkler solid set systems are used to avoid frost damages during the nights when temperatures fall below 0°C.

Two main regions produce more than 80% of the Argentinean production of blueberries (Fig. 6.1): Concordia region in northeast and the other in the northwest (Tucuman and Salta). Among the factors, that influence the production of blueberries in this region, are excessive drainage and soil salinity.[32,37,38,40]

This chapter discusses water footprint research in blueberries under drip irrigation in Concordia, Argentine.

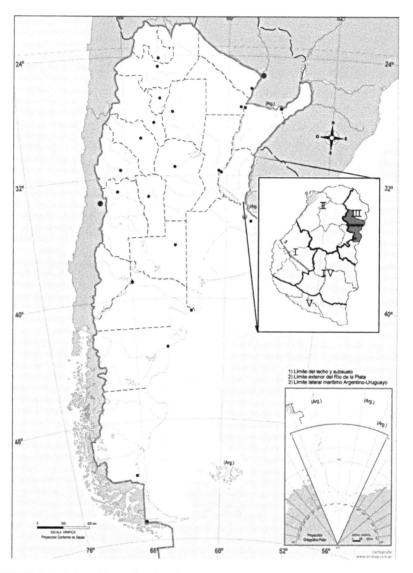

FIGURE 6.1   Map of Argentine showing Concordia area.

6.2   CLIMATIC CHARACTERIZATION OF CONCORDIA

The study area is located within 31°11′24″ S, 58°02′54″ W at 58 m above the sea level in "Berries del Sol" farm in Colonia Ayui, Concordia, Entre

Ríos Province. The Concordia climate is a temperate humid, with medium temperatures between 17°C and 20°C, and rainfall in the area varies from 1000 to 1300 mm/year spreading all over the year.[8] These values of rainfall are not enough to meet the demands of water use for blueberry crop, considering the sandy soil of the area and the shallow root system of the crop (Table 6.1). In this region, "El Niño" phenomenon is associated with water excess during November while dry seasons can be expected in the other periods.[16]

TABLE 6.1 Evapotranspiration with Penman–Monteith Method ($_0$) and Climatic Data: Concordia Region.

Month	Min. temp	Max. temp	Humidity	Wind	Sunshine	Radiation	ET$_0$
	°C	°C	%	m/s	Daily hours	MJ/m²/day	mm/day
January	8.4	42.2	63	3.6	8.9	24.6	8.69
February	7.0	41.6	68	3.4	7.7	21.5	7.75
March	4.7	39.9	72	3.2	7.7	19.0	6.71
April	1.5	35.0	75	3.0	6.1	13.8	4.89
May	−1,3	31.8	78	3.1	5.9	11.0	3.96
June	−5.3	29.0	80	3.2	4.6	8.6	3.42
July	−3.9	31.6	78	3.5	5.1	9.5	4.04
August	2.2	32.4	74	3.6	5.7	12.2	4.7
September	−3.0	34.4	78	3.8	6.5	16.1	5.92
October	0.9	37.8	72	3.9	7.8	20.7	7.31
November	2.3	39.0	69	3.7	8.5	23.6	7.96
December	5.4	41.0	64	3.6	9.3	25.6	8.61
Average	1.2	36.3	72	3.5	7.0	17.2	6.16

6.3 SOIL WATER POTENTIAL CHARACTERIZATION AND ITS IMPORTANCE TO DETERMINE WATER REQUIREMENTS

In Concordia region, just aside the Uruguay River, sandy soils over a strong clay subsoil can be found in some spotted areas, in which blueberry crop is cultivated. Main roots of the crop were found up to 40 cm, where soil's texture consisted of clay 9%, silt 10.5%, sand 80.5%, pH of ≈5, salinity of 0.3 dS·m⁻¹, and cation exchange capacity (CEC) of 8 meq/(100 g).

For research on crop production, it is important to know the crop response to different water potentials at which water is retained in the soil. It is useful to measure crop water requirements to know irrigation threshold for the crop.

To determine these values, soil samples from the plot were tested in laboratory. Water content was determined at potentials: 10, 33, 50, 75, 100, and 150 kPa. With these values and for 0–40 cm depth (root activity zone of blueberries), a curve was constructed and regression coefficients were found to obtain water content at different water potentials (Fig. 6.2).

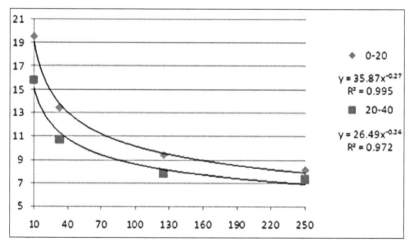

FIGURE 6.2   Soil moisture retention curve for two layers soils (Y-axis: % moisture and X-axis: tension in kPa): 0–20 cm and 20–40 cm.

Blueberry crop is very sensitive to water stress because of superficial root system.[39] About 80% of the water is extracted from first 20 cm soil depth with significant root activity, even though roots can extend up to 40 cm. Therefore, crop water requirements at 25 cm depth were studied in this research. For analyzing the water potential interval for irrigation scheduling, the range of 10–20 kPa was considered[15] because water potential greater than 20 kPa implies decrease in yield.[30] Based on these results, a strict range of water potential must be followed to obtain high yields. For this range and the water content of sandy soil, only a small quantity of water should be applied in each irrigation to maintain the expected water potential, avoiding deep percolation and also water stress. In this study, plants were spaced at 0.85 m with row spacing of 3.5 m. The wetted area per plant was observed to be 0.68 m²/plant. Considering 25 cm depth and a range of 10–20 kPa, the soil can hold only 1.92 L/plant.

6.4 WATER REQUIREMENTS FOR BLUEBERRY CROP IN CONCORDIA

The estimation of water requirements is a key factor to design and operate irrigation system. Evapotranspiration demand of crop was estimated using the Penman–Monteith formula, crop coefficient (Kc), and crop coverage factor:[23]

6.4.1 THE PENMAN–MONTEITH METHOD

The Penman–Monteith includes all parameters that govern energy exchange and corresponding latent heat flux (evapotranspiration, ET) from uniform expanses of vegetation. Most of the parameters are measured or can be readily calculated from weather data. The equation can be utilized for the direct calculation of any crop evapotranspiration as the surface and aerodynamic resistances are crop specific.

$$\lambda_v E = \frac{\overset{\text{Energy flux rate}}{\Delta(R_n - G) + \rho_a c_p (\delta e) g_a}}{\Delta + \gamma(1 + g_a / g_s)} \quad \text{or}$$

$$ET_o = \frac{\overset{\text{Volume flux rate}}{\Delta(R_n - G) + \rho_a c_p (\delta e) g_a}}{(\Delta + \gamma(1 + g_a / g_s))L_v}, \tag{6.1}$$

where λ_v = latent heat of vaporization (J g⁻¹), L_v = volumetric latent heat of vaporization (L_v = 2453 MJ m⁻³), E = mass water evapotranspiration rate (g s⁻¹ m⁻²), $_o$ = evapotranspiration (mm s⁻¹), Δ = rate of change of saturation specific humidity with air temperature (Pa K⁻¹), R_n = net irradiance (W m⁻²), G = ground heat flux (W m⁻²), c_p = specific heat capacity of air (J kg⁻¹ K⁻¹), ρ_a = dry air density (kg m⁻³), δe = vapor pressure deficit or specific humidity (Pa), g_a = conductivity of air (m s⁻¹), g_s = conductivity of stoma (m s⁻¹), and γ = psychrometric constant ($\gamma \approx 66$ Pa K⁻¹).

Crop evapotranspiration (ET_c)

$$ET_c = ET_o \times K_c \times F_c, \tag{6.2}$$

where ET_0 is the reference evapotranspiration, K_c is the crop coefficient and F_c is the coverage factor depending on the percentage of shaded area of the crop.

ET_0 is calculated using empirical formulas, Class A pan evaporating and lysimeter measurements. In this study, Penman–Monteith formula was applied using CROPWAT 8.0 with data from Table 6.1. The effective rainfall is required to know the effective water provided by rainfall and moisture storage in the root zone depth. One must also consider water percolation and erosion potential. To calculate the effective rainfall, the USDA Soil Conservation Service method was used. This method considers monthly rainfall and monthly crop evaporation crop (Table 6.2).

TABLE 6.2   The Effective Moisture Storage (mm) in the Soil.

Month	2010	2011	2012	2013	2014	2015	Average
January	31	101	15	31	85	96	58
February	64	100	100	82	100	29	80
March	27	50	60	50	60	14	44
April	32	38	0	26	48	8	29
May	6	20	16	20	20	20	17
June	16	12	0	0	20	20	11
July	12	20	8	14	18	0	15
August	24	18	36	13	6	40	22
September	60	45	0	14	21	40	30
October	0	22	0	29	58	32	24
November	120	9	0	38	85	96	47
December	62	76	121	0	140	48	73
Average	455	511	356	317	660	442	451

The correct determination of the crop coefficient (K_c) is basic to irrigation requirements and to manage irrigation systems scheduling.[34] Several researchers have indicated K_c value of blueberry crop between 0.2 and 0.97 for a crop of 1–3 years old,[35] while others indicate K_c of 0.2–1.1 for the same crop.[5] One must also consider density of the plantation.[6] In this study, plants per hectare were 3465. Research studies have shown high crop yield at 100% of water demand, while applying 150% of the requirements does not give high yield.[3,4]

6.5 DESIGN AND EVALUATION OF DRIP IRRIGATION SYSTEM

English et al.[12] mentioned that irrigated agriculture will need to provide two-thirds of the increase in food to feed the growing population. The drip irrigation system was designed considering two laterals of drippers at a dripper spacing of 30 cm and drippers with 1.34 L/h/emitter. Mini-sprinkler irrigation system is commonly used as antifrost system in this area. Partial wetting system with 15 m³/h or solid set system with 33 m³/h is also being used. Dug well at the site can produce up to 300 m³/h for antifrost irrigation purpose. The pH of water is 6.5 with electrical conductivity of 0.14 dS/m. The design and operation of the system are relevant in to improve water management and the economic profitability of irrigated agriculture.[22] The uniformity coefficient of Christiansen (UCC)[25] is a statistical coefficient to show the dispersion between all values and the average value.

$$\text{UCC (\%)} = \left[1 - \frac{\sum_{i=1}^{n} |x_i - \bar{x}|}{n\bar{x}} \right] \times 100, \tag{6.3}$$

where x_i is the emitter flow (L/h), $\bar{x}$ is the average flow of the evaluated emitters (L/h), and n is the number of emitters.

The UCC of 95.13% was obtained for the study areas. The uniform coefficient of the minor quarter (UCMQ) is given below:

$$\text{UCMQ} = \frac{x_{25}}{x_x}, \tag{6.4}$$

where x_x is the average flow of the evaluated emitters and x_{25} is the average flow emitted by 25% of the emitters of the minimum flow.

The UCMQ for the same emitters was 93.69%.

$$\text{EDT} = \left[1 - \frac{\sum_{i=1}^{n} |x_i - Xr|}{n * Xr} \right] \times 100, \tag{6.5}$$

where EDT is the total distribution efficiency, x_i is the water infiltrated for ith point (mm), Xr is the crop water need (mm), and n is the number of observations according to Holzapfel model.[20, 21]

The total distribution efficiency shows the way in which water distribution is compared with water requirements as well as the soil water holding capacity in the extracted root zone. This analysis is focused on the variations between water required and water infiltrated at each point and also between water required and water stored in the root zone in each point. The EDT was found to be 81.49%, while applying all the crops need in one irrigation per day.

6.6 BLUEBERRY YIELD (VAR. SNOWCHASER): CONCORDIA, ENTRE RÍOS, ARGENTINE

The crop yield is shown in Figure 6.3. The harvest for 2015 has not finished yet, and the expected yield will be around 1 kg/plant.

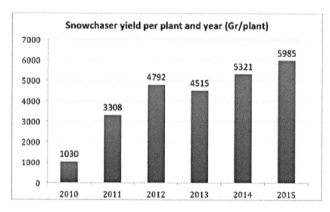

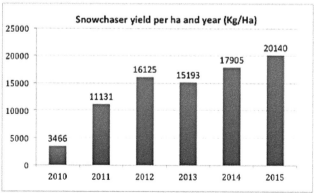

FIGURE 6.3   Blueberry yield: top—g/plant, bottom—kg/ha.

6.7 ESTIMATION OF WATER FOOTPRINT FOR BLUEBERRY (SNOWCHASER VAR.): CONCORDIA, ENTRE RÍOS, ARGENTINE

Blue water is the water applied by the drip irrigation system. Green water is the water received by rainfalls and represents effective storage in the soil volume explored by roots. The effective rain was calculated following the procedure of the Soil Conservation Service of the United States Agriculture Department. In case of gray water, authors considered the water applied by the sprinkler antifrost irrigation system. Considering different components of the water footprint and the yield obtained, water footprint of Snowchaser variety under Concordia conditions is shown in Figure 6.4 during 2010–2015.

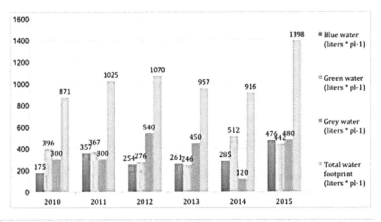

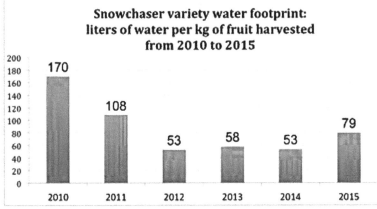

FIGURE 6.4 Crop water use: top—L/plant, bottom—L/kg of fruits.

6.8   SUMMARY

Water footprint in irrigated agriculture is an important issue to be developed and studied, not only by growers and farmers but also by irrigation planners. Irrigation designer must follow strict criteria to provide high efficient systems, in order to allow farmers to have a good tool. Irrigation scheduling must consider water threshold and water holding capacity of the soil, daily water requirements of the crop, soil water potential, and wetting depth. Efficient water management is one of the key elements in successful operation and management of irrigation projects. Water footprint involves aspects related to the basin, well ruled and legal monitoring of rights and obligations of all social actors, and it can give a fair frame to reach better results for the community without compromising the environment.

KEYWORDS

- Argentine
- Argentine blueberry
- drip irrigation
- evapotranspiration
- irrigation design criteria
- irrigation uniformity
- Snowchaser variety
- water footprint
- water potential

REFERENCES

1. Altieri, M.; Nicholls, C. A Rapid, Farmer—Friendly Agro Ecological Method to Estimate Soil Quality and Crop Health in Vineyard Systems. Agroecology and the Search for a Truly Sustainable Agriculture. *PNUMA* **2005,** *277*, 290.
2. Allan, J. Watersheds Explaining the Absence of Armed Conflict over Water in the Middle East. *MERIA Middle East Rev. Int. Aff.* **1998,** *2*(1), 1–3.

3. Bryla, D.; Yorgey, B.; Shireman, A. Irrigation Management Effects on Yield and Fruit Quality of Highbush Blueberry. *Acta Hortic.* **2009**, *810*, 649–656.

4. Bryla, D. Water Requirements of Young Blueberry Plants Irrigated by Sprinklers, Microsprays and Drip. *ISHS Acta Horti.* **2008**, *792*, 135–139.

5. Bryla, D.; Strik B. Effects of Cultivar and Plant Spacing on the Seasonal Water Requirements of High Bush Blueberry. *J. Am. Soc. Hortic. Sci.* **2007**, *132*, 270–277.

6. Bryla, D.; Strik, B. Water Requirements of High Bush Blueberry Planted at Normal and High Density Spacings. *HortScience* **2005**, *40*, 1058–1059.

7. Cañas, J.; Jiménez, M.; Arellano, R.; Moreno-Pérez, M. Improving Water Application Using Irrigation Indicators (Mejora de la gestión del agua de riego mediante el uso de indicadores de riego). *Rev. FCA UNCuyo* **2010**, *42*(1), 107–124.

8. Cifuentes, H.; Merino, F. *Hydraulic Footprints to Determine Water Use and to Manage Water Resources* (Spanish); Book Series 50 (ISSN 0717-8410), 2013; p 210.

9. Chapagain, A.; Hoekstra, A. Water Footprints of Nations. In *Value of Water Research. Report Series No. 16*, UNESCO-IHE. Delft, The Netherlands, 2004; pp 30–34.

10. Deuret, M.; Green, S.; Clothier, B.; Mowat, A. Can Product Water Footprints Indicate the Hydrological Impact of Primary Production? A Case Study of New Zealand Kiwi Fruit. *J. Hydrol.* **2011**, *408*, 246–256.

11. Duarte, O.; Díaz, E.; *Garcia, M. Water Footprint for Rice Cultivation in the Province of Entre Ríos (Spanish)*, XXIV Proceedings on National Congress on Water, Congreso Nacional, San Juan, 2013.

12. English, M. J.; Solomon, K. H.; Hoffman, G. J. A Paradigm Shift in Irrigation Management. *J. Irrig. Drain. Eng.* **2002**, *128*, 267–277.

13. Falkenmark, M. Freshwater as Shared Between Society and Ecosystems: From Divided Approaches to Integrated Challenges. *Philos. Trans. R. Soc. Lond. B. Biol. Sci.* **2003**, *358*, 2037–2049.

14. Garran, S. The Climate in the Concordia Region (Spanish). *XXII Jornadas Forestales de Entre Ríos*, Concordia, Argentine, 2007; p. III 1–III 3.

15. Haman, R.; Pritchard, T.; Smajstrla, A.; Zazueta, F.; Lyrene, P. Evapotranspiration and Crop Coefficients for Young Blueberries in Florida. *Appl. Eng. Agric.* **1997**, *13*(2), 209–216.

16. Heinzenknecht, G. *Impact of "El Niño or La Niña" on Rainfall*; Report by Office of Risks and Agriculture, Ministry of Agriculture, Ganadería y Pesca. http://www.ora.gov.ar/informes/enso_precipitaciones.pdf (accessed March 15, 2005).

17. Herath, I.; Clothier, B.; Green, S.; Horne, D.; Singh, R.; Marsh, A.; Buchanan, A.; Burgess, R. Measuring the Grey-water Footprint of Potato Production. http://www.massey.ac.nz/~flrc/workshops/12/Manuscripts/Herath_2012.pdf (accessed April 20, 2012).

18. Hoekstra, A.; Chapagain, A.; Aldaya, M.; Maite, M.; Mesfin, M. Water Footprint Manual. http://waterfootprint.org/media/downloads/WaterFootprintManual2009.pdf (accessed May 21, 2009).

19. Hoekstra, A.; Hung, Q. Virtual Water Trade: A Quantification of Virtual Water-flows Between Nations in a Relation to International Crop Trade. In *Value of Water Research Report Series 11,* UNEASCO—IHE, Delft, The Netherlands. www.waterfootprint.org/Reports/Report11 (accessed May 31, 2002).

20. Holzapfel, E. A. Selection and Design of Surface Irrigation Methods. Ph.D. Thesis, University of California, Davis, California, USA, 1984, p 221.
21. Holzapfel, E. A. In *Irrigation in Cranberry*, International Congress on *Vaccinium corymbosum*, Universidad de Concepción, Chillan, Chile, 2002; pp 110–112.
22. Holzapfel E.; Pannunzio, A.; Lorite, I. Design and Management of Irrigation Systems, Chapter 8, In *Research Advances in Sustainable Microirrigation Principle and Practices,* Goyal, M., Ed.; Apple Academic Press Inc.: New Jersey, 2014; Vol. 1, pp 147–160.
23. López-Martos, J. *Integrated Approach in Water Management* (Spanish). *Foresta N. 41,* Septiembre, Madrid, 2008; p 16–24.
24. Mekkonen, M.; Hoekstra, A. A Global and High-resolution Assessment of the Green, Blue and Grey Water Footprint of Wheat. *Hydrol. Earth Syst. Sci.* **2010,** *14*(7), 1259–1276.
25. Mekonnen, M.; Hoekstra, A. The Green, Blue and Grey Water Footprint of Crops and Derived Crop Products. *Hydrol. Earth Syst. Sci.* **2011,** *15,* 1577–1600.
26. Merriam, J. L.; J. Keller. *Farm Irrigation System Evaluation. A Guide for Management;* Utah State University, Logan, Utah, USA, 1978; p 18.
27. Lorite, I.; Holzapfel, E.; Pannunzio, A. Irrigation Systems (Spanish). In *Sustainable Water Management in South America;* Fernandez, C. A., Holzapfel, E., Iván del C., Billib, M., Eds.; 2009; p 183, ISBN 978-987-25074-1-1.
28. Pannunzio, A.; Texeira-Soria, P.; Borello, L. In *Impact of Climate Variability on Irrigation Requirements for Two Crops and Runoff Effects in Concordia; Entre Ríos, Argentine.* Poster Presentation at the International Conference of Agricultural Engineering, CIGR AGENG 2012, Valencia, Spain, July 8–12, 2012; p 210.
29. Pannunzio, A.; Vilella, F.; Texeira-Soria, P.; Premuzik, Z. Impact of Irrigation Systems in Cranberry. *AGRIAMBI—Revista Brasileira de Engenharia Agrícola e Ambiental,* **2011,** *15*(1), 20–25.
30. Pannunzio, A.; Texeira-Soria, P. In *Fertigation Trial in Four Blueberries Varieties in Northern Argentine.* Paper presentation at 28th International Horticultural Congress, Lisboa, Portugal, August 22–27, 2010.
31. Pannunzio, A. *Sustainability Study on Irrigation Systems in Cranberry. Editorial Orientación*; Buenos Aires, p 86, ISBN 978-987-9260-75-3.
32. Pengue, W. Excessive Exploitation of Natural Resources and Agricultural Export Market: Towards Determination of Ecological Imbalance in Argentine. Ph.D. Thesis, Universidad de Córdoba, España, Córdoba, 2006; p 132.
33. Pérez, A. *International Business and Environment*; Colección libros de investigación, Programa Editorial Universidad del Valle, Colombia, 2008; p 153.
34. Pritchard, R.; Haman, D.; Smajstrla, A.; Lyrene, P. In *Water Use and Irrigation Scheduling of Young Blueberries*, Proceedings of the Florida State Horticultural Society, Florida, USA, 1993; Vol. 106, pp 147–150.
35. Riveros, C. Response of Cranberry (*Vaccinium corymbosum* L. cv Bluetta) to Irrigation Rates Under Drip Irrigation and Microjets (Spanish). M.Sc. Thesis, Chillan: Facultad de Agronomía, Universidad de Concepción, 2006; p 99.
36. Ridoutt, G.; Juliano, P.; Sanguansri, P. The Water Footprint of Food Waste: Case Study of Fresh Mango in Australia. *J. Clean. Prod.* **2010,** *18*, 1714–1721.

37. Rodríguez Casado, R.; Garrido Colmenero, A.; Llamas Madurga, M.; Varela Ortega, C. *Hydrological Foot Print of Spanish Agriculture*; Reportes de Agua Virtual, Fundación Marcelino Botín, 2008; p 132.
38. Starr, G.; Seymour, R.; Olday, F.; Yarborough, D. Determination of Evapotranspiration and Drainage in Low Bush Blueberries (*Vaccinium angustifolium*) Using Weighing Lysimeter. *Small Fruits Rev.* **2004,** *33*, 273–283.
39. Undurraga, P.; Vargas, S. *Manual de Cranberry;* Instituto de Investigaciones Agropecuarias Centro Regional de Investigación Quilamapu, Chillán, Chile, 2013; p 120.
40. Wright, E.; Penzo, G.; Pérez. B.; Sánchez, J.; Carboni, G., Cruzáte, G. Factors that Favor Cranberry Production in Concordia, Entre Ríos. *XXXVI Congress ASAHO—* Tucumán, 2013; p 145.

PART III
Fertigation Technology

CHAPTER 7

EFFICIENT NUTRIENT MANAGEMENT THROUGH FERTIGATION

MAHESH RAJENDRAN[1,*] and ARCHANA IRENE[2]

[1]Central Sericultural Research & Training Institute, Central Silk Board, Ministry of Textiles, Govt. of India, Berhampore 742101, Murshidabad, West Bengal, India

[2]Department of Agronomy, Agricultural College and Research Institute, Tamil Nadu Agricultural University (TNAU), Coimbatore 625104, Tamil Nadu, India

*Corresponding author. E-mail: maheagri@gmail.com

CONTENTS

ABSTRACT

Most of farmers in India apply fertilizer through broadcasting method. This practice leads to heavy loss of nutrients through volatilization and denitrification of N and fixation of P and K in the soil, which ultimately reduces the nutrient use efficiency (NUE). This type of fertilizer application adversely affects soil health and production capacity. Therefore, efficient use of fertilizer and water is critical to sustained agricultural production. High fertilizer use efficiency (FUE) is possible through fertigation, where plant nutrients are placed around the plant roots as per crop nutrient demand at different growth stages. This chapter explains the importance of fertigation for efficient management of nutrients.

7.1   INTRODUCTION

Agricultural development has provided much evidence that adequate supply of nutrients through fertilizer application is the most efficient practice for increasing crop yield and quality and that sustained yield growth is almost impossible without supply of fertilizer nutrients. At the global scale, crop yields have increased by at least 30–50% as a result of fertilization.[20] Due to intensive cultivation of crops year after years in the same field, the soil fertility was depleted in most of the Indian soils. Therefore, for essential supply, it requires quantity of nutrients (N, P, and K) through fertilization to obtain optimum yield on sustainable basis. In India, most of the farmers apply more fertilizer as splits through broadcasting on surface soil. This practice leads to heavy loss of nutrients through volatilization and denitrification of N and fixation of P and K in the soil, which ultimately reduces the nutrient use efficiency (NUE).[3] This imbalanced and inappropriate fertilizer application adversely affects soil health and limits the capacity of the soil to produce optimally and sustainably at spatial and temporal scales. Therefore, efficient use of fertilizer and water is critical to sustained agricultural production. High fertilizer use efficiency (FUE) is possible through fertigation, where plant nutrients are placed around the plant roots as per crop nutrient demand at different growth stages.

The research study in this chapter discusses the effects of efficient nutrient management on growth, yield, and NUE of different crops.

7.2 FERTIGATION TECHNOLOGY

Fertigation is the application of plant nutrients through drip irrigation system. Fertigation enables adequate supplies of water and nutrients with precise timing and uniform distribution to meet the crop nutrient demand.[2] Further, fertigation ensures substantial saving in fertilizer usage and reduces leaching losses. In fertigation, both water and fertilizers are delivered to crop simultaneously through an irrigation system. Fertigation controls precisely the time and rate of both water and fertilizer application to meet the requirements of a crop at each physiological growth stage.[14] This improves water and NUE, and minimizes leaching and volatilization losses as well as groundwater contamination. Various scientists indicated that drip irrigation is used to apply water-soluble fertilizer (WSF) or normal fertilizer (NF) in precise amounts, as and when required to match the plant needs[1,2,4,13] directly to the root zone of the crop. In fertigation, all the three major nutrients (nitrogen, phosphorus, and potassium) are applied in one solution thus, helping in increased uptake and efficient use by plants. Fertigation is a viable technology mainly for commercial and high-value horticultural crops, such as fruits, vegetables, flowers, sugarcane, cotton, and mulberry.

7.2.1 WHY FERTIGATION?

In the surface flood irrigation method, alternate wetting and drying will reduce the NUE. Compared to drip irrigation (where water application is restricted around the root zone only), the proper air water ratio is maintained the loss of nutrients will be minimum and FUE will be higher.[11,17] According to Satisha,[17] nitrogen use efficiency was 95% in drip fertigation compared to 65% in fertigation + banded method and 30% in banded method, phosphorous use efficiency was 45% in drip fertigation compared to 30% in fertigation + banded method and 20% in banded method, and potassium use efficiency was 80% in drip fertigation compared to 60% in fertigation + banded method and 50% in banded method.

7.2.2 ADVANTAGES OF DRIP FERTIGATION

- Regular and uniform supply of both water and nutrients.

- Nutrient applied to meet specific crop demands according to the crop stages.
- Improved nutrient availability and their uptake by their roots.
- Reduction in fertilizer application resulting in increased FUE.
- Higher crop yield by 25–100% and better quality.
- Saving in fertilizers by 25–30%.
- Nutrient losses through leaching reduced.
- Small quantities of fertilizers can be applied at close intervals.
- Micronutrients can also be effectively applied by fertigation.
- Groundwater pollution reduced.
- Offers simpler and more convenient application thus, saving time, labor, and energy

7.2.3 FERTILIZER SUITABLE FOR FERTIGATION

WSFs are high-analysis fertilizers (more than 25% nutrients), which are totally soluble in water. WSFs are available in double and multinutrient combinations with or without secondary elements or micronutrients. WSFs are manufactured mainly for administering through drip irrigation.[7] These imported WSFs have high solubility and nutrient content and could supply all major nutrients along with micronutrients (Table 7.1). Apart from the solubility, the other major factors to be considered for selection of the soluble fertilizers are compatibility and corrosiveness.

TABLE 7.1   The Solubility of Some Common Fertilizers for Fertigation.

Name	Chemical formula	N:P_2O_5:K_2O	Solubility (g/L) at 20°C
Ammonium nitrate	NH_4NO_3	34:0:0	1185
Ammonium sulfate	$(NH_4)_2SO_4$	21:0:0	700
Di-ammonium phosphate	$(NH4)_2HP_2O_5$	18:46:0	413
Monoammonium phosphate	$NH_4H_2PO_4$	12:61:0	225
Potassium chloride	KCL	0:0:60	277
Potassium nitrate	KNO_3	13:0:45	135
Potassium sulfate	K_2SO_4	0:0:50	67
Urea	$CO(NH_2)_2$	46:0:0	1190

Adopted from Goyal, M. R. *Management of Drip/Trickle or Micro Irrigation* (Chapter 8); Apple Academic Press Inc.: Oakville, ON, 2014; p 172.

7.2.4 SPECIAL WATER-SOLUBLE FERTILIZER

Water-soluble specialty fertilizers specifically meant for fertigation are available as polyfeed (19:19:19, 20:20:20, 11:42:11, 16:8:24, 15:15:30), UP (17:44:00), MAP (12:61:00), Multi-K (13:00:46), MKP (00:52:34), and SOP (00:00:50) at present and most of them are imported in India and marketed by irrigation systems and fertilizer dealers.

7.2.5 FERTIGATION DEVICES

Fertilizers can be injected into drip irrigation system by selecting appropriate equipment from a wide assortment of available pumps, valves, tanks, venturi, and aspirators. There are two types of fertigation, namely, quantitative fertigation and proportional fertigation. The advantages of proportional fertigation are precision and accurate injection of nutrients and are not affected by water pressure changes and it can be easily automated.

7.2.6 REQUIREMENT FOR FERTIGATION

- Test the native soil fertility status.
- Fix the correct fertilizer dose for selected crop.
- Develop appropriate fertigation schedule.
- Select suitable fertilizer grade according to crop stage.
- Install appropriate fertigation device along the main line.

7.2.7 STEPS IN FERTIGATION

- Calculate the required fertilizer quantity for the actual cropped area and prepare the nutrient stock solution (dissolve the solid fertilizer with water at 1:5 ratio).
- Operate the drip system for 10–20 min for wetting (first process).
- Regulate the valves and initiate fertigation (second process) at 95 or 186 L/h injection rate based on the selection of a fertigation device.
- Complete the fertigation and finally flush (third process) for 10–15 min for removal of left out fertilizers in pipe network.
  CAUTION: Fertigate during middle of irrigation.

7.3   EFFICIENT NUTRIENT MANAGEMENT THROUGH DRIP FERTIGATION: CROPS

7.3.1   SUGARCANE

Sugarcane is the world's primary sugar crop. In India, sugarcane is cultivated over an area of 5.31 million hectare with annual production of 361 million tons of cane with an average productivity of 66.36 t/ha. Currently, India consumes about 18.5 million tons of sugar but to meet the future demands of an increasing population, 28 million tons of sugar is to be produced by 2020. Hence, there will be more and more stress on the sugarcane ecosystem to meet this growing demand in future. In recent years, sugarcane yield is declining mainly due to inappropriate water and nutrient management practices. Therefore, fertigation is essential in order to make efficient nutrient management for higher cane productivity and NUE in sugarcane. Soil type, water quality, and type of fertilizer govern the adoption of fertigation for sugarcane. Sugarcane removes substantial amount of plant nutrients because it is a long duration and high-biomass-accumulating crop. On an average, a crop of 100 t/ha cane yield might use 205 kg N, 55 kg P, 275 kg K, and 30 kg S.

Table 7.2 gives a view of the efficiency of fertilizers applied under different sources of fertilizer and fertilizer levels. In an experiment studying the effects of different fertilizer sources and fertilizer levels under subsurface drip fertigation (SSDF) and surface drip fertigation (SDF) on yield and NUE of sugarcane var. Co-86032 in Tamil Nadu, application of 100% recommended dose of fertilizer (RDF) through SSDF with WSF has resulted significant cane yield compared to control with soil application of 100% RDF with NF. It was observed from this investigation that 97% higher cane yield was registered under SSDF at 100% RDF with WSF compared to control. Further research clearly indicated that NUE was higher in SSDF with WSF compared to soil application of NF. Experimental results from Sugarcane Breeding Institute, Coimbatore also indicate 25% saving in nitrogen and potash fertilizer when supplied through drip compared to the conventional application of the RDFs.[5,11]

TABLE 7.2 Effect of WSF and NF on Cane Yield and NUE of Sugarcane Under Fertigation.

Treatments		Cane yield (t/ha)	Increased cane yield over control (%)	NUE (kg/ kg of NPK applied)	Increased NUE over control (%)
Subsurface drip fertigation					
T_1	75% RDF (WSF)	167.63	70.39	372.50	127.13
T_2	100% RDF (WSF)	193.94	97.13	323.20	97.07
T_3	75% RDF (WSF)	172.22	75.06	382.70	133.35
T_4	100% RDF (NF)	148.67	51.12	247.80	51.10
T_5	75% RDF (NF)	132.58	34.76	294.60	79.63
T_6	75% (WSF) + 25% (NF)	184.41	87.45	307.40	87.44
T_7	50% (WSF) + 50% (NF)	163.48	66.17	272.50	66.16
T_8	25% (WSF) + 75% (NF)	156.52	59.10	260.9	59.09
Surface drip fertigation					
T_9	100% RDF (WSF)	175.14	78.02	291.90	77.99
T_{10}	100% RDF (NF)	139.53	41.83	232.50	41.77
Control (surface irrigation + soil application of fertilizer)					
T_{11}	100% RDF (NF)	98.38	–	164.00	–
SED		8.08	–	–	–
CD ($P = 0.05$)		16.86	–	–	–

T_1 alone—urea phosphate; all 19, SOP, and urea were used as WSF

WSF: all 19, MAP, and potassium nitrate.

WSF, water-soluble fertilizer; NF, normal fertilizer.

Adapted from Mahesh, R. Effect of Water-Soluble Fertilizers and Normal Fertilizers at Different Levels on Chlorophyll Content, Leaf Area Index and Cane Yield of Sugarcane Under Subsurface Drip Fertigation. Unpublished Ph.D. Thesis, For Department of Agronomy at Tamil Nadu Agricultural University, Coimbatore, TN; 2015.

7.3.2 BANANA

Banana production needs a shift from the present peasantry farming system to large-scale corporate cultivation to meet worldwide consumer demands.[10] This cannot be achieved with current cultural operations, namely annual replanting, manual fertilizers application, irrigation and weeding etc., which are highly labor intensive and constitute a major portion of the input costs. Adoption of a new system for an easy, efficient, and cost-effective cultivation of banana with considerably minimum labor involvement for increasing productivity at lesser cost is essential. One such improved technology is fertigation. In this context, fertigation experiment conducted with cv. Robusta (AAA) revealed that fertigation treatment with 50–75% of recommended NPK (200:30:300 g NPK/plant) registered the maximum bunch weight, more number of hands and fingers both under normal and high-density planting (HDP) system (Table 7.3).[4]

TABLE 7.3   Influence of Fertigation on Performance of Banana cv. Robusta (AAA).[4]

Treatments	Bunch weight (kg)	Yield (t/ha)	% increase over conventional	No. of hands	No. of fingers	TSS (%)
Normal planting system (single plant/pit)						
Plant crop 25 LPD + 100:30:150 g NPK	38.00	95.00	61.07	9.34	163.94	19.29
Ratoon crop 25 LPD + 150:30:225 g NPK	44.42	111.05	61.07	13.47	261.27	20.10
High-density planting system (three plants/pit)						
50 LPD + 450:90: 675 g NPK	34.99	174.88	196.51	10.22	173.38	21.20
Conventional (single plant/pit)						
50 LPD + 450:90:675 g NPK	23.59	58.98	–	8.12	118.01	22.13

Adapted from Kavino, M. Standardization of Fertigation Technique for the Ratoon Crop of Banana cv. Robusta (AAA) Under High Density Planting System. Unpublished M.Sc. Thesis, Tamil Nadu Agricultural University, Coimbatore, 2001.

The effect of fertigation on the growth, yield quality, and physiological parameters of banana cv. red banana (AAA) revealed that planting of one plant/pit along with 100% of RDF (110:35:300 g NPK) through fertigation resulted in higher bunch weight (22.55 kg), number of fingers (98.92), and finger weight (255.36 g).[22] Conventional fertilizers are equally effective as that of WSFs for fertigation in banana besides reducing the cost.[8]

7.3.3 MANGO

India is the largest producer of mango, grown on about 2.3 million hectare with an annual production of 15.02 million tons. However; its productivity is only 6.5 t/ha as against 30 t/ha in Israel. Low yield may be mainly attributed to wider trees spacing (10 × 10 m or even more). However, to enhance the productivity, HDP/ultra-high-density planting (UHDP) coupled with drip fertigation system has been recently recommended for mango in India particularly for commercial choice cultivars.[19] Recently, drip fertigation has been standardized for cv. Alphonso under UHDP.[9] Results indicated that 100% RDF (i.e.,120:75:100 g NPK/tree/year) has registered significantly higher number of panicles unit area, fruit weight, number of fruits/tree, and fruit yield/tree (Table 7.4) especially in the second year and this was on par with 125% of RDF through fertigation.

TABLE 7.4   Effect of Drip-Fertigation Levels on Yield and Yield Components in Mango.[19]

Fertigation levels	Number of panicles/m² canopy area		Mean fruit weight (g)		Number of fruits/tree		Fruit yield (kg/tree)	
	2009–10	2010–11	2009–10	2010–11	2009–10	2010–11	2009–10	2010–11
50% RDF	6.12	10.77	241.30	230.78	28.73	27.66	6.96	6.91
75% RDF	6.35	11.77	251.70	235.58	30.82	30.48	7.8	7.69
100% RDF	6.55	14.41	264.80	262.63	34.83	38.50	9.30	10.69
125% RDF	6.92	14.92	260.80	265.39	33.53	39.92	8.77	11.14
SED	0.460	0.427	1.545	1.84	0.287	1.55	0.092	0.34
CD (0.05)	NS	0.898	3.247	3.86	0.603	3.25	0.192	0.72

Adapted from Sivakumar, V. Studies on Influence of Drip Irrigation Regimes and Fertigation Levels on Mango (*Mangifera indica* L.) cv. Ratna Under High Density Planting. Unpublished Ph.D. (Hort.) Thesis, Submitted to Tamil Nadu Agricultural University, 2007.

7.3.4 DRIP FERTIGATION UNDER TAMIL NADU PRECISION FARMING PROJECT (TNPFP) VERSUS CROP YIELD

TNPFP envisages key technologies, namely remote sensing, Chisel plough, hi-tech community nursery, and drip fertigation system. Tremendous increase in yield was obtained in many horticultural crops as indicated in Table 7.5.

TABLE 7.5   Effects of Fertigation on Yield of Various Crops Under TNPFP.

Crop	Yield (t/ha)		Remarks % increase
	National average	**Precision protocol**	
Banana	28.58	110	284.88
Beetroot	16.75	35	108.95
Bitter gourd	6.23	15	140.77
Bottle gourd	12.21	66	440.54
Cabbage	14.38	120	734.49
Cauliflower	14.22	33	132.06
Chili	12.02	35	191.18
Chrysanthemum	8–15	25	117.39
Cotton	15–20 Q	30 Q	111.43
Cucumber	6.48	20	208.64
Eggplant	10.46	150	1334.03
French beans	5.8	12	106.89
Marigold	10	25	150.00
Muskmelon	21.95	40	82.23
Okra	6.28	16	154.78
Onion	11.32	21	85.51
Pumpkin	11.91	50	319.81
Ribbed gourd	15.85	34	114.51
Rose	10 lakh stems	25 lakh stems	150.00
Salidago	12,000 bunches	25,000 bunches	108.33
Sugarcane	80–100	250	177.77
Tapioca	25.52	52	103.76
Tomato	17.35	150	764.55
Turmeric	4.95	9	81.81
Watermelon	12.71	60	372.06

Q = quintal = 100 kg; lakh = 100,000.

Adapted from unpublished data, WTC, Annual Reports by Water Technology Center at Tamil Nadu Agricultural University, Coimbatore, 1985–2015.

7.3.5   DRIP FERTIGATION AND FARM PROFITABILITY: A CASE STUDY IN ANDHRA PRADESH

Rain-fed districts (Kurnool, Cuddapah, Chittoor, Anantapur, etc.)—where the precipitation is very low and water source is very minimal and farmers are purely dependent on rain-fed crops, such as groundnut, sunflower, and millets—were giving very nominal income. In these regions, by adoption of drip irrigation technology, farmers have started using WSFs for horticulture crops like banana, papaya, and melons through fertigation technique thereby enhancing their income levels. The incremental cost benefit ratios (ICBR) (Table 7.6) were worked out to be highly encouraging and were 1:6.5, 1:16, 1:8.4, and 1:5.2 for banana, muskmelon, pomegranate, and sweet orange, respectively.

TABLE 7.6   Impact of Fertigation with Water-soluble Fertilizers on Fruit Crops with ICBR.[21]

Crop	Farmers expenditure (Rs./ha)		Crop yield (t/ha)		Revenue (Rs./ha)	ICBR
	Without fertigation	With fertigation	Without fertigation	With fertigation		
Banana	134,800	151,838	45	75	111,000	1:6.5
Muskmelon	93,750	101,250	25	35	120,000	1:16
Pomegranate	191,250	215,000	15	25	200,000	1:8.4
Sweet orange	60,000	81,250	15	25	110,000	1:5.2

US$ 1.00 = Rs. 60.
Source: Subrahmanyam, S. V. S. In *Global Scenario of 100% Water-Soluble Fertilizers*, Proceedings of CAFTA, Department of Agronomy, TNAU, Coimbatore, 2013. Reprinted with permission.
http://www.brijj.com/dr-svs-subrahmanyam

7.3.6   TOMATO

A comparative study conducted by Hebber[6] in tomato hybrid Arka Abhijit on furrow and drip irrigation revealed that drip irrigation with soil application of fertilizer registered higher yield (71.92 t/ha) of 19.9% compared with furrow irrigation with soil application of fertilizers (59.50 t/ha). This yield increase can be attributed to higher number of flowers and fruits per plant in drip irrigation over furrow irrigation. This might be due to the maintenance of favorable soil moisture in the root zone which in turn helped the plants to utilize moisture as well as nutrients more efficiently from the limited wetted area.

Drip fertigation treatments with WSF resulted in higher fruit yield (79.27 t/ha) over drip irrigation with soil application of fertilizers, which accounted for 10.20% increase in yield. Fertigation with normal fertilizers (73.27 t/ha) gave significantly lower yield compared to fertigation with WSF (79.27 t/ha). WSF fertigation had higher concentration of available plant nutrients in top layers over normal fertilizers thus, increasing the marketable fruit yield of tomato. In fertigation, use of 100% WSF is recognized to safeguard the drip system in a long run. Ascorbic acid content was significantly higher in WSF fertigation (19.33 mg/100 g) compared to furrow irrigation (16.00 mg/100 g) and drip fertigation (16.67 mg/100 g). Researchers concluded that water-soluble K increased the ascorbic acid concentration in tomato fruits and similar trend was also observed for the titrable acidity.

Singh[18] conducted a study to find out minimum water use, optimum spacing, and compact use of nutrients for tomato and capsicum for maximum yield through drip irrigation under polyhouse in north Indian plains.[16] A pressurized drip system was installed on 50,000 L capacity water tank with filtration system with dripper and water discharge capacity of 2 L/h. The pressurized drip irrigation had significantly increased the yield 11.38 and 12.50 kg/m^2 and net income 65.40 and 67.78 Rs./m^2 in tomato and capsicum as compared to flood irrigation in all the years. The crop yield was improved by 65.40% in tomato and 67.70% in capsicum in drip irrigation over flood irrigation. Fertigation resulted in maximum yield of tomato (10.85 kg/m^2) and capsicum (12.15 kg/m^2), net income, minimal disease, weed incidence and saved water (tomato—43.23%; capsicum—42%), and total irrigation time (71.43%) as compared to top-dressing method in both tomato and capsicum crops under polyhouse condition. Fertigation of nitrogen recorded 25.14% and 26.82% higher yield in tomato and capsicum, respectively, as compared to top-dressing method.

7.3.7   CHILI

Field experiments were conducted during 2001–2004 at Agricultural College and Research Institute, Madurai on drip irrigation and fertigation in chilies to evaluate the comparative merits of drip fertigation over surface irrigation. Surface irrigation 0.90% IW/CPE ratio + entire NPK as soil application was compared with drip irrigation at 100%, 75%, and 50% of PE each in combination with 75%, 100%, and 125% of recommended N and K (120:60:30 kg NPK/ha) through fertigation in six splits at 15 days interval from the establishment of seedlings. The results revealed that drip irrigation at 100% PE + 100% N and K through fertigation recorded 2.3 t/ha as dry pod yield (3 years pooled data). The treatment recorded 67% increased yield over surface irrigation at 0.90 IW/CPE ratio + entire NPK as soil application. Irrespective of the fertilizer dose, there was a marked increase in pod yield under fertigation. Fertigation permits a slow and controlled rate of fertilizer application at the root zone of the crop, when the crop requires nutrients. In turn, fertigation activates mechanisms, such as enhanced mobility, availability and uptake of available nutrients at high soil moisture content, prevention of leaching, volatilization, and denitrification due to controlled rate of application in more splits and matching of application rate and time with crop demand through increased splits. The investigators recommended drip irrigation at 50% PE along with fertigation of RDF (entire 30 kg K/ha as basal and 120:30 kg N and K/ha) in six splits at 15 days interval to maximize the yield and conserve moisture in chilies.[12]

7.3.8   CAPSICUM ANNUM

An experiment with polyhouse-grown early capsicum (*Capsicum annum* var. grossum) was conducted with an objective to find out economic dose of N and K through fertigation. Results revealed that drip irrigation at 100% evaporation replenishment along with 100% recommended N and K through fertigation recorded significantly highest growth attributes and yield attributes. Pooled data over 3 years (2005–2008) revealed that fertigation with 100% recommended N and K (120:60 kg/ha) recorded 61.09% increased yield over conventional fertilization. Cost economics study revealed that 100% recommended N and K through fertigation recorded the highest cost–benefit ratio of 1:1.72. It was inferred that early season crop grown—inside the naturally ventilated polyhouse irrigation scheduling at 100% evaporation replenishment through drip irrigation coupled with 100% RDF (N and K 120:60 kg/ha) as fertigation—improved the growth, yield, and quality of the crop with the highest cost–benefit ratio.

7.3.9 OKRA (BHENDI)

Field study on sandy loam soil at Hyderabad conditions was conducted to investigate the effects of drip irrigation and resource use efficiency in okra var. Mahyco Hybrid-10 during summer season. Results revealed that yield and water use efficiency differed significantly. Consistently high yields (4.19 and 4.15 t/ha in consecutive years) and water use efficiency (8.23 and 8.10 kg/ha/mm) were observed when the crop was drip irrigated at 1.0 Epan and fertigated with 120 kg N/ha. Furrow-irrigated crop showed 54% and 57% lower yield than drip irrigated at 1.0 Epan and fertigated with 120 kg N/ha in consecutive years. Increase in water use efficiency under drip system was mainly due to controlled water release near the crop zone.[15]

7.4 SUMMARY

Precise and efficient use of water and nutrients are prime concerns of sustainable crop production. Fertigation is a sophisticated and efficient method of applying fertilizers, in which the irrigation system is used as the carrier and distributor of crop nutrients. Correct design of micro irrigation is an essential prerequisite for efficient distribution of nutrients thus, avoiding deficiency in some pockets and excess in other areas. Various types of fertilizers meant for open-field and greenhouse conditions may be chosen for appropriate use for achieving maximum FUE. Fertigation scheduling should be developed before stating fertigation, considering essential criteria, such as native soil fertility status, targeted yield, variety or hybrid, growing conditions, nutrient uptake pattern, actual soil, and plant nutrient concentrations. Fertigation with WSFs is a costly process due to high cost of WSF. Hence, fertigation with imported grades of WSF has to be targeted in high-value crops for getting greater net return and to have shorter payback periods. The synergism and combination of water and nutrient lead to an efficient use by the plant. On the basis of studies conducted on different field and horticultural crops, it was found that adoption of this technology improves the yield and NUE of crops. It is also highly beneficial to farming community to reduce the cost of production. Further, it helps in maintaining the soil health for better productivity and reducing environmental pollution.

KEYWORDS

- **broadcasting**
- **NUE**
- **drip irrigation**
- **efficient use**
- **fertigation**
- **nutrient uptake**
- **FUE**

REFERENCES

1. Bafna, A. M.; Dafter, S. Y.; Khade, K. K.; Patel, P. V.; Dhotre, R. S. Utilization of Nitrogen and Water by Tomato Under Drip Irrigation System. *J. Water Manag.* **1993,** *1*(1), 1–5.
2. Bar-Yosef, B.; Sagiv, B. Response of Tomato to N and Water Applied via Trickle Irrigation. *Water Agron. J.* **1982,** *74*, 633–637.
3. Cassel, S. F.; Sharmasakar, S.; Miller, S. D.; Vance, G. F.; Zhang. R. Assessment of Drip and Flood Irrigation on Water and Fertilizer Use Efficiencies for Sugar Beets. *Agric. Water Manag.* **2001,** *46*, 241–251.
4. Clothier, B. E.; Saucer, Y. Nitrogen Transport During Drip Fertigation with Urea. *Soil Sci. Soc. Am. J.* **1988,** *52*, 345–349.
5. Esther-Shekinath, D.; Gupta, C.; Sundara, B.; Rakkiyappan, P. Effect of Drip Irrigation, Planting Methods and Fertigation on Yield, Quality and Water Use Efficiency of Sugarcane (*Saccharum* species hybrid). *Int. J. Agric. Stat. Sci.* **2013,** *8*(2), 691–696.
6. Hagin, J.; Sneh, M.; Lowengart-Aycicegi, A. *Fertigation: Fertilization Through Irrigation.* IPI Research Topics No. 23, International Potash Institute (IPI), Horgen, Switzerland, 2002; p 81.
7. Hebbar, S. S.; Ramachandrappa, B. K.; Nanjappa, H. V.; Prabhakar. M. Studies on NPK Drip Fertigation in Field Grown Tomato (*Lycopersicon esculentum* Mill.). *Eur. J. Agron.* **2004,** *21*, 117–127.
8. Kavino, M. Standardization of Fertigation Technique for the Ratoon Crop of Banana cv. Robusta (AAA) Under High Density Planting System. M.Sc. Thesis, Tamil Nadu Agricultural University, Coimbatore, 2001.
9. Kumar, N. In *Drip-fertigation to Increase the Productivity of Mango cv. Alphonso Raised Under Ultra-high Density Planting (UHDP) System*, XVI IPNC, Turkey, 2013.
10. Mahalakshmi, M. Water and Fertigation Management Studies in Banana cv. Robusta (AAA) Under Normal Planting and High Density Planting System. Ph.D. Thesis, Tamil Nadu Agricultural University, Coimbatore, 2000.

11. Mahesh, R.; Asoka-Raja, N. Effect of Water Soluble Fertilizers and Normal Fertilizers at Different Levels on Chlorophyll Content, Leaf Area Index and Cane Yield of Sugarcane Under Subsurface Drip Fertigation. *Trends Biosci.* **2015,** *8*(4), 1091–1094.

12. Muralikrishnasamy, K.; Veerabadran, V.; Krishnasamy, S.; Kumar, V.; Sakthivel, S. In *Drip Irrigation and Fertigation in Chilies (Capsicum annum L.)*, 7th International Micro Irrigation Congress Held at PWTC, Kuala Lumpur, Malaysia, 2006.

13. Or, D.; Coelho, F. E. Soil Water Dynamics Under Drip Irrigation: Transient Flow and Uptake Models. *Trans. ASAE* **1996,** *39*(6), 2017–2025.

14. Papodopoulos, K. Constant Feeding of Field Grown Tomatoes Irrigated with Sulphate Water. *Plant Soil* **1985,** *88*, 213–236.

15. Rekha, K. B.; Reddy, M. G.; Mahavishnan, K. Nitrogen and Water Use Efficiency in Bhendi as Influenced by Drip Fertigation. *J. Trop. Agric.* **2005,** *43*(1–2), 43–46.

16. Sanchita, B.; Luchon, S.; Pankaj, B.; Tridip, H.; Bhaskaryoti, D. Studies on Effect of Fertigation with Different Levels of N and K Fertilizers on Growth, Yield and Economics of Early Season Capsicum Under Cover. *Veg. Sci.* **2010,** *37*(2), 160–163.

17. Satisha, G.C. (1997). Fertigation new concept in Indian agriculture. Kisan World, **1997,** *24,* 29-30.

18. Singh, A. K.; Chandra, P.; Srivastava, R. Response of Micro-irrigation and Fertigation on High Value Vegetable Crops Under Control Conditions. *Indian J. Hortic.* **2010,** *67*(3), 418–420.

19. Sivakumar, V. Studies on Influence of Drip Irrigation Regimes and Fertigation Levels on Mango (*Mangifera indica* L.) cv. Ratna Under High Density Planting. Ph.D. (Hort.) Thesis, Submitted to Tamil Nadu Agricultural University, 2007.

20. Stewart, W. M.; Dibb, D. W.; Johnston, A. E.; Smyth. T. J. The Contribution of Commercial Fertilizer Nutrients to Food Production. *Agron. J.* **2005,** *97*, 1–6.

21. Subrahmanyam, S. V. S. In *Global Scenario of 100% Water Soluble Fertilizers*, Proceedings of CAFTA, Department of Agronomy, TNAU, Coimbatore, India, 2013.

22. Suganthi, L. Fertigation Management Studies in Banana cv. Red Banana (AAA) Under Different Planting Densities. M.Sc. Thesis, Tamil Nadu Agricultural University, 2002.

APPENDIX
Glossary of Terms

Broadcasting refers to spreading fertilizers in surface of the soil uniformly all over the field.

Fertigation refers to the application of fertilizers along with irrigation water through drip irrigation system.

Fertilizer use efficiency is defined as quantity of economic produce produced from 1 kg of fertilizer.

Leaching refers to the loss of water-soluble plant nutrients from the soil due to rain and irrigation.

Soil fertility refers to the ability of soil to sustain plant growth, that is, to provide plant habitat and result in sustained and consistent yields of high quality.

Volatilization refers to the excess nutrients that are lost through volatilization (when nitrogen vaporizes in the atmosphere in the form of ammonia), surface runoff and leaching to groundwater.

CHAPTER 8

SUBSURFACE DRIP FERTIGATION TO ENHANCE YIELD AND NET INCOME OF SUGARCANE

MAHESH RAJENDRAN[1,*] and ARCHANA IRENE[2]

[1]Central Sericultural Research & Training Institute,Central Silk Board, Ministry of Textiles, Government of India, Post Berhampore 742101, Murshidabad, West Bengal, India

[2]Department of Agronomy, Agricultural College and Research Institute, Tamil Nadu Agricultural University (TNAU), Coimbatore 625104, Tamil Nadu, India

*Corresponding author. E-mail: maheagri@gmail.com

CONTENTS

In this chapter: US$ 1.00 = Rs. 60.00 (Indian Rupees)

ABSTRACT

The present study clearly indicated that adopting wider lateral spacing at 180 cm with DSP under SSDF was economically viable as evidenced by higher economic net returns and agronomically feasible for mechanization, particularly mechanized cane harvesting and is recommended for sugarcane growers for realizing higher yields and economic returns.

8.1 INTRODUCTION

Sugarcane is commercially an important crop in terms of its contribution to the national economy and livelihood support to millions of farmers. India ranks second after Brazil among sugarcane producing countries of the world and contributes to an extent of 25.0% and 22.2% of area and production of the world, respectively. In India, sugarcane is cultivated over an area of 5.31 million hectares producing around 361.0 million tons of cane with an average productivity of 66.36 t/ha. In India, sugar crop is worth Rs. 550 billion industry, and more than 50 million sugarcane farmers and a large mass of agricultural laborers are involved in sugarcane cultivation, harvesting, and ancillary activities. There is a growing demand for sugar in India, the largest sugar-consuming country in the world.

Sugarcane cultivation is facing a rough path ahead due to depleting water resources, increasing input and labor costs, and lack of alternate and viable innovative technologies to boost the productivity. Sugarcane is a high water-consuming crop and it requires about 200 t of water to produce 1 t of cane and with an average of 20 ML of water/year.[4] The traditional surface method of irrigation widely practiced by the sugarcane farmers directly leads to inefficient use of irrigation water and fertilizers owing to enormous losses in evaporation and distribution.[22] Besides, the cane farmers face new challenges and threats due to the nonavailability of labor for cane cultivation, particularly for harvesting. Sugarcane production is highly labor-intensive, requiring about 250–400 labor man-days per ha for different operations and cane harvesting alone requires 70 labor man-days. Harvesting operation alone accounts for 30% of the cost of production; and added to it, this acute labor scarcity results in escalating the cost of cultivation and thus pulls down the net profitability.[39] Out of the total cost of cultivation, 60% of the cost of cultivation wages is toward labor wages alone. Even though the cane price has increased two folds, the profit margins remain more or less static for the cane farmers.

Hence, to make the cane cultivation a more remunerative one, the introduction of mechanization has become essential and subsurface drip irrigation (SSDI) will offer the best scope for mechanized cultivation, particularly for harvesting besides favoring higher cane productivity with less water use. Conventionally, sugarcane is planted at 90–105 cm spacing and this row spacing is not ideally suitable for mechanized operations. Therefore, planting at different row distance technique was primarily thought to introduce mechanization in sugarcane cultivation. Cultivation of sugarcane at a distance of 120, 150, 165, or 180 cm is often referred as "wider row planting technique." Wider row planting is conceived to facilitate and introduce mechanization in sugarcane to reduce the cost of production in contrast to conventional methods of planting. To achieve this, there is a strong need to plant sugarcane at wider spacing. Wide-row spaced planting helps to provide abundant sunlight for increasing cane yield, and provides adequate space for intercropping and intercultural operations and also the proper adoption of mechanization thereby increasing the per unit profitability.[8,21] Early and short-duration varieties perform well under closer spacing, while late-maturing varieties require wider spacing.[11] Harvesting machines currently available in the international market can operate at wider spaced sugarcane.[18] In order to facilitate mechanical harvesting, sugarcane needs to be grown at a row spacing of 140–180 cm.[7] Mechanical harvesting brought clearly triple benefits to the farmers. It saves labor and cost, and ensures timely harvesting of the crop.

SSDI is the application of water below the soil surface by buried laterals with drip emitters.[17] It has many benefits over conventional drip irrigation.[34] SSDI system can contribute to maximizing water use efficiency (WUE) due to negligible soil evaporation, percolation, runoff,[24] and also improve crop yield and quality.[41] Providing optimum soil moisture conditions throughout the crop-growing period through drip irrigation is therefore of paramount importance to obtain higher yields.[37]

Further, SSDI offers application of water and nutrients at optimum amounts to the most active part of the crop root zone, with timing appropriate for maximum plant response, while minimizing the potential for nutrient leaching. Drip fertigation offers great scope to enhance cane productivity,[22,27,31] saves 40–50% irrigation water, and enhances nutrient efficiency by 40%[36] and low mobility of nutrients into the soil,[35] and also offers the possibility to optimize the water and nutrient distribution over time and space.[19]

Cane yield was increased by 23% in drip cultivation compared to conventional method of irrigation.[20] Hence, the introduction of appropriate technologies that paves way for mechanization particularly for cane harvesting will be a boon to the cane growers besides increasing the profitability by way of reducing the cost of cultivation. Information on the influence of crop geometry on growth, development, and productivity of cane under subsurface drip fertigation (SSDF) is rather scanty. It is necessary to optimize the lateral spacing for mechanized cane harvesting.

This book chapter reveals the effect of SSDF on yield, quality, and economics of sugarcane under different lateral spacing and methods of planting.

8.2   MATERIALS AND METHODS

8.2.1   EXPERIMENTAL SITE

A field experiment was conducted in Agronomy Research Block of the Central Farm at Agricultural College and Research Institute of Tamil Nadu Agricultural University, Madurai during midseason (March–April) of 2008–09. The site is geographically located at 9°54′N latitude and 78°54′E longitude at an elevation of 147 m above mean sea level. The region is semiarid with a mean annual rainfall of 853 mm. The daily mean maximum and minimum temperatures range between 33.6°C and 23.5°C, respectively. The daily mean pan evaporation per day was 4.1 mm with a relative humidity of 73.6% during the cropping period. A mean annual rainfall of 761.3 mm was received with 46 rainy days during the cropping period.

The soil of the experimental site was sandy clay loam in texture and taxonomically called as *Typic Haplustalf*. The soil has a pH of 7.5 and EC of 0.4 dS/m, having 0.4% organic carbon, 220 kg/ha available N (low), 19 kg/ha available P (medium), and 425 kg/ha available K (high). The experimental soil has a bulk density of 1.43 g/cc with a hydraulic conductivity of 4.2 cm/h and a permanent wilting point of 12.3%. The field capacity and infiltration rate are 25.4% and 10.6 mm/h, respectively.

8.2.2 EXPERIMENTAL TREATMENTS

The experiment was laid out in a randomized block design with 11 treatments and 4 replications. The treatments consisted of lateral spacing and methods of planting as follows:

- SSDF with 120 cm lateral spacing as single-side planting (SSP) (T_1)
- SSDF with 120 cm lateral spacing as double-side planting (DSP) (T_2)
- SSDF with 135 cm lateral spacing as SSP (T_3)
- SSDF with 135 cm lateral spacing as DSP (T_4)
- SSDF with 150 cm lateral spacing as SSP (T_5)
- SSDF with 150 cm lateral spacing as DSP (T_6)
- SSDF with 165 cm lateral spacing as SSP (T_7)
- SSDF with 165 cm lateral spacing as DSP (T_8)
- SSDF with 180 cm lateral spacing as SSP (T_9)
- SSDF with 180 cm lateral spacing as DSP (T_{10})
- Surface irrigation with soil application of fertilizers (T_{11}).

8.2.3 METHODS OF PLANTING

The experimental field was plowed with tractor-drawn disc plow followed by two plowings with cultivator and the clods were broken with rotavator. The field was uniformly leveled and the trenches were dug to a depth of 25 cm. The cane setts were planted in the trench with two methods as SSP and DSP under SSDF. In DSP, setts were planted on both sides of the trench having a width of 40 cm. The drip tape lateral was placed at the center of the trench and maintained 15 cm distance on both sides of the drip tape from the setts.

In SSP, setts were planted at one side of the trench having a width of 30 cm and the drip tape lateral was placed on the opposite side. The setts were planted by an overlapping method in control as done by the farmers under surface method of irrigation. Setts at the rate of eight per running meter were planted under SSDF in both SSP and DSP methods. Sugarcane variety Co 86032 with two-bedded setts was planted in April 2008 and harvested in December 2008. The crop attained maturity earlier and hence, harvest was also made earlier in 9 months.

8.2.4 SSDI SYSTEM

The water pump of 7.5 HP was placed in the system for delivering of water from the open well. SSDI system had water filtration unit at the base of the system with sand filter (20 m³) and disc filter (2.0″) with 200 mesh. After filtration unit, PVC main line (75 mm) and submain line (63 mm) were installed to take water from filtration unit to experimental field. Inline drip tape laterals were fixed to submains at different spacing of 120, 135, 150, 165, and 180 cm and were placed manually in the trench at a depth of 20 cm from the soil surface. One inline drip lateral was placed in every trench. Inline drip tapes had 16 mm diameter, 15 mil wall thickness, 20 cm emitter distance, and 1.29 L/h emitter discharge with emission uniformity 95.31%. The operating pressure was maintained 1 KSC at the end of laterals.

8.2.5 FERTIGATION TECHNOLOGY

The sugarcane was fertilized with 275:62.5:112 of $N:P_2O_5:K_2O$ kg/ha. From this recommendation, 50% of P and K was applied as single super-phosphate, muriate of potash as basal application, and remaining 100% of N and 50% of P and K were applied in 30 splits at 6 days interval starting from 7th to 210th day through subsurface drip as water-soluble fertilizers, namely urea (46:00:00), polyfeed (13:40:13) and potassium nitrate (00:00:45).[5,40] Each plot consisted of three laterals for irrigating three rows of cane crop. A drip tape was provided at beginning of each lateral for giving controlled fertigation. For each lateral, either side of drip tape with 4 mm diameter microtube and end of microtubes was attached with 5 L capacity plastic cane. At the time of fertigation, fertilizer solution was controlled by tape. The required quantity of N, P, and K fertilizers for each treatment was dissolved separately in the bucket. The required quantity of fertilizer solution was filled with each plastic cane and then fertilizer solution was injected through subsurface drip system by adjusting the length of drip tape. Apart from this, calcium nitrate at 62.5 kg/ha, humic acid at 2.5 L/ha at 30 and 60 DAP (days after planting) and liquid biofertilizers, namely Azosphi, Phosphofix, and Potash activa at 750 mL/ha at 30, 60, and 90 DAP and liquid bioinoculants at 20 L/ha at 70 DAP were applied through subsurface drip to all the treatments as common dose.

8.2.6 IRRIGATION

Irrigation was applied uniformly after planting until the crop establishment, then SSDIs were scheduled once every 2 days based on the 100% crop evapotranspiration (ET_c) and surface method of irrigation was scheduled whenever the cumulative pan evaporation reached 50 mm after previous irrigation.

8.2.7 AFTER PLANTING

Crokran at 370 mL/ha and chloripyrifos at 2.5 L/ha were injected through drip system against internode borer and termite. The micronutrients, namely magnesium sulfate, borax and zinc sulfate at 12.5, 5.0, and 12.5 kg/ha were applied on 180 DAP and 210 DAP through the drip system. *Tricogramma chilonis* at 2.5 cc/ha was released on 180 DAP for the control of internode borer. Carbofuran 3G at 10 kg/ha was applied to control rat attack. Trifluralin was injected through drip system at the rate of 625 mL/ha for controlling root intrusion in drip tape emitters on the sixth month and 10 days prior to harvest. Random samples of cane stalks from each plot were collected at harvest. The cane yield and juice quality parameters, namely brix percentage, sucrose percentage, commercial cane sugar (CCS) percentage, purity percentage, and sugar yield were recorded at harvest-adopting standard procedures.

8.2.8 COST ECONOMICS

Cost of laying out SSDI system and cultivation charges were worked out for 1 ha of sugarcane crop. Cost of production and gross return for all the treatments were worked out on the basis of the prevailing input costs and price of sugarcane at the time of experimentation. Net returns were estimated by deducting the total cost of cultivation invested in each treatment from the respective gross returns. The cost of cultivation was computed by adding the cost of the sugarcane cultivation with annualized fixed cost (installation of drip system and fertigation). The gross return per rupee investment was calculated by dividing the gross return of each treatment with a total cost of cultivation.

8.3 RESULTS AND DISCUSSION

8.3.1 BRIX, SUCROSE, AND PURITY PERCENTAGE

The data on the effect of different lateral spacing and methods of planting under SSDF on the quality of cane juice, as expressed by brix, pol, and purity percentage are presented in Table 8.1. The results showed that different lateral spacing and methods of planting under SSDF had no significant influence on brix, pol, and purity percentage. The data related to brix, pol, and purity were not significantly varied among the treatments and numerically higher brix (18.2%), pol (15.6%), and purity (88.56%) values were obtained under SSDF at wider spacing. Similar results were obtained with the sugarcane cultivation under drip fertigation by other workers.[12,26,39] Juice quality mainly depends on genetic nature of the variety.[39] Quality parameters such as pol, brix, and purity percentage of juice were not significantly influenced by row spacing due to early harvesting.[2,3]

TABLE 8.1 Influence of Lateral Spacing and Planting Methods on Brix, Pol, and Purity of Cane Juice Under SSDF.

Treatments	Brix (%)		Pol (%)		Purity (%)	
	SSP	DSP	SSP	DSP	SSP	DSP
Lateral spacing under SSDF						
120 cm	17.60	17.40	14.80	15.20	84.09	87.36
135 cm	17.70	18.00	14.90	15.30	84.18	85.00
150 cm	18.00	18.10	15.40	15.60	85.56	86.19
165 cm	18.20	17.90	15.50	15.60	85.16	87.15
180 cm	17.48	17.40	15.30	15.40	87.53	88.56
Surface irrigation	17.40		14.30		82.18	
SED	0.58		0.50		2.83	
CD ($P = 0.05$)	NS		NS		NS	

SSDF = subsurface drip fertigation; SSP = single-side planting; DSP = double-side planting; NS= nonsignificant.

8.3.2 CCS, JUICE EXTRACTION, AND WEIGHT PERCENTAGE

SSDF significantly influenced CCS and juice extraction percentage, juice weight, and fiber content compared to surface irrigation. Methods of planting had no significant variation among the treatments, but all DSP treatments recorded numerically higher juice quality compared to SSP. Increasing lateral spacing from 120 to 180 cm as DSP under SSDF significantly increased juice weight and juice extraction percentage (Table 8.2).

TABLE 8.2 Influence of Lateral Spacing and Planting Methods on Juice Extraction, Juice Weight, and Fiber Content of Sugarcane Under SSDF.

Treatments	Juice extraction (%)		Juice weight (kg/cane)		Fiber content (%)	
	SSP	DSP	SSP	DSP	SSP	DSP
Lateral spacing under SSDF						
120 cm	74.07	75.16	1.13	1.18	12.80	12.70
135 cm	74.51	76.10	1.14	1.21	10.20	12.40
150 cm	80.89	82.14	1.27	1.38	12.40	11.60
165 cm	84.34	84.83	1.40	1.51	11.20	10.80
180 cm	89.22	89.62	1.49	1.58	10.60	10.10
Surface irrigation	65.48		0.73		13.90	
SED	2.62		0.04		0.37	
CD ($P = 0.05$)	5.36		0.08		0.76	

SSDF = subsurface drip fertigation; SSP = single-side planting; DSP = double-side planting.

The juice weight and juice extraction percentage were also significantly higher with 180 cm lateral spacing as DSP and this was on par with 165 cm lateral spacing as DSP under SSDF. Significantly, lesser fiber content was recorded in SSDF with wider spacing at 180 cm as DSP (10.10%) compared to closer spacing at 120 cm as DSP (12.70%) and surface method of irrigation (13.90%).

The highest CCS percentage was recorded with SSDF at 180 cm lateral spacing as DSP and this was at par with all other lateral spacing as DSP under SSDF but significantly higher than SSDF at 120 and 135 cm lateral spacing as SSP and surface method of irrigation.

Fertigation of water soluble fertilizers also had a positive effect on CCS value and it was improved to 10.9% in SSDF at 180 cm as DSP over the surface method of irrigation treatment (Table 8.3). The similar results have been reported by Gurusamy.[13] Dhotre reported that paired row planting at wider row spacing had higher CCS than single-row planting.[10] Higher CCS of 13.0% was recorded at wider spacing (120 + 30 + 30 + 30 cm).[28] The uniform maturity of cane at harvest consequently resulted in higher quality characters under drip irrigation compared to the surface method of irrigation.[30]

TABLE 8.3 Influence of Lateral Spacing and Planting Methods on CCS Percentage, Sugar, and Cane Yield of Sugarcane Under SSDF.

Treatments	CCS (%)		Sugar yield (t/ha)		Cane yield (t/ha)	
	SSP	DSP	SSP	DSP	SSP	DSP
Lateral spacing under SSDF						
120 cm	10.26	10.74	17.24	19.33	168.0	180.0
135 cm	10.34	10.67	17.00	18.98	164.5	178.0
150 cm	10.77	10.95	16.54	18.62	153.6	170.0
165 cm	10.82	11.01	16.22	18.94	150.0	172.0
180 cm	10.82	10.96	15.97	18.63	147.6	170.0
Surface irrigation	9.80		9.40		96.0	
SED	0.35		0.56		5.32	
CD ($P = 0.05$)	0.71		1.16		10.86	

SSDF = subsurface drip fertigation; SSP = single-side planting; DSP = double-side planting.

8.3.3 CANE YIELD

A significant progressive increase in cane yield was observed under SSDF compared to the surface method of irrigation (96.0 t/ha) and data are presented in Table 8.3. Significantly superior sugarcane yield of 180.0 t/ha was recorded at 120 cm lateral spacing as DSP and this was comparable with 135 cm (178.0 t/ha) and 165 cm (172.0 t/ha) lateral spacing as DSP under SSDF. The yield obtained with 165 cm was at par with 135 and 120 cm lateral spacing under DSP, indicating no significant yield variation among the lateral spacing from 120 to 165 cm under DSP. These findings are in conformity with Kumari.[16] The higher cane yield under wider spacing with DSP was mainly due to the availability of sufficient sunlight with better aeration coupled with increased nutrients and WUE.[8,21] Wider row spacing of 150 cm gave significantly higher cane yields than the conventional row spacing.[14] The increased cane yield was noticed with DSP compared to SSP in all SSDF treatments which account 7.0% at 120 cm and 15.2% at 180 cm lateral spacing.

Drip irrigation with paired row planting increased cane yield by 13.9% with better juice quality over normal furrow irrigation.[15] Wider row spacing gave a higher yield of 20–30 t/ha over normal row spacing.[29] The increased cane yield was around 53.75–87.75% in SSDF compared to the surface method of irrigation. Those findings are in conformity with those of other researchers.[12,20,33] The yield improvement under drip fertigation was mainly due to the maintenance of soil near field capacity throughout the growth period in the active root zone, leading to low soil-suction, thereby facilitated better water utilization, higher nutrients uptake, and excellent maintenance of a soil–water–air relationship with a higher oxygen concentration in the root zone.[25] The reason for low yield in surface irrigation might be water stress between irrigations. These results agree with those reported by Dalri and Cruz.[9]

8.3.4 SUGAR YIELD

There was marked effect of SSDF on sugar yield and data are presented in Table 8.3. The significantly increased sugar yield (19.33 t/ha) over the surface method of irrigation (9.4 t/ha) was observed with 120 cm lateral spacing in DSP and this was on par with rest of all the treatments under SSDF. It was found that SSDF registered almost more than double the sugar yield over the surface method of irrigation. In SSDF, the increase in sugar yield was around 105.63% and 98.19% at 120 and 180 cm lateral spacing under DSP, respectively over the surface method of irrigation. Increased sugar yield as obtained under SSDF was mainly due to improved juice characters, such as brix, pol, purity, and CCS percentage, as a result of uniform millable cane production under SSDF treatments.[16] Marginal improvements in juice quality were also observed in drip irrigation,[23] while Ahluwalia[1] found that the sugar yield with drip irrigation was higher than surface method.

8.3.5 COST ECONOMICS

SSDF is an innovative technology for maximizing the yield of cane crop. Though the cost of SSDF unit was very high, considering the longer life period of SSDF system, the benefit obtained out of SSDF will be for a longer period.

The economics of different lateral spacing and methods of planting in sugarcane cultivation under SSDF and surface method of irrigation were worked out and discussed in Tables 8.4 and 8.5. SSDF was found to be more profitable than surface irrigation due to a higher yield of sugarcane. In SSDF, closer lateral spacing at 120 cm exhibited significantly higher cost of cultivation due to higher lateral cost (33.33%) and other variable costs, such as field preparation, seed material, planting, weeding than the wider lateral spacing of 180 cm (Table 8.4). The maximum gross return obtained from the lateral spacing of 120 cm with DSP (Rs. 225,000) being at par with 135 cm with DSP (Rs. 222,500) were significantly superior to other treatments under SSDF. The increased gross return in closer spacing was mainly due to higher cane yield than wider spacing. Significantly, the highest net return was achieved in the wider lateral spacing of 180 cm with DSP (Rs. 106,956) being at par with T_2, T_4, and T_8 under SSDF.

TABLE 8.4 Total Cost of Sugarcane Cultivation and Gross Return as Affected by Different Lateral Spacing and Planting Methods Under SSDF.

Treatments	Cane yield increase over surface (%)		Total cost of cultivation Rs./ha		Gross return Rs./ha	
	SSP	DSP	SSP	DSP	SSP	DSP
Lateral spacing under SSDF						
120 cm	75.00	87.50	120,095	123,095	210,000	225,000
135 cm	71.35	85.41	114,743	118,118	205,625	222,500
150 cm	60.00	77.08	108,226	112,326	192,000	212,500
165 cm	56.25	79.16	103,968	109,468	187,500	215,000
180 cm	53.75	77.08	98,344	103,944	184,500	212,500
Surface irrigation	–		70,369		120,000	
SED			3228		5957	
CD ($P = 0.05$)			6593		12,167	

Prevailing market price of sugarcane: Rs. 1250/t.
SSDF = subsurface drip fertigation; SSP = single-side planting; DSP = double-side planting.

In general, net return was not significantly decreased with increasing lateral spacing from 120 to 180 cm with DSP under SSDF and numerically higher net return registered in wider spacing at 180 cm as DSP. Besides this, DSP registered significantly higher net return of 13.34%, 14.85%, 19.57%, 26.33%, and 26.49% than SSP at 120, 135, 150, 165, and 180 cm lateral spacing, respectively under SSDF. Significantly, lesser net return obtained in the surface method of irrigation (Rs. 49,361) which was 116.68% lesser than 180 cm lateral spacing with DSP under SSDF.

TABLE 8.5 Net Return of Sugarcane as Affected by Different Lateral Spacing and Planting Methods Under SSDF.

Treatments	Net return Rs./ha		Increased net return over surface irrigation (%)	
	SSP	DSP	SSP	DSP
	Lateral spacing under SSDF			
120 cm	120,095	123,095	71	75
135 cm	114,743	118,118	63	68
150 cm	108,226	112,326	54	60
165 cm	103,968	109,468	48	56
180 cm	98,344	103,944	40	48
Surface irrigation	70,369		–	
SED	2736			
CD ($P = 0.05$)	5588			

Prevailing market price of sugarcane: Rs. 1250/t.
SSDF = subsurface drip fertigation; SSP = single-side planting; DSP = double-side planting.

SSDF performed well at 180 cm lateral spacing as DSP with respect to net returns and gross return per rupee investments due to fewer cultivation expenses particularly reduced drip laterals cost. Arvind and Tripathi[3] also noted that paired row planting (40:110 cm) registered higher net return and benefit–cost ratio followed by the conventional planting of sugarcane at 75 cm spacing. Torres[38] also reported higher economic results for drip irrigation system than gravity irrigation. The drip fertigation with water-soluble fertilizers significantly increased the cane yield, net seasonal income, and benefit–cost ratio.[4,6,32]

KEYWORDS

- double-side planting
- mechanical harvesting
- nutrient use efficiency
- polyfeed
- potassium nitrate
- purity
- quality
- single-side planting
- water use efficiency

REFERENCES

1. Ahluwalia, M. S.; Sing, K. J.; Sharma, K. P. Influence of Drip Irrigation on Water Use and Yield of Sugarcane. *Int. Water Irrig. Rev.* **1998,** *18*, 12–17.
2. Ahmed, A. Z.; Khaled, K. A. M. *Detection of Genetic Similarity of Sugarcane Genotypes;* Sugar Crops Research Institute, Agricultural Research Center: Giza, Egypt, 2008; p 112.
3. Misra, A.; Tripathi, B. K. Feasibility of Mechanical Harvesting of Sugarcane (*Sachharam* spp. Hybrid). *Indian J. Agron.* **2006,** *51*(1), 65–67.
4. Bangar, A. R.; Chaudhari, B. C. In *Nutrient Mobility in Soil, Uptake, Quality and Yield of Sugarcane as Influenced by Fertigation Through Drip in Medium Vertisols*, Proceedings of International Conference on Micro and Sprinkler Irrigation System, Jalgaon, Maharashtra, India, Feb 8–10, 2001; pp 480–488.
5. Bhattacharya, A. *Bio-fuel Production and Water Constraint: An Effective Utilization of Water Shed Projections in India;* ISSF: Mumbai, 2010.
6. Bhoi, P. G.; Pawar, D. D.; Raskar, B. S.; Bangar, A. R.; Shinde, S. H. *Growth, Yield and Quality of Suru Sugarcane as Influenced by Water Soluble Fertilizer Applied Through Drip*, Extended Summary: International Conference on Managing Natural Resources for Sustainable Agricultural Production 21st Century, Feb 14–18, 2000; pp 1342–1344.
7. Blackburn, F. *Sugarcane*; Longman: New York, 1984; p 388.
8. Chaudari, S.; Dorge, J. T.; Tilekar, S. N. Impact of Agricultural Technologies and Development on Cane Productivity of Sugarcane in Western Maharashtra. *Coop. Sugar* **2010,** *41*(11), 69–74.
9. Dalri, A. B.; Cruz, R. L. Productivity of Sugarcane Fertigation with NK by Subsurface Drip. *Eng. Agricola* **2008,** *28*(3), 516–524.

10. Dhotre, R. S.; Gadge, S. B.; Gorantiwar, S. D. In *Yield Response of Sugarcane (Saccharum officinarum L) to Porus Pie Subsurface Irrigation Under Medium Black Soil*, Proceedings of National Seminar on Fertigation in Sugarcane, Vasantdada Sugar Institute, Pune, Dec 19–20, 2007; pp 53–59.

11. Gopalasundaram, P. Cane Agronomy. In *Sugarcane Production Technology*; Rajula Shanthy, T., Nair, N. V., Eds.; NFCSF & SBI Kalaikathir Printers: Coimbatore, 2009; pp 134–145.

12. Gouri, V.; Chitkala, D. T.; Prasada Rao, K.; Ankaiah, R. Evaluation of Drip Fertigation in Sugarcane in North Coastal Zone of Andhra Pradesh. *J. Sugarcane Res.* **2012,** *2*(2), 54–57.

13. Gurusamy, A.; Mahendran, P. P.; Krishnasamy, K.; Prabhu, R. Multispecialty Water Soluble Fertilizers and Sulphur Enhances the Yield and Quality of Sugarcane Under Subsurface Drip Fertigation System. *Int. J. Chem. Environ. Biol. Sci.* **2013,** *1*(2), 387–390.

14. Khandagave, R. B.; Hapase, D. G.; Somaiya, S. S. *Maximization of Sugarcane Yields and Reduction of Production Costs a Participatory Rural Appraisal,* Silver Jubilee Congress, Guatemala, Agriculture Commission, January 30 to February 4, 2005.

15. Kittad, C. H.; Jagtap, B. K.; Ghugare, R. V.; Pharande, A. L. Effect of Drip and Surface Irrigation Methods on Yield and Quality of Seasonal Sugarcane (Co 7219). *Indian Sugar* **1995,** *45*(8), 613–616.

16. Kumari, M. B. G. S.; Maheswara Reddy, M. P.; Srinivasulu Reddy, D.; Bharatha Lakshmi, M. Growth and Yield of Sugarcane as Affected by Planting Geometry and Intercropping. *Andhra Agric. J.* **2008,** *55*(1), 11–13.

17. Lamm-Freddie, R. In *Managing the Challenges of Subsurface Drip Irrigation*, Proceedings of Irrigation Association Technical Conference, San Antonio, Texas, Dec 2–5, 2009 (Available from the Irrigation Association, Falls Church, Virginia).

18. Murali, P.; Balakrishnan, R. Labor Scarcity and Selective Mechanisation of Sugarcane Agriculture in Tamil Nadu, India. *Sugar Tech.* **2012,** *14*(3), 223–228.

19. Nanda, R. S. Fertigation to Enhance Farm Productivity. *Indian J. Fertil.* **2010,** *6*(2), 13–16.

20. Narayanamoorthy, A. Can Drip Method of Irrigation be Used to Achieve the Macro Objectives of Conservation Agriculture? *Indian J. Agric. Econ.* **2010,** *65*(3), 428–438.

21. Pangghal, S. S. Can Production Mechanization a Solution for Labor Problems. *Indian Sugar* **2010,** *45*, 27–32.

22. Patil, V. G.; Dixit, R. M.; Bhoite, D. S.; Bhot, P. G. *Effect of Planting Patterns Under Drip and Surface Irrigation in Pre-season and Ratoon Sugarcane*, Proceedings of the 67th Annual Convention of STAI, 2006; pp 183–190.

23. Pawar, D. D.; Thawal, D. W.; Gaikwad, C. B. *Water Management in Sugarcane*, Souvenir, Group Meeting of All India Coordinated Research Project on Sugarcane, Held at College of Agriculture, Shivajinagar, Pune, India, Oct 9–11, 2007; pp 30–33.

24. Phene, C. J.; Hutmacher, R. B.; Ayars, J. E.; Davis, K. R.; Mead, R. M.; Schoneman, R. A. *Maximizing Water Use Efficiency with Subsurface Drip Irrigation*, International Summer Meeting of the American Society of Agricultural Engineers, St. Joseph Michigan; Paper # 922090.

25. Raina, J. N.; Thakur, B. C.; Verma, M. L. Effect of Drip Irrigation and Polyethylene Mulch on Yield, Quality and Water-use Efficiency of Tomato. *Indian J. Agric. Sci.* **1992**, *69*, 430–433.

26. Rajanna, M. P.; Patil, V. C. Effect of Fertigation on Yield and Quality of Sugarcane. *Indian Sugar* **2003**, *52*(12), 1007–1011.

27. Rajendra, G.; Yada, R. L.; Prasad, S. R. Comparison of Planting Methods and Irrigation Techniques for Water Use Efficiency, Yield and Quality of Sugarcane in Semi-arid Sub Tropics of India. *Indian J. Sugarcane Technol.* **2004**, *19*(1 & 2), 1–6.

28. Rajendran, B.; Paneerselvam, R. In *Studies on Planting Systems and Varieties Under Drip Fertigation for Sugarcane*, Proceedings of the National Seminar on Fertigation in Sugarcane, Vasantdada Sugar Institute, Pune, Dec 19–20, 2007; pp 35–42.

29. Rajula-Shanthy, T.; Muthusamy, G. R. Wider Row Spacing in Sugarcane: A Socio-economic Performance Analysis. *Sugar Tech.* **2012**, *14*(2), 126–133.

30. Selvaraj, P. K.; Asokaraja, N.; Manicksundaram, P. Drip Irrigation for Sugarcane. *Indian Farming*, **1997**, 17–21.

31. Senthil, K. Feasibility of Drip Irrigation in Sugarcane as Influenced by Planting Techniques and Sources of Fertigation Under Drip Irrigation. *Indian Sugar.* **2009**, *50*(11), 801–810.

32. Shinde, P. P.; Deshmukh, A. S. In *Crop Geometry and Fertigation Through Drip Irrigation for Sugarcane*, Proceedings of 71st Annual Convention of STAI, 2012; pp 149–157.

33. Singandhupe, R. B.; Bankar, M. C.; Anand, P. S. B.; Patil, N. G. Management of Drip Irrigated Sugarcane in Western India. *Arch. Agron. Soil Sci.* **2008**, *54*(6), 629–649.

34. Singh, D. K.; Rajput, T. B. S. Response of Lateral Placement Depths of Subsurface Drip Irrigation on Okra (*Abelmoschus esculentus*). *Int. J. Plant Prod.* **2007**, *1*(1), 73–84.

35. Solomon, K. Subsurface Drip Irrigation: Product Selection and Performance. In *Subsurface Drip Irrigation: Theory, Practices and Application;* Jorgennsen, G. S., Norum, K. N., Eds.; CATI Publication No. 921001: California, 1993; p 81.

36. Solomon, S. *Cost Effective and Input Efficient Technologies for Productivity Enhancement in Sugarcane*, 25th Meeting of Sugarcane Research and Development Workers of A.P., Visakhapatnam, July 20–21, 2012; pp 1–10.

37. Sundara, B. *Sugarcane Cultivation;* Vikas Publishing House: New Delhi, 1998; p 223.

38. Torres, S. G.; Prado-Vazquez, V. H.; Rivera-Espinoza, M. P. Sensibility Analysis of Sugar Cane Productions with Two Irrigation Technologies (Drip and Gravity) in Zapotiltic Jalisco Mexico. *Rev. Mex. Agronegocios* **2010**, *14*(26), 193–201.

39. Yadav, R. N. S.; Yadav, S.; Tejra, R. K. Labor Saving and Cost Reduction Machinery for Sugarcane Cultivation. *Sugarcane Technol.* **2003**, *1–2*, 7–10.

40. Yanam, I.; Emtryd, O.; Dinguing, L. Effect of Organic Manures and Chemical Fertilizers on Nitrogen Uptake and Nitrate Leaching in an *Enmerthicanthrosollos* Profile. *J. Appl. Entomol.* **1997**, *132*, 778–788.

41. Zhou, J. B.; Cheng, Z. J.; Li, S. X. Fertigation: An Efficient Way to Apply Water and Fertilizer. *Agric. Res. Arid Areas* **2001**, *19*(4), 16–27.

APPENDIX
Glossary of Terms

Double-side planting refers to the planting of setts on both sides of the trench having a width of 40 cm and the drip tape laterals are to be placed at the center of the trench and maintained 15 cm distance on both sides of the drip tape from the setts.

Fertigation refers to the application of fertilizers along with irrigation water through drip irrigation system

Single-side planting refers to the planting of setts on one side of the trench having a width of 30 cm and the drip tape lateral to be placed on the opposite side.

Subsurface drip irrigation is the irrigation of crops through buried plastic tubes containing embedded emitters located at regular spacings.

CHAPTER 9

IRRIGATION SCHEDULING USING CROPWAT 8.0 AND LEVEL OF FERTILIZERS FOR MAIZE UNDER FULLY AUTOMATIC DRIP IRRIGATION AND FERTIGATION SYSTEM

T. ARTHI[1,*] and K. NAGARAJAN[2]

¹Department of Land and Water Management, Agricultural Engineering College & Research Institute (AEC & RI), Pallapuram Post, Kumulur, Trichy 621712, Tamil Nadu, India

²Water Technology Centre, Tamil Nadu Agricultural University, Coimbatore 641003, Tamil Nadu, India

*Corresponding author. E-mail: arthi.elakia@gmail.com

CONTENTS

ABSTRACT

This chapter evaluates the yield and other biometric parameters of hybrid maize under fully automatic drip irrigation and fertigation. The CROPWAT 8.0 gave accurate CWR. Results indicated 20–30% of irrigation water saving.

9.1 INTRODUCTION

Maize is one of the world's leading crops cultivated over an area of about 167 million hectares with a production of about 860 million tons according to Directorate of Economics and Statistics, TNAU.[3] In India, it occupies third place among the cereals after rice and wheat, grown in area of 8.49 million hectares with the production of 21.28 million tons with an average productivity of 2507 kg ha[-1].[3,6] By 2020, the requirement of maize for various sectors will be around 100 million tons of which the poultry sector alone will demand 31 million tons.

The present field research was undertaken to assess the feasibility of drip fertigation (fully automatic) in hybrid maize with following objectives:

- To find the soil moisture content at different depths and distances from the emitters.
- To optimize the fertigation level for hybrid maize.[4,5]
- To calculate the crop water requirement (CWR) and to provide irrigation scheduling for hybrid maize NK 6240 based on the climatic data using CROPWAT 8.0.
- To find the uniformity of the fertigated water in head, middle, and tail in fully automatic condition.[7,8]
- To calculate the yield of hybrid maize under fully automatic irrigation and fertigation.

9.2 MATERIALS AND METHODS

Experiments were conducted at the Irrigation Cafeteria of Water Technology Centre of TNAU, Coimbatore, to evaluate the yield effects under optimized fertilizer levels in hybrid maize in fully automatic drip irrigation system. Irrigation scheduling was done by calculating the CWR using the climatic data of the cropping period. CWR was calculated by using formula and also by CROPWAT 8.0 provided by Food and Agricultural Organization (FAO) (http://www.fao.org/nr/water/infores_databases_cropwat.html).

The analysis was made to compare the uniformity of the fertigated water in head, middle, and tail through automatic condition. Water samples were collected in head, middle, and tail reaches of the laterals connected with the head, middle, and tail positions of the submain, when irrigation and fertigation were given through automatic fertigation.

The experiment was conducted to optimize the irrigation requirement and fertilizer levels for hybrid maize during *rabi* season. The CWR, water use efficiency, and fertilizer efficiency were calculated. The fertilizer dosage was scheduled according to the Crop Production Guide for 2012 by the Directorate of Extension of Tamil Nadu Agricultural University. On this basis, irrigation scheduling and fertilizer scheduling were programmed in the fully automatic drip irrigation and fertigation setup provided by NETAFIM.

The yield and other biometric parameters of the hybrid maize under fully automatic irrigation and fertigation were observed.

9.3 RESULTS AND DISCUSSION

The clogging study was conducted for 4 lph discharge capacity drippers in the experimental field after the harvest. It was found that dripper discharge was comparatively less when maize was irrigated through drippers in different treatments compared to control drippers. However, the discharge rate was uniform in each case separately.

9.3.1 COMPARISON OF IRRIGATION SCHEDULING

On the basis of the climatic data at the site, irrigation scheduling was calculated both manually and also by using CROPWAT 8.0 by FAO. The solid lines show the best fit for linear equation ($Y = Ax + B$) with regression coefficient of 0.99.

Figure 9.1 shows that the linear regression equation, $Y = 0.215x + 3.1593$, has a R^2 value of 0.9289 for the FAO method, while the equation $Y = 0.1768x + 0.2752$ has a R^2 value of 0.632 in case of manually calculated data (Fig. 9.2). The FAO method was more accurate.

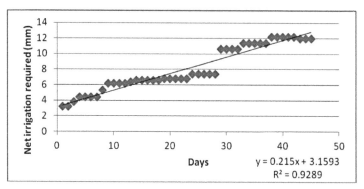

FIGURE 9.1   Irrigation requirement calculated by CROPWAT 8.0 method.

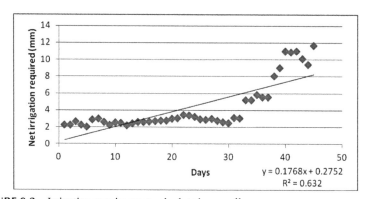

FIGURE 9.2   Irrigation requirement calculated manually.

9.3.2   UNIFORMITY OF FLOW

During fertigation, the data revealed that the dripper flow was uniform in head, middle, and tail reaches of the laterals connected with the head, middle, and tail positions of the submain. After harvest, the discharge from the emitter was not uniform, resulting in clogging of the emitters.

9.3.3   EFFECT ON BIOMETRIC PARAMETERS AND YIELD OF HYBRID MAIZE

The yield was higher in T5 (125% RDF) and showed a significant difference in the plant growth and other biometric characters, which revealed that the fertilizers applied at root zone at correct intervals were efficiently used. Following this, T4 (100%) and T3 (75%) showed similar readings showing that fertilizer applied at root zone is effectively used by the plant giving 20–25% saving of fertilizers when treated with minute variations under automatic drip irrigation (Table 9.1 and Fig. 9.3). These results agree with those reported by other investigators.[1,2,9]

TABLE 9.1 Effect of Fully Automatic Drip Irrigation System and Fertigation on Plant Height.

Treatments	30 days	60 days	90 days
T1	76.02	154.84	220.12
T2	75.164	186.96	227.55
T3	79.72	197.66	232.20
T4	80.06	202.16	233.73
T5	87.148	212.06	239.47
Mean	79.66	191.936	230.61
SED	4.40	4.89	3.67
CD (0.05)	10.16	11.28	8.45

Fertigation T5 = 125% RDF, T4 = 100% RDF, T3 = 75% RDF, T2 = 50% RDF, and T1 = 100% RDF conventional method.

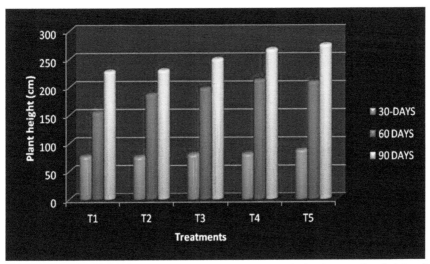

FIGURE 9.3 Effect of fully automatic drip irrigation and fertigation on plant height of maize.

9.3.4 SOIL MOISTURE DISTRIBUTION

The soil moisture was determined by taking soil samples at different depth and distances away from the dripper. The observations were plotted using a software surfer. Results are shown in Figure 9.4.

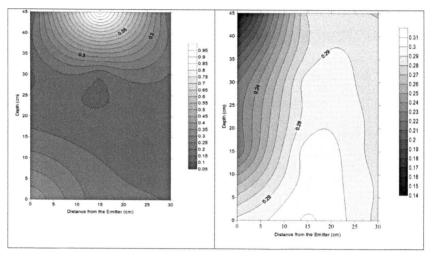

FIGURE 9.4   Moisture distributions immediately after 0, 3, 7, and 14 h irrigation.

Usage of fully automatic irrigation and fertigation reduced the labor input due to weeding, as the area between the plant rows was dry. It was found that there was complete reduction of the weeds in the experimental field, since the irrigation and fertilizer were limited only to the root zone. The control plot showed 2–3 kg of weeds per plot during the growing period.

9.4 SUMMARY

In this chapter, the yield and other biometric parameters of the hybrid maize under fully automatic drip irrigation and fertigation were observed. Using of CROPWAT 8.0 gave accurate CWR, which is useful for programming in fully automatic irrigation. Results indicated 20–30% of irrigation water saving. Leaching effect of fertilizer was also reduced in the soil.

KEYWORDS

- **crop water requirement**
- **drip irrigation**
- **FAO**
- **irrigation scheduling**
- **maize**
- **RDF**
- **soil moisture**

REFERENCES

1. Akbari, M.; Dehghanisanij, H.; Mirlatifi, S. M. Impact of Irrigation Scheduling on Agriculture Water Productivity. *Iran. J. Irrig. Drain.* **2009,** *1*, 69–79.

2. Bhunia, S. R.; Singh, V. Scheduling of Irrigation to Wheat (*Triticum aestivum*) Under Shallow Water Table Condition. *Indian J. Agric. Sci.* **2000,** *70*(7), 494–495.

3. Directorate of Economics and Statistics (DES). *Agricultural Statistics for Tamil Nadu* [Online], 2011. DES, Government of India (GOI). http://www.agritech.tnau.ac.in (accessed on Dec 12, 2016).

4. Fanish, A. S. Influence of Drip Fertigation and Intercropping on Yield, Agronomic Efficiency and Partial Factor Productivity of Maize. *Madras Agric. J.* **2013,** *100*(1–3), 102–106.

5. Fanish, S.; Muthukrishnan, K. Nutrient Distribution Under Drip Fertigation System. *World J. Agric. Sci.* **2013,** *9*(3), 273–283.

6. http://agricoop.nic.in/Divisions.aspx (accessed Sept 30, 2016).

7. Vijayakumar, G.; Tamilmani, D.; Selvaraj, P. K.; Ramaswamy, K. Moisture Movement and Nitrogen Distribution in Onion Under Drip Fertigation in Sandy Loam Soil. *Indian J. Soil Conserv.* **2011,** *39*(1), 37–43 (Dept. SWCE, TNAU, Coimbatore).

8. Vijayakumar, G.; Tamilmani, D.; Selvaraj, P. K. Irrigation and Fertigation Scheduling Under Drip Irrigation in Brinjal (*Solanum melongena* L.) Crop. *IJBSM* **2010,** *1*, 72–76.

9. Wasif, E. A.; Magdy, B.; El-Tantawy, T.; Yasser, E. Impact of Fertigation Scheduling on Tomato Yield Under Arid Ecosystem Conditions. *Res. J. Agric. Biol. Sci.* **2009,** *5*(3), 280–286.

CHAPTER 10

COST ECONOMICS OF CUCUMBERS GROWN UNDER A SHADE NET HOUSE WITH DIFFERENT FERTIGATION LEVELS

M. A. PATIL[1,*], S. B. GADGE[2], SUNIL D. GORANTIWAR[2], and ARUN D. BHAGAT[2]

[1]*Department of Irrigation and Drainage Engineering, College of Technology & Engineering, Maharana Pratap University of Agriculture & Technology, Udaipur 313001, India*

[2]*Department of Irrigation and Drainage Engineering, Dr. Annasaheb Shinde College of Agricultural Engineering (Dr. ASCAE), Mahatma Phule Agricultural University (MPKV), Rahuri 413722, India*

*Corresponding author. E-mail: mangalpatil43@gmail.com

CONTENTS

In this chapter: US$ 1 = INR 60.00 (Indian Rupees)

ABSTRACT

The field experiment was conducted at the Instructional Farm of Department of Irrigation and Drainage Engineering, Mahatma Phule Krishi Vidyapeeth, Rahuri during the period from January 2012 to May 2012. The cost economics of cucumber (Vvar. Gypsy) production per m^2 under shade net house with 35%, 50%, and 75% per cent shading with open-field trial and different fertigation levels were worked out. While working out the cost economics, cost of production, gross monetary returns, and net income were considered to solve away the benefit–cost ratio. The study expressed that, the maximum cost of production was recorded under the shade net with 75% per cent shading with the application of NPK ratio as per the growth stage of cucumber with 125% per cent NPK of Rs. 74.34/m^2, the maximum gross monetary returns and net returns recorded under shade net with 75% per cent and application of 125% per cent NPK through the drip of Rs. 125. 2/- and Rs. 51.28/m^2, respectively, with benefit–cost ratio of 1.69.

10.1 INTRODUCTION

Agriculture is the backbone of India's economic activity and the experience during the last 50 years has demonstrated the strong correlation between agricultural growth and economic prosperity. The present agricultural scenario is a mix of outstanding achievements and missed opportunities. If India has to emerge as an economic power in the world, the agricultural productivity should be equal to those countries, which are currently rated as the economic power of the world. India needs a new and effective technology, which can improve continuously the productivity, profitability, and sustainability of the major farming systems. One such technology is the protected cultivation technology. About 95% of plants, either food crops or cash crops, are grown in open field; therefore, for higher and qualitative yield, cultivation under shade net with low cost is affordable.

Cucumber (*Cucumis sativus* L.) is an important and commercially popular cucurbitaceous vegetable crop, which is native to India, and is one of the most nutritive vegetables, rich in vitamins, and minerals such as phosphorus, potassium, calcium, and iron. It is mainly grown for its fruits both in tropics and subtropics of the world and produces tender fruits continuously. Growing plants under cover improves the quality of the produce. This, in turn, is helpful in getting higher price that becomes remunerative to the grower. It is also possible to make the produce available in the market when it is in great demand, provided the grower takes the action of protected cultivation.[1] The growers can be offered to cultivate a crop in any season under protected environment, as he can provide the temperature, humidity, and light, as required by the plant species.[7]

Economic analysis permits to identify the strengths and weaknesses of technical and economic results, to take immediate and decisive action, at any time, and to solve problems affecting the agricultural activity, helping the grower to manage and use the available resources more efficiently, favoring their maximization and increasing the level of the production system with a simultaneous reduction in costs. Thus, an economic analysis of cucumber under a protected environment was carried out with the aim of increasing the grower's profitability.[2–6,8]

The aim of this research study is to determine the investment and operation—maintenance expenses for cucumber under shade net house with different shading percent and fertigation levels.

10.2 MATERIAL AND METHODS

10.2.1 STUDY AREA

The investigation on economics of cucumber under shade net house with different fertigation levels was carried out at the Instructional Farm of Department of Irrigation and Drainage Engineering which is situated in the transitional tract 74°38′00″ E longitude and 19°20′00″ N latitude at 557 m above the mean sea level, in the central campus of Mahatma Phule Krishi Vidyapeeth, Rahuri. The experiment was carried during January–May 2012 under four different shading percentages of 288 m² area each. Figures 10.1 and 10.2 show the general and internal view of the shade net house. "Gypsy" variety of cucumber was selected for the study under shade net with different fertigation levels as given below:

Main Treatment Details

S_1: 35% shading
S_2: 50% shading
S_3: 75% shading
S_4: 0% shading

Subtreatments

T_1 = soil application of recommended dose of NPK (control).

T_2 = application of 100% N through drip and soil application of P and K.

T_3 = application of 125% N through drip and soil application of P and K.

T_4 = application of 100% NPK through drip.

T_5 = application of 125% NPK through drip.

T_6 = application of NPK ratio as per the crop growth stages with 125% N.

FIGURE 10.1   General view of shade net house.

FIGURE 10.2   Internal view of shade net house.

10.2.2   COST ECONOMICS

Cost economics of the cucumber per shade net house and per hectare was worked out by adopting following procedure:

10.2.2.1   COST OF CULTIVATION

The total cost of cultivation for cucumber grown under shade net house included labor charges, fertilizer, water charges, seeds, insecticide and pesticide, and miscellaneous, etc.

10.2.2.2   COST OF PRODUCTION

The cost of production was worked out for each treatment. The cost included paid out cost on hired labor, seeds, fertilizers, water charges, interest on working capital, interest on fixed capital, depreciation, repair and maintenance for drip irrigation system, and shed net house.

10.2.2.3 GROSS MONETARY RETURNS

The gross monetary returns per hectare were worked out by considering the fruit yield from different treatments and the prevailing market price of cucumber.

10.2.2.4 NET INCOME

The net income was worked out by subtracting the cost of production from the gross momentary returns in each treatment.

10.2.2.5 BENEFIT–COST RATIO

The benefit–cost ratio (BCR) was worked out by dividing the cost of production to the gross returns in each treatment under study. The data were statistically analyzed to check its suitability.

10.3 RESULTS AND DISCUSSION

The cost economics of cucumber production per meter square under shade net house with different shading percentage and different fertigation levels were worked out. While working out the cost economics, cost of production and gross monetary returns were considered to work out the BCR and are presented in Tables 10.1–10.5 and in Figure 10.3.

TABLE 10.1 Benefit–Cost Ratio for Cucumber Under Shade Net House with 35% Shading (288 m²).

Sr. no.	Particulars		T_1	T_2	T_3	T_4	T_5	T_6
1		Fixed cost						
	a.	Cost of structure (excluding fogger and cladding material) Rs. 180/m²	51,840	51,840	51,840	51,840	51,840	51,840
	b.	Life of structure (years)	25	25	25	25	25	25
	c.	Depreciation/year (a/b)	2074	2074	2074	2074	2074	2074

TABLE 10.1   *(Continued)*

Sr. no.	Particulars	T_1	T_2	T_3	T_4	T_5	T_6
	d. Cost of cladding material at Rs. 13/m²	3744	3744	3744	3744	3744	3744
	e. Life of cladding material	5	5	5	5	5	5
	f. Depreciation/year (d/e)	748	748	748	748	748	748
	g. Weed mat Rs. 10/m²	2880	2880	2880	2880	2880	2880
	h. Life of weed mat	8	8	8	8	8	8
	i. Depreciation/year (g/h)	360	360	360	360	360	360
	j. Drip irrigation/288 m²	1285	1285	1285	1285	1285	1285
	m. Fogging system Rs. 25/m²	7200	7200	7200	7200	7200	7200
	n. Life of system (years)	7	7	7	7	7	7
	o. Depreciation (m/n)	1028	1028	1028	1028	1028	1028
	p. Trellis system Rs. 35/m²	10,080	10,080	10,080	10,080	10,080	10,080
	q. Life of system (years)	20	20	20	20	20	20
	r. Depreciation (p/q)	504	504	504	504	504	504
	Total	**75,744**	**75,744**	**75,744**	**75,744**	**75,744**	**75,744**
2	Repair and maintenance (2% of total cost)	1515	1515	1515	1515	1515	1515
3	Interest rate (10% of total cost)	7574	7574	7574	7574	7574	7574
4	Total operational cost/m²	52.03	52.03	52.03	52.03	52.03	52.03
5	Cost of cultivation	19.61	19.72	19.82	20.36	20.61	21.03
6	Total cost of production/m² (4 + 5)	71.63	71.74	71.84	72.38	72.64	73.06
7	Average yield of produce (kg/m²)	2.03	2.42	2.68	2.25	2.9	2.02
8	Average market price	40	40	40	40	40	40
9	Gross monetary returns (7 × 8)	81.2	96.8	107.2	90.0	116.0	80.8
10	Net income/m²	9.57	25.06	35.36	17.62	43.36	7.74
11	**B:C ratio**	**1.13**	**1.35**	**1.49**	**1.24**	1.60	1.11

TABLE 10.2 Benefit–Cost Ratio for Cucumber Under Shade Net House with 50% Shading (288 m²).

Sr. no.		Particulars	T_1	T_2	T_3	T_4	T_5	T_6
1		**Fixed cost**						
	a	Cost of structure (excluding fogger and cladding material) Rs. 180/m²	51,840	51,840	51,840	51,840	51,840	51,840
	b	Life of structure (years)	25	25	25	25	25	25
	c	Depreciation/year (a/b)	2074	2074	2074	2074	2074	2074
	d	Cost of cladding material Rs. 15/m²	4320	4320	4320	4320	4320	4320
	e	Life of cladding material	5	5	5	5	5	5
	f	Depreciation/year (d/e)	864	864	864	864	864	864
	g	Weed mat Rs. 10/m²	2880	2880	2880	2880	2880	2880
	h	Life of weed mat	8	8	8	8	8	8
	i	Depreciation/year (g/h)	360	360	360	360	360	360
	j	Drip irrigation/288 m²	1180	1180	1180	1180	1180	1180
	m	Fogging system Rs. 25/m²	7200	7200	7200	7200	7200	7200
	n	Life of system (years)	7	7	7	7	7	7
	o	Depreciation (m/n)	1028	1028	1028	1028	1028	1028
	p	Trellis system Rs. 35/m²	10,080	10,080	10,080	10,080	10,080	10,080
	q	Life of system (years)	20	20	20	20	20	20
	r	Depreciation (p/q)	504	504	504	504	504	504
		Total	**76,320**	**76,320**	**76,320**	**76,320**	**76,320**	**76,320**
2		Repair and maintenance (2% of total cost)	1526	1526	1526	1526	1526	1526
3		Interest rate (10% of total cost)	7632	7632	7632	7632	7632	7632
4		Total operational cost/m²	52.67	52.67	52.67	52.67	52.67	52.67
5		Cost of cultivation	19.61	19.72	19.82	20.36	20.61	21.03
6		Total cost of production/m² (4 + 5)	72.27	72.39	72.48	73.03	73.28	73.70
7		Average yield of produce (kg/m²)	2.25	2.51	2.74	2.44	2.38	2.39
8		Average market price	40	40	40	40	40	40
9		Gross monetary returns (7 × 8)	90.0	100.4	109.6	97.6	95.2	95.6
10		Net income/m²	17.73	28.01	37.12	24.57	21.92	21.90
11		**B:C ratio**	**1.25**	**1.39**	**1.51**	**1.34**	**1.30**	1.30

TABLE 10.3 Benefit–Cost Ratio for Cucumber Under Shade Net House with 75% Shading (288 m²).

Sr. no.	Particulars		T_1	T_2	T_3	T_4	T_5	T_6
1		**Fixed cost**						
	a	Cost of structure (excluding fogger and cladding material) Rs. 180/m²	51,840	51,840	51,840	51,840	51,840	51,840
	b	Life of structure (years)	25	25	25	25	25	25
	c	Depreciation/year (a/b)	2074	2074	2074	2074	2074	2074
	d	Cost of cladding material Rs. 17/m²	4896	4896	4896	4896	4896	4896
	e	Life of cladding material	5	5	5	5	5	5
	f	Depreciation/year (d/e)	979	979	979	979	979	979
	g	Weed mat Rs. 10/m²	2880	2880	2880	2880	2880	2880
	h	Life of weed mat	8	8	8	8	8	8
	i	Depreciation/year (g/h)	360	360	360	360	360	360
	j	Drip irrigation/288 m²	1180	1180	1180	1180	1180	1180
	m	Fogging system Rs. 25/m²	7200	7200	7200	7200	7200	7200
	n	Life of system (years)	7	7	7	7	7	7
	o	Depreciation (m/n)	1028	1028	1028	1028	1028	1028
	p	Trellis system Rs. 35/m²	10,080	10,080	10,080	10,080	10,080	10,080
	q	Life of system (years)	20	20	20	20	20	20
	r	Depreciation (p/q)	504	504	504	504	504	504
		Total	**76,896**	**76,896**	**76,896**	**76,896**	**76,896**	**76,896**
2		Repair and maintenance (2% of total cost)	1538	1538	1538	1538	1538	1538
3		Interest rate (10% of total cost)	7690	7690	7690	7690	7690	7690
4		Total operational cost/m²	53.31	53.31	53.31	53.31	53.31	53.31
5		Cost of cultivation	19.61	19.72	19.82	20.36	20.61	21.03
6		Total cost of production/m² (4 + 5)	72.91	73.03	73.12	73.66	73.92	74.34
7		Average yield of produce (kg/m²)	2.31	2.69	2.86	2.56	3.13	2.84
8		Average market price	40	40	40	40	40	40
9		Gross monetary returns (7 × 8)	92.4	107.6	114.4	102.4	125.2	113.6
10		Net income/m²	19.49	34.57	41.28	28.74	51.28	39.26
11		**B:C ratio**	**1.27**	**1.47**	**1.56**	**1.39**	**1.69**	**1.53**

TABLE 10.4   Benefit–Cost Ratio for Cucumber in Control Treatment.

Sr. no.	Particulars	T_1	T_2	T_3	T_4	T_5	T_6
1		**Fixed cost**					
	a Cost of structure (excluding fogger and cladding material) Rs. 180/m²	51,840	51,840	51,840	51,840	51,840	51,840
	b Life of structure (years)	25	25	25	25	25	25
	c Depreciation/year (a/b)	2074	2074	2074	2074	2074	2074
	g Weed mat Rs. 10/m²	2880	2880	2880	2880	2880	2880
	h Life of weed mat	8	8	8	8	8	8
	i Depreciation/year (g/h)	360	360	360	360	360	360
	j Drip irrigation/288 m²	1180	1180	1180	1180	1180	1180
	m Fogging system Rs. 25/m²	7200	7200	7200	7200	7200	7200
	n Life of system (years)	7	7	7	7	7	7
	o Depreciation (m/n)	1028	1028	1028	1028	1028	1028
	p Trellis system Rs. 35/m²	10,080	10,080	10,080	10,080	10,080	10,080
	q Life of system (years)	20	20	20	20	20	20
	r Depreciation (p/q)	504	504	504	504	504	504
	Total	**72,000**	**72,000**	**72,000**	**72,000**	**72,000**	**72,000**
2	Repair and maintenance (2% of total cost)	1440	1440	1440	1440	1440	1440
3	Interest rate (10% of total cost)	7200	7200	7200	7200	7200	7200
4	Total operational cost/m²	48	48	48	48	48	48
5	Cost of cultivation	19.61	19.72	19.82	20.36	20.61	21.03
6	Total cost of production/m² (4 + 5)	67.47	67.59	67.68	68.23	68.48	68.90
7	Average yield of produce (kg/m²)	0.40	0.38	0.36	0.34	0.33	0.36
8	Average market price	40	40	40	40	40	40
9	Gross monetary returns (7 × 8)	16.00	15.20	14.40	13.60	13.20	14.40
10	Net income/m²	–	–	–	–	–	–
11	B:C ratio	0.24	0.22	0.21	0.20	0.19	0.21

TABLE 10.5   Cost Economics of Cucumber Under Different Treatments.

Items	Cost of production (Rs./m²)	Gross monetary returns (Rs./m²)	Net income (Rs./m²)	B:C ratio
	S₁: Shade net with 35% shading			
T₁	71.63	81.2	7.57	1.13
T₂	71.74	96.8	25.60	1.35
T₃	71.84	107.2	35.36	1.49
T₄	72.38	90.0	17.62	1.24
T₅	72.64	116.0	43.36	**1.60**
T₆	73.06	80.8	7.74	1.11
	S₂: Shade net with 50% shading			
T₁	72.27	90.0	17.73	1.25
T₂	72.39	100.4	28.01	1.39
T₃	72.48	109.6	37.12	**1.51**
T₄	73.03	97.6	24.57	1.34
T₅	73.28	95.2	21.92	1.30
T₆	73.70	95.6	21.90	1.30
	S₃: Shade net with 75% shading			
T₁	72.91	92.4	19.49	1.27
T₂	73.03	107.6	34.57	1.47
T₃	73.12	114.4	41.28	1.56
T₄	73.66	102.4	28.74	1.39
T₅	73.92	125.2	51.28	**1.69**
T₆	74.34	113.6	39.26	1.53
	S₄: Open field			
T₁	67.47	16.0	–	0.24
T₂	67.59	15.2	–	0.22
T₃	67.68	14.4	–	0.21
T₄	68.23	13.6	–	0.20
T₅	68.48	13.2	–	0.19
T₆	68.90	14.4	–	0.21

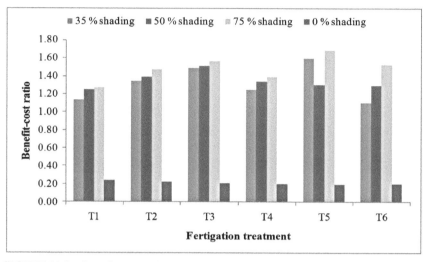

FIGURE 10.3   Benefit–cost ratio observed in different treatments.

10.3.1   TOTAL YIELD OF CUCUMBER

The maximum yield of fruit per plot was observed under 75% shading (21.31 kg), significantly superior to 50% shading (19.13 kg), which was at par to 35% shading (18.60 kg). Minimum yield (2.83 kg) was observed in open field conditions. The total fruit yield recorded from shade net with 35%, 50%, and 75% shading were 23.84, 24.52, and 27.32 t/ha, respectively that were 8–10 times more than open field condition, that is, 3.63 t/ha.

10.3.2   COST OF CULTIVATION

The cost of cultivation for cucumber grown under shade net house ranged from Rs. 19.61 to Rs. $21.03/m^2$ and average total cost of cultivation was Rs. $20.19/m^2$.

10.3.3   COST OF PRODUCTION

The maximum cost of production Rs. $74.34/m^2$ was observed under treatment combination of shade net of 75% shading and with the application of NPK ratio as per the crop growth stages with 125% N (Rs. $74.34/m^2$) and minimum under control treatment.

10.3.4   GROSS MONETARY RETURNS

Maximum gross monetary returns of Rs. $125.2/m^2$ was reported under shade net of 75% shading and 125% NPK through drip and minimum (Rs. $80.8/m^2$) was reported under shade net of 35% shading and 100% NPK soil application when compared only under shade net besides, minimum gross monetary returns were obtained from control treatment (Rs. $13.2/m^2$).

10.3.5   NET INCOME

Maximum net income was gained from treatment combination of shade net of 75% shading and 125% NPK through drip (Rs. $51.28/m^2$), whereas minimum was reported in open field condition (Rs. $7.57/m^2$).

10.3.6   BENEFIT–COST RATIO

The calculated BCR data were analyzed statistically (Table 10.6). The BCR was significantly influenced by different shading percentage of shade net. Maximum BCR was observed under shade net with 75% shading (1.49) at par to shade net with 35% shading (1.32) and shade net with 50% shading (1.35). BCR was not significantly influenced by different shading percentage and fertigation treatment.

TABLE 10.6   Statistical Analysis of Benefit–Cost Ratio.

Treatment	S_1	S_2	S_3	S_4	Mean
T_1	1.13	1.25	1.27	0.24	**0.97**
T_2	1.35	1.39	1.47	0.22	**1.11**
T_3	1.49	1.51	1.56	0.21	**1.20**
T_4	1.25	1.34	1.39	0.20	**1.04**
T_5	1.60	1.30	1.69	0.19	**1.20**
T_6	1.11	1.30	1.53	0.21	**1.03**
Mean	**1.32**	**1.35**	**1.49**	**0.21**	**1.09**
Interaction	S. E. ±		C. D. (5%)		
Level A	0.10		NS		
Level B	0.11		NS		

Note: Level A—between subplots means at the same level of main plot mean.
Level B—between main plots means at the same level of sub plot mean.

Table 10.6 reveals that the BCR was found to be maximum under shade net with 75% shading and application of 125% NPK through drip, that is, 1.69. Minimum BCR was observed under shade net with 35% shading application of NPK ratio as per the crop growth stages (1.11). The economic analysis of cucumber production under open field with different fertigation levels revealed that the production is not economically viable as the B:C ratio was less than 1.0.

10.4   SUMMARY

The net income was found to be maximum in shade net with 75% shading with the application of 125% NPK through drip system (Rs. 125.20/m²). The BC ratio (1.69) was found to be maximum in shade net with 75% shading with the application of 125% NPK through the drip system. Though the effect of interaction between shading percentage and fertigation levels was found statistically nonsignificant in case of yield, yet the individual effect of shading percentage and fertigation levels gave significant results.

KEYWORDS

- benefit–cost ratio
- cost economics
- cucumber
- fertigation
- lettuce
- NPK
- shade net house
- shading percentage

REFERENCES

1. Agarwal, K. N.; Satapathy, K. K. In *Potential of Using Low Cost Polyhouse in NEH Region*, Proceedings of All India on Seminar Potential and Prospects for Protective Cultivation, The Institute of Engineers, Ahmednagar, Dec 12–13, 2003; pp 47–53.
2. Agasimani, A. D.; Harish, D. K.; Imamsaheb, S. J.; Patil, V. S. *Anthurium* Varieties Performance and Economics Under Greenhouse. *Res. J. Agric. Sci.* **2011**, *2*(2), 226–229.
3. Braulio, I. A.; Rezende, A. B., Filho, C.; Barros, A. P. Economic Analysis of Cucumber and Lettuce Intercropping Under Greenhouse in the Winter-spring. *Ann. Braz. Acad. Sci.* **2010**, *83*(2), 705–717.
4. Gajanana, T.; Singh, K. P.; Subrahamnyam, K. V. Economic Analysis of Gerbera Cultivation Under Protected Cultivation. *Indian J. Hortic.* **2003**, *60*, 104–107.
5. Naik, H. B.; Chauhan, N.; Patil, A. A.; Patil, V. S.; Patil, B. C. Comparative Performance of Gerbera (*Gerbera jamesonii* Bolus ex Hooker F.) Cultivars Under Naturally Ventilated Polyhouse. *J. Orna. Hortic.* **2006**, *9*(3), 204–207.
6. Pattanshetti, C. N. Evaluation of Gerbera Cultivars Under Protected Conditions. M. Tech. Thesis, Submitted to University of Agricultural Sciences, Dharwad, Karnataka, India, 2009.
7. Singh, R.; Asrey R.; Nangare D. D. In *Studies on the Performance of Tomato and Capsicum Under Medium Cost Greenhouse,* Proceedings of All India Seminar on Potential and Prospects for Protective Cultivation, The Institute of Engineers, Ahmednagar, Maharashtra, India, 2003.
8. Stachowiak, M. Costs of Integrated Plant Protection of Greenhouse Cultivation of Tomato and Cucumber. *Prog. Plant Prot.* **2009**, *49*(4), 1656–1663.

CHAPTER 11

FISHPOND WASTEWATER: THE POTENTIAL IRRIGATION SOURCE

LALA I. P. RAY[1,*], B. C. MAL[2], S. MOULICK[3], and
P. K. PANIGRAHI[4]

[1]*School of Natural Resource Management, College of Postgraduate
Studies (Central Agricultural University, Imphal), Umiam, Barapani
793103, Meghalaya, India*

[2]*Department of Agricultural and Food Engineering, Indian Institute
of Technology, Kharagpur 721302, West Bengal, India and
Chhattisgarh Swami Vivekananda Technical University, Bhilai,
Chhattisgarh, India*

[3]*Department of Civil Engineering, Kalinga Institute of Industrial
Technology (KIIT) University, Bhubaneswar, Odisha, India*

[4]*West Bengal and Fisheries Development Officer, Government of
Odisha, India*

Corresponding author. E-mail: lalaipray@rediffmail.com

CONTENTS

ABSTRACT

With dwindling global water resources, attempts have been made by several sectors to establish alternate sources of irrigation. Under this scenario, an effort has been made to assess the potential of aquacultural wastewater for irrigation. In an integrated agri-aquaculture system when the cultured fishery is being practiced under a semi-intensive or intensive manner, fishpond water needs to be exchanged along with all the established management practices. This exchanged water is supposed to be harmful for fish growth; however, this so-called polluted water is nutrient rich and can be made use for irrigation in growing crops. In the present investigation, tomato has been grown as a test crop to study the efficacy of the fishpond wastewater.

11.1   INTRODUCTION

Reuse of freshwater is being advocated to overcome its exploitation in the agricultural sector. Under these scenarios, several potential water sources of irrigation (municipal water, brackish water, industrial wastewater, wastewater from agricultural and allied processed industries, and wastewater from aqua cultural firms) have emerged.[8,13,14,16,17,32,34,36,38] Irrigation water from these sources with some treatment has been used suitably for irrigating agricultural crops. The reuse of water for irrigation is often viewed as a positive means of recycling, the advantage being a constant, reliable source and reduction in the amount of water extracted from the environment.[3,37] The practice of wastewater reuse for landscape irrigation in Saudi Arabia was a success story.[1] The use of wastewater can save up to 50% application of inorganic nitrogen fertilizer when it contains 40 mg of N L^{-1}.[14] The feasibility study was conducted in Brazil for using fishpond effluent to irrigate cherry tomatoes, grown with different types of organic fertilizers.[7] Higher productivity was observed in effluent treatments. Researchers have reported that water reclamation, recycling, and reuse address the challenges of water scarcity by resolving water resource issues, creating new sources of high-quality water supplies in an integrated way.[19]

   This chapter discusses vegetable-based remunerative cropping integrated with a semi-intensive aquaculture system. The polluted water exchanged from the aquacultural fishponds was used to irrigate tomato during the winter season. The efficacy of exchanged water from fishponds stocked with high densities of three species of Indian Major Carps (IMC) as an irrigation source was monitored for three consecutive growing seasons.

11.2   MATERIALS AND METHODS

Field experiments were conducted at the experimental farm of the Department of Agricultural and Food Engineering, Indian Institute of Technology, Kharagpur, West Bengal in eastern India for 3 consecutive years (2006, 2007, and 2008). The site is located at a latitude of 22°19′ North and longitude of 87°19′ East with an altitude of 48 m above the mean sea level with an average annual rainfall of 1200 mm. The experimental site is shown in Figure 11.1. The average soil type of this region is light textured, acidic lateritic with pH ranging from 4.0 to 6.8. Soil at the experimental site is lateritic with sandy loam texture and very low fertility. The physical and chemical properties of the soil at the site are presented in Tables 11.1 and 11.2.

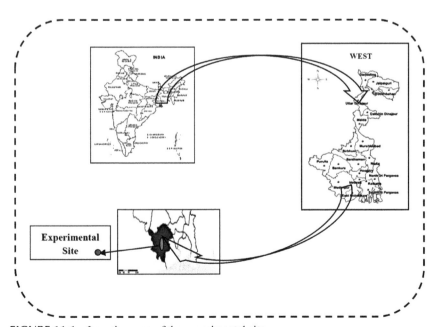

FIGURE 11.1   Location map of the experimental site.

TABLE 11.1   Physical Properties of Soil at the Experimental Site.

Soil depth (cm)	Particle size distribution (%)			Bulk density (g cm^{-3})	FC (mm cm^{-1})	WP (mm cm^{-1})	K_s (cm hr^{-1})
	Clay	Silt	Sand				
15	14.5	26.2	59.3	1.61	2.0	0.9	0.487
30	21.2	19.4	59.4	1.56	2.2	0.9	0.375
45	27.8	20.1	52.1	1.59	2.2	1.1	0.278
60	28.2	19.2	52.6	1.63	2.4	1.2	0.162
90	29.6	24.8	45.6	1.69	2.6	1.6	0.107

TABLE 11.2   Chemical Properties of Soil at the Experimental Site.

Parameters	Values
pH (1:2.5::soil:water)	5.2
Electrical conductivity (1:1::soil:water)	0.56 dS m^{-1} at 25°C
Cation exchange capacity	6.00 meq per 100 g soil
Organic carbon	0.28%
Available nitrogen	0.025%
Available phosphorus	0.004%
Available potassium	0.015%
Total nitrogen	0.035%
Total phosphorus	0.045%
Total potassium	0.420%

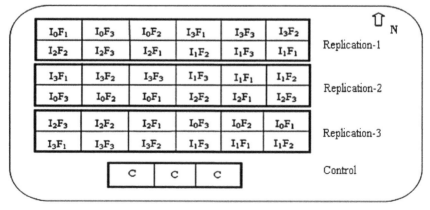

FIGURE 11.2   Schematic layout of the experimental plot: I = main treatment, the source of irrigation; F = subtreatment, fertilizer; and C = control plot.

Dugout ponds with a depth of 1.5 m and average water spreading area of 150 m² were constructed. The ponds were lined with suitable polythene sheets (Silpauline, blue colored, 250 gauge thickness). Fishponds were stocked with IMC with three different stocking densities (SD: 2.0, 3.5, and 5.0 numbers per square meter of water spread area), with a stocking ratio of 4:3:3 for *Catla, Rohu,* and *Mrigala,* respectively. Depending on the fish stocking density and the supply of enriched feed, the fishpond water gets polluted with time. Fishpond water needs to be exchanged, which could be used as nutrient-rich irrigation water. This exchanged water was used as a source of irrigation for tomato crop in this study.

Thirty-six plots (6 m × 5 m size each) were prepared adjacent to the fishponds along with three control plots. The plots were separated by 60 cm bunds. A field study was taken up with tomato (*Lycopersicum esculentum* L.) cultivar MHTM-256 (Suparna). The seedlings were raised in the nursery inside a polyhouse. Three weeks old seedlings were planted in the experimental plots with 0.75 m row to row and 0.6 m plant to plant spacings. Irrigation was based on the volume of water available from the fishponds. Split-plot experimental design was followed with irrigation as main treatments and suboptimal doses of fertilizer as subtreatments.

Four different sources of irrigation water (I_0 with direct tube well water, I_1 from fishpond with stocking density of 5.0/m², I_2 from fishpond with stocking density of 3.5/m², and I_3 from fishpond with stocking density of 2.0/m²) constituted the main treatment, three reduced doses of nitrogen fertilizers constituted the subtreatments and three replications were followed. Recommended full dose of fertilizer for the experiment was 80:40:40 kg ha⁻¹ (NPK). However, for nitrogen application, three subtreatments [90% N (F_1), 80% N (F_2), and 70% N (F_3)] were followed. All recommended doses of fertilizers except nitrogen were applied as a basal dose. Nitrogen fertilizer was applied in all cases at 20% of the treatment amount as a basal dose, and rest in two equal splits during the crop growth period. The schematic layout of the field experiment is shown in Figure 11.2.

The volume of water supplied to a given plot was known by measuring the discharge obtained from the pump and the time of application. Weather parameters (temperature, solar radiation, wind speed, rainfall, and evaporation during the crop growing seasons) are presented in Table 11.3.

a. Lined fishpond under construction   b.   A view of pond for fish rearing

c. Water filling in a fishpond   d.   Water exchange from a fishpond

e. Junction box showing aeration pipe   f.   Water quality analysis in the
    line   laborotory

g. Preparation of the experimental crop h   View of the experimental field
    field

i. Installation of a Symon Rain-gauge   j.   Installation of a USWB Class-A Pi
    at the experimental site   Evaporimeter at the experimental site

FIGURE 11.3   Details of field layout and site.

TABLE 11.3 Rainfall, Temperature, Solar Radiation, and Wind Speed During Crop Growing Seasons.

Parameter	Experiment-1 (2005–06)	Experiment-2 (2006–07)	Experiment-3 (2007–08)
Rainfall (mm)			
R_{max}	1.02	38.3	0
Total	2.8	148.9	0
Temperature (°C)			
T_{max}	38.5	35.5	33.4
T_{min}	9.2	9.0	8.6
Mean	25.1	24.6	23.54
Solar radiation ($kW\ m^{-2}\ h^{-1}$)			
Average	0.18	0.21	0.15
Wind speed ($m\ s^{-1}$)			
Maximum	6.48	7.26	7.44
Mean	0.36	0.28	0.26

A total of 2.8, 148.9, and 0 mm of rainfall were received during each winter growing seasons for tomato crop. The details of experimental layout and the site are shown in Figure 11.3. The yield obtained from the control plot with 100% recommended dose of fertilizer and tube well water was compared with the yield from the treatment plots. The nutrient supplementing potential of the fishpond water was also studied in a framework of proper experimental design.

11.3   RESULTS AND DISCUSSION

11.3.1   WATER QUALITY

Average values of various water quality parameters from two different sources, namely, tube well water and fishpond wastewater along with their permissible limits are listed in Table 11.4. The average values of the water quality parameters show wide variations between the tube well water and fishpond wastewater. The values for ammoniacal N (NH_3–N), orthophosphate (PO_4–P), and nitrite-N (NO_2–N) were almost zero for tube well water. The physicochemical characteristics of water in fishponds are one of the deciding factors in optimizing the conditions for fish productivity in small fishponds. The fertility status of the fishponds is known to be directly related to the water quality.[22] Water exchange has direct influence over the water quality of the pond, growth of fish, and economy of the fish culture.

Repeated water exchange during the later stage of fish growth could reduce the total ammonia nitrogen (TAN) concentration and other nitrogenous parameters in the fishpond water for all the three SD and in all the 3 years of the study. More frequent water exchange was needed for the higher stocking density ponds than the lower density ponds. For example, fishpond water was exchanged eight times in S.D-2.0 fishpond compared to 13 times in S.D-3.5 and 17 times in S.D-5.0 during 2005–06. Similar requirements were noticed in the remaining 2 years of the study.

TABLE 11.4 Mean Values of Water Quality Parameters of Tube Well Water and Wastewater from Fishpond and Their Ideal Values.

Parameter	Tube well water	Wastewater from fishpond	Ideal value (range)[a]
Temperature (°C)	28.64 ± 4.26	26.55 ± 4.85	25–32
pH	6.65 ± 0.87	7.24 ± 0.56	6.7–8.5
DO (mg L^{-1})	2.2 ± 1.35	5.66 ± 1.05	5–10
TSS (mg L^{-1})	74.65 ± 18.66	88.65 ± 22.5	30–200
NO_3^-–N (mg L^{-1})	0.021 ± 0.004	0.55 ± 0.42	0.1–3.00
Ammonia nitrogen (NH_3–N) (mg L^{-1})	–	<0.1	
Unionized		<1.0	0–0.1
Ionized			0–1.0
NO_2^-–N (mg L^{-1})	–	0.07 ± 0.05	0–0.5
Total N (mg L^{-1})	–	2.12 ± 0.74	0.05–4.5
PO_4^{3-}–P (mg L^{-1})	–	0.12 ± 0.06	0.05–0.4

[a]The ideal values of water quality parameters (physical and chemical) for aquacultural practices in freshwater, prescribed by Central Institute of Freshwater Aquaculture (CIFA).

Boyd et al.[6] suggested that water exchange is an effective measure in improving the water quality in small fishponds. Out of 11 months of the culture period, no water exchange is required during the initial 3 months of culture (i.e., June–August). It may be due to low biomass of fish in the culture ponds and dilution effect of pond water due to heavy rainfall during these months.[25,27–29] From the month of September onward, monthly water exchange ranging from 10% to 40% is needed based on the degree of pollution of the fishpond water. The total amount of water exchange during the culture period varied from 80% to 170%.

It may be noted that about 2.08 M m^3 of exchanged water is available from fishpond of 1 ha area with a stocking density of 5/m^2. For semi-intensive IMC culture, with stocking density ranging from 2.0 per m^2 to 5.0 per m^2, about 1.13–2.08 M m^3 of water is required for exchange in a year. In a recirculatory aquaculture system, the values of TAN or other parameters are easily controlled by filtering the water through a suitable filtration system. In case of intensive pond culture system, water exchange is considered as a better option for controlling TAN and other parameters. From the present study, it is estimated that large volume of water should be exchanged for maintaining the TAN values within its permissible range. Disposal of the huge volume of polluted water from an intensive fish farm to the adjacent environment is of increasing concern.[9,21,28–30,33,41] On the other hand, the scarcity of fresh water is found as a limitation of water exchange.[4] The introduction of integrated aquaculture-cum-irrigation (IAI) has been recommended as a solution to the problem by many researchers. In this chapter, the water was used twice, first for aquaculture and then for irrigation. The water with high concentration of different inorganic nutrients is considered to be polluted water for fish culture. However, it is enriched with different nutrients for agricultural crop production.

11.3.2   SOIL MOISTURE DYNAMICS

The variation of soil moisture content during the cropping season in tomato crop is shown in Figure 11.4. The variation was less from January 15 to January 29 and from February 5 to harvest date in 2006–07 due to rainfall received during that period. But large variations in moisture level between irrigation treatments were observed during 2005–06 and 2007–08 throughout the cropping seasons as there was no rainfall. In 2005–06, the maximum variation (about 4%) was found in case of I_1 during the latter part of flowering because of less number of water exchanges from the SD of 2.0 for the fishpond. The soil moisture variation in case of I_0 was not conspicuous during all the seasons due to the application of water at the time of crop need directly from the tube well. The variation in the treatments was narrowed down during the later stage of the crop due to frequent water exchange resulting in the availability of more irrigation water. The least variation in soil moisture content with I_3 treatment was observed due to the frequent availability of exchanged water from S.D-5.0 fishpond.

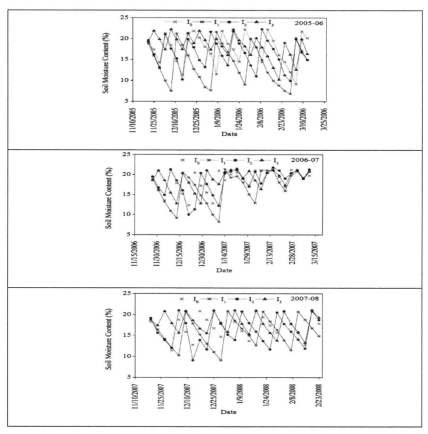

FIGURE 11.4   Variation of soil moisture in different treatments during the winter season for tomato.

11.3.3 NUMBER OF IRRIGATIONS

Irrigation was provided to the crop based on the availability of water from fishpond due to exchange. The stage of application with numbers of irrigation was monitored regularly and the data are presented in Table 11.5. In 2005–06, very less amount of rainfall (2.8 mm) was received during the latter part of the crop when it was about to be harvested. During 2007–08, the crop received no rainfall, whereas in 2006–07 and 149.3 mm of rainfall was received during the crop growth. Since the rainfall was received during harvesting time, there was rotting of tomato in the field resulting in low yield during that year.

TABLE 11.5 Number of Irrigation to Tomato Crop in Different Crop Stages and Years.

Crop seasons	Irrigation sources	Number of irrigation in different stages of crop					Total number of irrigation
		Initial estab-lishment	Vegetative	Flowering	Fruit initiation/ fruiting	Fruit harves-ting	
2005–06	I_0	2	1	2	1	–	6
	I_1	2	1	1	1	–	5
	I_2	2	2	2	1	–	7
	I_3	2	2	2	2	2	10
2006–07	I_0	1	1	2	1	–	5
	I_1	1	1	1	1	–	4
	I_2	2	1	2	1	–	6
	I_3	2	2	2	1	1	8
2007–08	I_0	2	1	2	1	–	6
	I_1	2	1	1	1	–	5
	I_2	2	1	2	1	–	6
	I_3	2	2	2	1	1	8

I_0—Irrigation with tube well water; I_1—Irrigation from SD-2.0; I_2—Irrigation from SD-3.5; I_3—Irrigation from SD-5.0.

The treatment I_3 received the highest number of irrigation (10 in 2005–06 and 8 both in 2006–07 and 2007–08). The number of irrigation provided under I_1 treatment was the least (4–5) due to the availability of less water from fishpond water exchange. Irrigation from the treatment I_0 was provided by tube well water based on the moisture depletion pattern of the soil in the field. The total number of irrigation provided under I_0 varied from 5 to 6. As the water of the fishpond (SD: 5) was polluted early, more number of irrigation was applied under I_3 and at late fruiting stage of the crop, this excess irrigation water was not utilized properly by the crop. Irrigation from source I_2 was almost uniform in different crop growth stages.

11.3.4   NUTRIENT RECOVERY FROM FISHPOND

The amount of nutrient recovered from the fishpond was estimated from the nutrient loads of exchanged wastewater. It is found that as the number of water exchange was more in high stocking density fishpond (I_3), the amount of recovered nutrient was also more with I_3 as compared to the two other SD. The amount of nutrient recovered from different fishponds is presented in Table 11.6. It may be noted that the nutrient recovery (N) is higher in case of SD of 5.0 fishpond (33.27 kg ha^{-1}) compared to the other two treatments. The total nutrient recovered from the fishponds through water exchange was not fully utilized during the cropping season due to mismatch of crop stage with the water exchange calendar. However, about 65–75% of the total nutrients were utilized by the crop. The recovery of phosphate from the exchanged water of fishpond was estimated to be about 70%. Higher recovery (1.70 kg ha^{-1}) occurred with SD of 5.0 fishpond as compared to the other two treatments.

11.3.5   CROP PERFORMANCE

The application of suboptimal and full doses of fertilizer favored the growth and yield of tomato. The crop yield was found to be statistically different under different treatments. An increase in yield was also observed with the increase in fertilizer dose. The control plot yield of tomato with reduced doses of N fertilizer was found to be statistically at par with the S.D-5.0 treatment.

There was no significant effect of N levels on fruit yield. The yield was low during 2006–07 due to unseasonal high rainfall at the maturity stage of the crop. Many fruits were damaged due to rotting. The maximum yield (66.85–70.19 t ha^{-1}) was recorded with I_3 and the minimum yield (53.07–61.29 t ha^{-1}) was from I_1. The mean data indicated that I_3 produced the highest yield of 68.27 t ha^{-1} which was 5.6–20.8% more than the other treatments. I_1 gave the lowest yield of 56.53 t ha^{-1}. It is revealed from the mean data that F_1 produced the highest yield of 63.03 t ha^{-1} and F_3 gave the lowest yield of 62.69 t ha^{-1}. The interaction effect of irrigation and N was not significant. However, the maximum yield of 69.3 t ha^{-1} was recorded with I_3F_2 followed by I_3F_3 (67.8 t ha^{-1}) and I_3F_1 (67.7 t ha^{-1}).

TABLE 11.6 Nutrient Recovery (kg ha⁻¹) from Fishpond.

Fishpond	Nitrogen (recovered)			Mean ± SD	Nitrogen (utilized)			Mean ± SD
	2005–06	2006–07	2007–08		2005–06	2006–07	2007–08	
SD-2.0	12.45	12.09	15.54	13.36 ± 1.55	6.70	6.49	9.75	7.65 ± 1.49
SD-3.5	26.33	15.06	22.01	21.13 ± 4.64	11.98	10.82	13.95	12.25 ± 1.29
SD-5.0	37.40	27.34	35.07	33.27 ± 4.30	21.5	16.39	21.55	19.81 ± 2.42
	Phosphate (recovered)				Phosphate (utilized)			
SD-2.0	0.79	0.40	0.92	0.70 ± 0.22	0.43	0.4	0.92	0.58 ± 0.24
SD-3.5	2.01	1.15	2.33	1.83 ± 0.50	1.00	0.89	1.40	1.10 ± 0.22
SD-5.0	2.75	1.88	3.60	2.74 ± 0.70	1.57	1.24	2.29	1.70 ± 0.44

TABLE 11.7   Effect of Irrigation Source and Nitrogen Level on Fruit Yield (t ha^{-1}) of Tomato.

Treatment	2005–06	2006–07	2007–08	Mean
	Irrigation sources			
I_0	63.25	56.82	65.53	61.87
I_1	55.22	53.07	61.29	56.53
I_2	67.39	59.98	66.52	64.63
I_3	67.77	66.85	70.19	68.27
SEM±	1.57	1.13	1.80	0.88
CD (0.05)	5.44	3.92	NS	2.62
	Nitrogen levels			
F_1	64.32	58.89	65.87	63.03
F_2	62.88	59.67	65.70	62.75
F_3	63.02	58.98	66.08	62.69
SEM±	1.76	1.70	1.86	1.03
CD (0.05)	NS	NS	NS	NS
	Interaction (I × F)			
SEM±	3.28	3.01	3.53	1.89
CD (0.05)	NS	NS	NS	NS
Control Yield	–	–	–	–
Control plot	70.52	68.26	69.67	69.48

I_0—Irrigation with tubewell water; I_2—Irrigation from SD-3.5; I_1—Irrigation from SD-2.0; I_3—Irrigation from SD-5.0; F_1—90% of recommended N; F_2—80% of recommended N; F_3—70% of recommended N.

The average yield obtained from the control plot was 70.48 t ha^{-1}. The comparison of yield between I_3 and control plot shows that both the yields are at par. Therefore, it can be inferred that even with the reduced dose of fertilizer application, the yield of tomato can be at par with that of 100% fertilizer dose use if the crop is irrigated with fishpond wastewater (Table 11.7; Fig. 11.5a and b).

The yield of tomato was the highest with irrigation from S.D-5.0 treatment fishpond and application of suboptimal fertilizer dose. This highest yield was almost at par with the result obtained from control plot with 100% of recommended doses of fertilizer. There was a significant variation of the yield of tomato irrigated with tube well water and with the wastewater from S.D-2.0 fishpond. It may be attributed to the less number

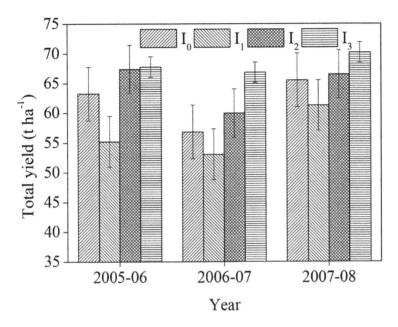

FIGURE 11.5a Effect of source of irrigation on fruit yield.

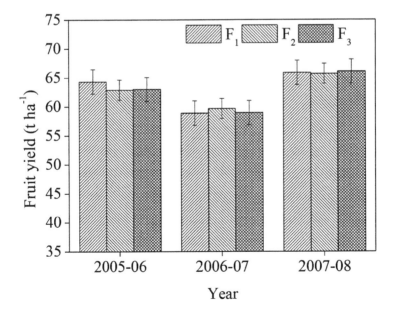

FIGURE 11.5b Effect of nitrogen level on fruit yield.

of irrigation that could be possible from S.D-2.0 fishpond wastewater and without any scientific scheduling. Ray et al.[24] reported a tomato yield of 64.5 t ha^{-1} irrigated with tube well water, whereas it increased to 95.8 t ha^{-1} due to irrigation with fishpond effluent. Similar findings were also reported by Castro et al.[7] Pinto[23] observed an increase in tomato productivity in the range of 19.5–21.8% when fertilizer application was changed from conventional method to fertigation. This is supported with the findings by other investigators.[10,15,35,39] Research conducted with aqua effluent irrigation claimed to have reduced the recommended fertilizer by almost 50% in a field trial in Saudi Arabia with wheat as the trial crop. However, in this literature no mention has been made of the type of fish reared and stocking details.[13] On the contrary, aqua effluent irrigation needs to be provided along with the recommended doses of fertilizer, as it contains the least amount of nutrients.[40] Similar findings were obtained in the present field investigation with IMC stocked fishpond wastewater carrying low nutrient value.[40] Castro et al.[7] also reported that there was nonsignificant interaction between the types of irrigation and fertilizers on fruit mean weight. Only types of irrigation had a significant effect, as the plants irrigated with well water had higher fruit mean weight than the plants irrigated with fish effluent. The increase of tomato fruit yield with fertilizer intervention was reported by several researchers. But the research work was done mostly on drip fertigation. There was an increase in tomato yield with an increase in N level as reported by several investigators.[2,5,10–12,18,20,26,31,32,35]

11.4   SUMMARY

Wastewater from fishponds cultured at three SD of IMC was evaluated to find out its efficacy as irrigation source with tomato as a test crop. A field trial was conducted in sandy loam soil of Indian Institute of Technology (IIT) Kharagpur, West Bengal, India during 2006–08. Water quality parameters such as temperature, pH, dissolved oxygen, nitrite, nitrate, TAN, orthophosphate, and total suspended solid for fishpond wastewater were monitored on every alternate date for the entire growth period of IMC. Water exchange was performed before the fishpond water attained the critical level of TAN value as it is harmful to the pond ecosystem and fish growth. The exchanged water was used for irrigation. The maximum numbers of water exchange for fishponds for a culture period of 300 days were 10, 13, and 17 for S.D-2.0, 3.5, and 5.0, respectively. The highest and lowest yields for tomato were 68.27 t ha^{-1} and of 61.87 t ha^{-1} with irrigation from S.D-5.0 and tube well water, respectively. There was a recovery of inorganic nitrogen to the tune of 13.36–33.27 kg ha^{-1} and phosphate to the tune of 0.70–2.74 kg ha^{-1} from the fishpond wastewater. The ratio of pond area to crop area for an integrated agri-aquaculture system was estimated to be 35:65 for S.D-2.0; 30:70 for S.D-3.5, and 22:78 for S.D-5.0.

ACKNOWLEDGMENT

Research funds available from the Indian Council of Agricultural Research (ICAR), New Delhi, for conducting the field research at Indian Institute of Technology, Kharagpur, West Bengal, India are thankfully acknowledged by the authors.

KEYWORDS

- alternate irrigation sources
- aqua effluent
- crop performance
- fishpond
- fishpond wastewater exchange
- Indian Major Carp
- integrated agri-aquaculture
- wastewater
- water exchange

REFERENCES

1. Al-Ama, M. S.; Nakhla, G. F. Waste Water Reuse in Jubail, Saudi Arabia. *Water Res.* **1995,** *29*(6), 1579–1584.
2. Al-Mutaz, I. S. Treated Wastewaters as a Growing Water Resource for Agriculture Use. *Desalination* **1989,** *73*, 27–36.

3. Al-Shammiri, M.; Al-Saffar, A.; Bohamad, S.; Ahmed, M. Waste Water Quality and Reuse in Irrigation in Kuwait Using Micro-filtration Technology in Treatment. *Desalination* **2005,** *185,* 213–225.

4. Avnimelech, Y.; Kochba, M. Evaluation of Nitrogen Uptake and Excretion by Tilapia in Biofloc Tanks, Using ^{15}N Tracing. *Aquaculture* **2009,** *287*(1), 163–168.

5. Bafna, A. M.; Daftardar, S. Y.; Khade, K. K.; Patel, V. V.; Dhotre, R. S. Utilization of Nitrogen and Water by Tomato Under Drip Irrigation System. *J. Water Manag.* **1993,** *1*(1), 1–5.

6. Boyd, C. E.; Pillai, U. K. *Water Quality Management in Aquaculture*; Special Publication No. 22, Central Marine Fisheries Research Institute, Cochin, India, 1984; p 76.

7. Castro, R. S.; Borges Azevedo, C. M. S.; Bezerra-Neto, F. Increasing Cherry Tomato Yield Using Fish Effluent as Irrigation Water in Northeast Brazil. *Sci. Hortic.* **2006,** *110,* 44–50.

8. David, D. J.; Williams, C. H. Effects of Cultivation on the Availability of Metals Accumulated in Agricultural and Sewage-treated Soils. *Progress. Water Technol.* **1979,** *11,* 257–264.

9. *FAO Statistics Database 2006;* FAO (Food and Agriculture Organization). http://faostat. fao.org, (accessed June 10, 2008).

10. Hebbar, S. S.; Ramachandrappa, B. K.; Nanjappa, H. V.; Prabhakar, M. Studies on NPK Drip Fertigation in Field Grown Tomato (*Lycopersicon esculentum* Mill.). *Eur. J. Agron.* **2004,** *21,* 117–127.

11. Herrera, F.; Castillo, J. E.; Chica, A. F.; Lopez-Bellido, L. Use of Municipal Solid Waste Compost (MSWC) as a Growing Medium in the Nursery Production of Tomato Plants. *Bioresour. Technol.* **2008,** *99*(2), 287–296.

12. Hussain, Z. Problems of Irrigated Agriculture in Al-Hassa, Saudi Arabia. *Agric. Water Manag.* **1982,** *5*(4), 359–374.

13. Hussain, G.; Al-Jaloud, A. A. Effect of Irrigation and Nitrogen on Water Use Efficiency of Wheat in Saudi Arabia. *Agric. Water Manag.* **1995,** *27,* 143–153.

14. Hussain, G.; Al-Saati, A. Wastewater Quality and Its Reuse in Agriculture in Saudi Arabia. *Desalination* **1999,** *123,* 241–251.

15. Ismail, S. M.; Ozawa, K.; Khondaker, N. A. Influence of Single and Multiple Water Application Timings on Yield and Water Use Efficiency in Tomato (var. First power). *Agric. Water Manag.* **2008,** *95,* 116–122.

16. Li, T. Y.; Baozhong, P.; Houquin, H.; Korong, J.; Monggae, C.; Disheitane, K. *Experimental Irrigation Project at Glen Valley Water Care Works: Final Report*; Sino-Botswana Government, 2001; p 100.

17. Leeper, G. W. *Managing the Heavy Metals on the Land;* Marcel Dekker: New York, 1978; p 233.

18. Mahajan, G.; Singh, K. G. Response of Greenhouse Tomato to Irrigation and Fertigation. *Agric. Water Manag.* **2006,** *84,* 202–206.

19. Miller, G. W. Integrated Concepts in Water Reuse: Managing Global Water Needs. *Desalination* **2006,** *187,* 65–75.

20. Mofoke, A. L. E.; Adewumi, J. K.; Babatunde, F. E.; Mudiare, O. J.; Ramalan, A. A. Yield of Tomato Grown Under Continuous-flow Drip Irrigation in Bauchi State of Nigeria. *Agric. Water Manag.* **2006,** *84,* 166–172.

21. Naylor, R. L.; Goldburg, R. J.; Primavera, J. H.; Kautsky, N.; Beveridge, M. C. M.; Clay, J.; Folke, C.; Lubchenco, J.; Mooney, H.; Troell, M. Effect of Aquaculture on World Fish Supplies. *Nature* **2000**, *405*, 1017–1024.

22. Ntengwe, F. W.; Edema, M. O. Physicochemical and Microbiological Characteristics of Water for Fish Production Using Small Ponds. *Phys. Chem. Earth* **2008**, *33*, 701–707.

23. Pinto, J. M. Doses and Period of Application of Irrigation on Tomato Crop (Portuguese). *Hort. Bras.* **1997**, *15*(1), 15–18.

24. Prinsloo, J. F.; Schoonbee, J. H. Investigations into the Feasibility of a Duck-fish-vegetable Integrated Agriculture: Aquaculture System for Developing Areas in South Africa. *Water Sci.* **1987**, *13*(2), 109–118.

25. Ray, L. I. P.; Bag, N.; Mal, B. C.; Das, B. S. In *Feasibility of Irrigation Options with Aquaculture Waste Water,* Proceedings of 2nd International Conference on Hydrology and Watershed Management, 2006; Vol. 2, pp 873–885.

26. Ray, L. I. P.; Panigrahi, P. K.; Moulick, S.; Bag, N.; Mal, B. C.; Das, B. S. In *Multiple Usage of Fresh Water in Aquaculture and Olericulture—A Case Study,* National Workshop on Sustainability of Aquacultural Industry, IIT Kharagpur, Sept 28–29, 2007; pp 88–95.

27. Ray, L. I. P.; Panigrahi, P. K.; Mal, B. C. In *Adequacy of Aquaculture Effluent as an Irrigation Source*, Paper #SWE-28 at XLII (42nd) ISAE Annual Convention and Symposium at CIAE, Bhopal, Feb 1–3, 2008.

28. Ray, L. I. P.; Moulick, S.; Mal, B. C.; Panigrahi, P. K. In *Effect of Intensification on Water Quality and Performance of Aqua-effluent as an Irrigation Source*, Proceedings of 2nd World Aqua Congress, Nov 26–28, 2008; Vol. 2, pp 402–410.

29. Ray, L. I. P.; Mal, B. C.; Das, B. S. In *Temporal Variation of Water Quality Parameters in Intensively IMC Cultured Lined Pond,* E-Proceedings of International Conference on Food Security and Environmental Sustainability (FSES-2009), Dec 17–19, 2009; pp 1–10.

30. Ray, L. I. P.; Moulick, S.; Mal, B. C.; Panigrahi, P. K. In *Effect Water Quality on Irrigation*, Proceedings of 2nd World Aqua Congress, Nov 26–28, 2008; Vol. 2, 397–402.

31. Ray, L. I. P.; Panigrahi, P. K.; Moulick, S.; Mal, B. C., Das, B. S.; Bag, N. Aquaculture Waste Water—An Irrigation Source. *Int. J. Sci. Nature (IJSN)* **2010**, *1*(2), 148–155.

32. Ray, L. I. P. Techno-economic Feasibility of Integrated Agri-aquaculture System. Unpublished Ph.D Thesis, Submitted to Indian Institute of Technology, Kharagpur 721302, West Bengal, 2014; p 211.

33. Read, P.; Fernandez, T. Management of Environmental Impacts of Marine Aquaculture in Europe. *Aquaculture* **2003**, *226*(1), 139–163.

34. Shuval, H. Water Pollution Control in Semi-arid and Arid Zones. *Water Res.* **1967**, *2*, 297–308.

35. Singandhupe, R. B.; Rao, G. G. S. N.; Patil, N. G.; Brahmanand, P. S. Fertigation Studies and Irrigation Scheduling in Drip Irrigation System in Tomato Crop (*Lycopersicon esculentum* L.). *Eur. J. Agron.* **2003**, *19*, 327–340.

36. Sopper, W. Disposal of Municipal Wastewater Through Forest Irrigation. *Environ. Pollut.* **1971**, *1*, 263–284.

37. Toze, S. Reuse of Effluent Water-benefits and Risks. *Agric. Water Manag.* **2006**, *80*, 147–159.

38. *Process Design Manual for Land Treatment of Municipal Wastewater*, USAEPA (United States Environmental Protection Agency), 1981; p 254 (EPA 625/1-81-013).
39. Wang, D., Kang, Y.; Wan, S. Effect of Soil Matric Potential on Tomato Yield and Water Use Under Drip Irrigation Condition. *Agric. Water Manag.* **2007,** *87,* 180–186.
40. Wood, C. W.; Meso, M. B.; Karanja, N.; Veverica K. *Use of Pond Effluent for Irrigation in an Integrated Crop/Aquaculture System;* Ninth Work Plan, Effluents and Pollution Research (9ER1) Final Report, 2005; p 112.
41. Yokoyama, H. Environmental Quality Criteria for Fish Farms in Japan. *Aquaculture* **2003,** *226*(1), 45–56.

PART IV
Performance of Agricultural Crops

CHAPTER 12

PERFORMANCE OF TOMATO UNDER BEST MANAGEMENT PRACTICES

K. NITHA and D. TAMILMANI*

Department of Soil and Water Conservation Engineering, Agricultural Engineering College and Research Institute (AEC & RI), Tamil Nadu Agricultural University, Coimbatore 641003, Tamil Nadu, India

Corresponding author. E-mail: swc@tnau.ac.in

CONTENTS

A modified version of "Nitha, K. *Effect of Drip Fertigation and Mulching on Tomato Crop.* Thesis submitted in partial fulfillment of the requirements for the award of the degree of Master of Technology (Agricultural Engineering) in Soil and Water Conservation Engineering, Department of Soil and Water Conservation Engineering, Agricultural Engineering College and Research Institute (AEC & RI), Tamil Nadu Agricultural University, Coimbatore 641003, Tamil Nadu, India, 2012."

In this chapter: US$ 1.00 = Rs. 60.00 (Indian Rupees)

ABSTRACT

Land and water are the indispensable resources of life system. Water is the most limiting natural resource in arid and semiarid areas for the economic development of the country. In most of the areas, the only water available is the rain that falls on the area; hence, for successful agriculture, proper utilization of water is very essential which means to increase the water use efficiency of a crop by adopting water conservation measures. The water loss takes place in nature due to evaporation, transpiration, and percolation. The percolation losses can be avoided by applying water to the plant properly. The evaporation losses can be minimized by the use of mulches such as crop wastes, polyethylene plastics, and chemicals.

In many countries and regions, fresh water is relatively scarce, but there are considerable resources of saline water which could be utilized for irrigation if proper crops, soil, and water management practices are established. During the last three decades, micro irrigation systems owing to their capability to apply water efficiently, low labor and energy requirement, and increase in quantity and quality of crop yield/produce have made a breakthrough in many countries around the globe. Micro irrigation encompasses drip/trickle systems, surface and subsurface drip tapes, micro sprinklers, sprayers, microjets, spinners, rotors, and bubblers. It is a concept where water is applied at low rates frequently near the root zone of the plant and is successfully applied to the vegetable crops. In recent years, fertilizer is also applied along with water through drip irrigation to get higher fertilizer use efficiency besides increased yields.

The experiment was conducted in PFDC research and demonstration plot of TNAU, Coimbatore during December 2011 to April 2012 to study the effect of drip fertigation and mulching on tomato hybrid variety COTH-2. The experiments were laid out in factorial randomized block design with nine treatments which included three mulching levels such as 25 μm thickness plastic mulch, 50 μm thickness plastic mulch, and control and three fertigation levels including 80%, 100%, and 120% RDF which were replicated thrice.

The observations were recorded on biometric parameters, yield, soil temperature, wetted zone diameter, moisture distribution, and also soil physical and chemical properties. Based on the observations, the water use efficiency, fertilizer use efficiency, and benefit–cost ratio were worked out and statistical analysis was carried out for each observation to find out the significant effect on the treatment.

Soil samples were collected before transplanting and after final harvest and were analyzed to find the physical and chemical properties of soil. The bulk density decreased from initial to final stage in the mulching treatments. In total, 50 μm thickness mulching treatments have observed greater reduction in bulk density, that is, from 1.4 to 1.34 g/cc. Particle density also showed similar results as that of bulk density. Porosity increased from initial to final. Maximum porosity is found in 25 μm thickness mulching treatment. The porosity of initial soil was 42%. After harvest, it changed to 45%.

After final harvest, it was observed that among most of the treatments, nutrient values were less in treatments with 80% recommended N, P, and K. It indicated that the fertigation applied in those treatments was fully taken up by the plants for their growth and yield. Also, we can see that the final nutrient has increased than the initial content in all the cases due to the application of the fertilizer. The available N in the initial soil was 131.2 ppm. After final harvest, it gets reduced in the 80% RDF treatments, and was less (118 ppm) in control plot. The maximum N (146 ppm) was found in 50 μm thickness mulch with 120% RDF, T6. The nutrients P and K also showed similar result as that of N.

The discharge from the drippers at different points of emission was measured for a particular period of time at 0.5 kgf/cm^2 pressure and parameters such as coefficient of variation (Cv), the statistical uniformity (SU), and coefficient of uniformity (CU) were evaluated from the observed discharge. The Cv was obtained as 0.0198, SU as 98%, and CU as 96.87%.

Soil temperature was low in the early morning and gradually increased from 12 to 4 p.m. in all the treatments and then declined. Temperature under mulches was higher than that of the control plots for all the times. In mulching treatments, weeds were completely absent. Weeds were found only in the control plots and their numbers increased with respect to the increase in fertilizer application. Increased dry matter production of weeds was observed with advancement of crop growth.

The hybrid was assessed for the mean performance in respect of growth characters, namely, plant height, flowering traits, such as days to first flowering, first fruit set and first harvest, fruit characters, such as fruit weight, size, and yield.

The highest yield was recorded in T6 (85.96 t/ha), that is, in 50 μm thickness plastic mulching at 120% RDF followed by T5 and T3 (83.16 t/ha) and (81.22 t/ha), respectively, (Table 4.10, Fig. 4.13) and lowest yield was recorded in T7 (57.98 t/ha), that is, in control.

The highest water use efficiency of 386 kg/ha mm was recorded in treatment T6 which is 50 μm plastic mulch with 120% RDF. The least water use efficiency (261 kg/ha mm) was noted in control (T7). Increased fertilizer use efficiency with the decreased level of fertilizer dose through drip was observed.

The fixed cost of installation of drip irrigation system and mulching sheets for 25 μm thickness treatments was Rs. 54,910 per ha and Rs. 51,710 for 50 μm thickness treatments per year. The treatment T6 registered the highest gross income of Rs. 416,100 per ha. The benefit–cost ratio was also higher (4.17) in this treatment (T6) compared to all other treatments. In control plot (T7) with 80% RDF, the benefit–cost ratio was (2.04), which is less than the other treatments.

Among all the mulching and fertigation treatments, the best performance in terms of growth, yield, and quality were observed in 50 μm thickness mulching treatment with 120% RDF. The results have indicated that the plastic mulching has higher benefit in terms of yield, quality, water use efficiency, and benefit–cost ratio as compared to other treatments.

12.1 INTRODUCTION

Land and water are the indispensable resources of our life. Water is the most limiting natural resource in arid and semiarid areas for economic development of the country. The demand of water for agricultural purpose is estimated to increase from 50 million ha-m in 1985 to 70 million ha-m by 2050.[166] According to World Water Council (WWC), by the year 2020, we shall need 17% more water than presently available to feed the world.[171] In most of the areas, only water available is the rain that falls on the area. Hence, for successful agriculture requires proper utilization of water,[170] which implies to increase the water use efficiency (WUE) of a crop by adopting different water conservation measures. The existing methods of irrigation and the available facilities are not reliable and we are fore fronted with many problems regarding soil and water. The water loss takes place in nature due to evaporation, transpiration, and percolation. The percolation losses can be avoided by applying water to root zone of the plant in a proper way. The evaporation losses can be minimized by the use of mulches such as crop wastes, polyethylene plastics, and chemicals. The transpiration losses can be minimized to some extents by erecting tunnels over the crop.

In many countries and regions, fresh water is relatively scarce, but there are considerable resources of saline water which could be utilized for irrigation if proper crops, soil, and water management practices were established. Surface irrigation method, with an overall efficiency of only 20–50% usually causes erosion, salinization, and water logging problems. Two important aspects to be considered in this regard are uniform water distribution in the field and accurate amount of water application by permitting accurate delivery control. These requirements are accomplished by adopting the promising drip or micro irrigation techniques. During the last three decades, micro irrigation systems owing to their capability to apply water efficiently, low labor and energy requirement, and increase in quantity and quality of crop yield/produce have made a breakthrough in many countries around the globe. Micro irrigation encompasses drip/trickle systems, surface and subsurface drip tapes, microsprinklers, sprayers, microjets, spinners, rotors, and bubblers. Among these methods, drip irrigation is a concept where water is applied at low rates frequently near the root zone of the plant and is successfully employed in vegetable crops. In recent years, fertilizer is also applied along with water through drip irrigation to get higher fertilizer use efficiency (FUE) besides increased yields.[59–62]

Fertigation is a new concept recently practiced in several parts of the world in horticultural crops. It offers intensive and economical crop production where both water and fertilizers are delivered to the crop through the drip system. It provides essential elements directly to active root zone thus minimizing losses of expensive fertilizers which ensure higher and quality yield along with saving time and labor.[128] Experiments have already indicated that through fertigation 40–50% of nutrient could be saved.[59–62] In fertigation, fertilizer is applied in small and frequent doses that fit within scheduled irrigation intervals matching the plant water use to avoid leaching. Fertilizer can be injected into drip system by selecting appropriate applications for a wide assessment of available pumps, tanks, valves, ventures, and aspirators. Among these, venturi is the cheap and economic one though it creates a high-pressure loss. It is generally used in high-value crops, like tomato, capsicum, eggplant, strawberry, etc. Because soluble nutrients move with wetting front, precise management of irrigation quantity and rating and timing of N, P, and K application are critical for efficient vegetable production. With fertigation, nutrient use, efficiency is increased and the risk of loss of nutrients to the ground water is reduced. Farmers are gradually getting convinced of the added benefits of using fertigation along with water application by a drip system.

The practice of applying mulches for the production of vegetables is a thousand years old.[100,142] Mulching is the practice of covering the soil/ ground to make more favorable conditions for plant growth, development, and efficient crop production. Typically, mulching involves placing a layer of material on the soil around the crop of interest to modify the growing environment to improve crop productivity. The primary purpose for using mulches is for weed suppression in the crop to be grown. Mulches typically function by blocking light or creating environmental conditions which can prevent germination or suppress weed growth shortly after germination. However, numerous other benefits are often obtained including: increased earliness, moisture conservation, temperature regulation of the root zone and aboveground growing environment, reduced nutrient leaching, altered insect and disease pressures, and, in some instances, reduced soil compaction or improved soil organic matter.[98,99,120] The use of mulches typically results in higher yields and quality in vegetable crops, enhancing profitability for the grower. While natural mulches such as leaf, straw, dead leaves, and compost have been used for centuries, during the last 60 years the advent of synthetic materials has altered the methods and benefits of mulching. The research as well as field data available on the effect of synthetic mulches make a vast volume of useful literature. When compared to other mulches, plastic mulches are completely impermeable to water; it, therefore, prevents direct evaporation of moisture from the soil and thus limits the water losses and soil erosion over the surface.

Black plastic is the predominate mulch utilized in vegetable production today. Much of its popularity is due to low cost per acre compared to other mulches. However, black plastic mulch (BPM) also effectively warms the soil, improving early crop production and eliminates most in-row weed growth.

Tomato is the most important vegetable crop. Many tomato growers face challenges in producing the crops due to stricter environmental regulations and fewer chemicals available for weed control. There is a demand for cultural practices that reduce chemical inputs and synthetic materials.

Tomato crop is one of the most common vegetables belonging to the nightshade family, *Solanaceae*. Tomato (*Solanum lycopersicum*) is a popular vegetable in India. It is rich in vitamins A and B, and iron. The fruit is consumed in diverse ways, including raw, as an ingredient in many dishes and sauces, and in drinks. While it is botanically a fruit, it is considered a vegetable for culinary purposes. Tomatoes are also commonly classified as determinate or indeterminate. Determinate types are preferred by commercial growers and indeterminate varieties are preferred by home growers and local market farmers who want ripe fruit throughout the season. Tomato requires a relatively cool, dry climate for high yield and premium quality. However, it is adapted to a wide range of climatic conditions from temperate to hot and humid tropical. The optimum temperature for most varieties lies between 21°C and 24°C.

The objectives of the research study in this chapter are: to study the effect of different plastic mulches on the yield and growth parameters of tomato; to find out optimum fertigation scheduling under drip irrigation for tomato; and to work out the cost economics of the system.

12.2   REVIEW OF LITERATURE

Due to the decreasing availability of water resources and the increasing competition for water between different users, improving agricultural WUE is vitally important in many parts of the world that have limited water resources. Drip irrigation is an effective tool for conserving water resources and studies have revealed significant water saving ranging between 40% and 70% by drip irrigation compared with surface irrigation. Fertigation is a new concept recently practiced in several parts of the world in horticultural crops. Drip fertigation increases the efficiency of the applied fertilizers thus economizing the quantity of fertilizers and water, and cost of labor and energy resulting in reduced cost of cultivation. The conservation of soil moisture may help in preventing the loss of water through evaporation from the soil facilitating maximum utilization of moisture by the plants. Productivity can be increased by adopting an improved package of practices, particularly in situ moisture conservation by mulching.

12.2.1   DRIP IRRIGATION

Drip irrigation is popularly known as an ingenious method of irrigation, wherein water and soluble nutrients are delivered near the roots of the plants. The modern technology of drip irrigation is successfully practiced in many countries for orchards, vegetables, ornamental crops, and high-value field crops. It is gaining momentum and its prospects in the years to come are expected to be very bright.

Nakayama and Bucks[117] stated that drip irrigation was considered as an emerging technology with its application limited to some special crops; nowadays, it is used on a wide variety of crops which were initially considered unprofitable for management under drip irrigation. The drip irrigation method provides the best possible conditions of total soil water potential for a given quality of irrigation water (IW)[59–62] with a 50% water savings from use of drip irrigation when compared to furrow irrigation.[59–62]

Water conservation is an important issue in Western Colorado, which depends on limited water supplies for use in both urban and agricultural areas. Subsurface drip irrigation also gives growers the flexibility to inject liquid fertilizer at the time when plants need it and according to the amount needed. Drip irrigation can be used in row crops such as sugarcane, vegetables, mulberry, cotton, cassava, etc., in water scarcity areas. There can be considerable saving of IW by adopting drip method since water can be applied almost precisely and directly in the root zone without wetting the entire surface area.[20] Micro irrigation can be used to improve the irrigation efficiency of vegetable gardens by reducing evaporation and drainage losses by creating and maintaining soil moisture conditions that are favorable to crop growth. Micro irrigation that was evaluated includes low-head drip irrigation, pitcher irrigation, and subsurface irrigation using clay pipes. Good results were obtained with subsurface irrigation when irrigation was carried out with poor-quality water.[24]

At Bhavanisagar, results were encouraging with drip irrigation for closer spaced crops like turmeric. The fresh rhizome yield of turmeric was enhanced up to 76.3% with water saving of 53.1% besides 25% saving in nitrogen fertilizer when applied through drip system daily at the rate of 40% of surface irrigation level.[150]

Decreasing the frequency of irrigation reduced plant growth, height, and thickness of stem in greenhouse tomato.[30] Shelke et al.[157] reported that maintaining available soil moisture at low water tension and maintaining it almost constant level during the entire growth period through drip irrigation was beneficial in terms of the economics of yield and could save up to 50% of IW.

A study has been conducted on economic comparison of drip and furrow irrigation methods at New Mexico.[56] The results indicated that yield was 25% greater when employing drip irrigation. The mean values of sugar beet yield and sugar content were higher with drip irrigation than with furrow practices. Comparing three drip treatments, a higher irrigation frequency was found to be more efficient for sugar beet production.[155]

Shock[162] reported that drip irrigation can help to use water efficiently. A well-designed drip irrigation system loses practically no water to runoff, deep percolation, or evaporation. Irrigation scheduling can be managed precisely to meet crop demands, holding the promise of increased yield and quality. Tingwu et al.[175] conducted studies in China with four irrigation treatments [control, 30%, 60%, and 90% of evapotranspiration (ET) from Chinese evaporation pan]. The yield and quality of water melon were improved under drip irrigation, as compared with control. The highest increase in both yield and quality was found in 60% treatment. Most salts were leached out of the root zone in the 60% and 90% treatments. The results suggested that drip irrigation of watermelon with saline water was feasible.

Enda et al.[48] conducted the experiment and understood the effect of different irrigation methods and scheduling on morphological, biophysical, yield, and WUE of capsicum. The plants grown under drip irrigation had more number of branches and plant heights compared to that of surface irrigated plants.

The yield had significant positive correlation with total dry matter (TDM, 0.865) and net photosynthesis (0.840). Thus, drip irrigation at 100% cumulative pan evaporation (CPE) is beneficial for capsicum plant in terms of yield and better plant morphological characters, namely plant height number of branches, root finesses, and root length.

Aujla et al.[19] evaluated the effect of various levels of water and N application through drip irrigation on seed cotton yield and WUE. In this experiment, three levels of water (IW/CPE ratio of 0.4, 0.3, and 0.2) and three levels of N (100%, 75%, and 50% of recommended N, 75 kg/ha) through drip were compared with check-basin method of irrigation under two methods of planting (normal sowing and paired sowing). The results revealed that when the same quantity of IW and N was applied through drip irrigation system, it increased the seed cotton yield to 2144 from 1624 kg/ha (an increase of 32%) under the check-basin method of irrigation. When the quantity of water through drip was reduced to 75%, the increase in seed cotton yield was 12%. However, when water was reduced to 50%, it resulted in 2% lower yield than the check basin.

Tyson et al.[179] of Georgia University stated that drip irrigation is gaining popularity for production of some vegetable crops. It can be used with or without plastic mulch. One of the major advantages of drip irrigation is its WUE if properly managed. Studies in Florida have indicated that 40% less water was required for drip-irrigated vegetables than the sprinkler-irrigated vegetables. Weeds are also less of a problem since only the rows are watered and the middles remain dry. Also, some studies have indicated that drip irrigation enhances earlier yields and fruit size. Shankar et al.[153] conducted an experiment to study the effect of various irrigation methods, that is, drip, mini sprinkler, big sprinkler, and surface irrigation on the growth, yield, and storage of onion cv. N-2-4-1. The highest yield was recorded in drip irrigation (47.47 t/ha) followed by big sprinkler (31.21 t/ha). The lowest yield was recorded in surface irrigation (22.79 t/ha).

12.2.1.1 UNIFORMITY AND WETTING PATTERN OF DRIP IRRIGATION

Uniformity of drip irrigation shows the efficient distribution of water throughout the field. Field uniformity is helpful in improving the operation and management of the drip irrigation system. Soil moisture distribution is one of the most important factors involved in successful design and management of a drip system. Keller and Karmeli[92] introduced Cv as a statistical measure of discharge variation in irrigation emitters due to the manufacturing process. Howell et al.[78] stated that the distribution pattern of soil moisture resulting from the drip irrigation wetting of soil was bulb like auxiliary symmetric pattern and the pattern of wetting would be of two dimensional.

Bralts and Kesner[28] developed a statistical method for field uniformity estimation of drip irrigation submain units based on the coefficient of variation and the statistical uniformity coefficient. The method was based on the assumption that the emitter flow variation was normally distributed. Khepar et al.[94] reported that the moisture distribution in drip irrigation systems depends on rates of application, amount of water applied, and the initial moisture content. As the rate of application increased, the vertical component of the wetting zone increased in light textured soils. Hanson et al.[72,73] conducted experiment on row crops to investigate wetting patterns under drip irrigation under a variety of conditions. The conditions revealed the wetting pattern in a very fine textured soil, under different irrigation frequencies and at different installation depths of drip tape. Patterns were also developed for conditions of mild and severe deficit irrigation. Kataria and Michael[90] observed that under drip irrigation in tomato, the surface soil layer up to 10 cm depth had the maximum soil moisture content and it decreased with increasing depth. This coincided with the regions having the maximum number of effective roots, resulting in a better environment for higher yields. Goel et al.[58] reported that the lateral movement of water varied between 24.4% and 24.2% in 0–30 cm depth at 40 cm distance away from the dripper. Water movement and its distribution in the soil depending upon many parameters such as soil type, the rate of infiltration, the rate of emitter discharge, the quantity of water applied, antecedent moisture content, depth to water table, and certain climatic factors.

Mishra and Pyasi[112] determined that the moisture distribution under drip irrigation was more uniform within a 10 cm radius of the emitter with maximum uniformity at zero, while nonuniformity increased with distance from the emitter, and also the water front advanced rapidly in the beginning and the rate of advance decreased with time. Dani[42] reported that the wetting pattern varied from the moisture of 50 cm suction near dripper source to that of 80 cm suction in the form of concentric circles away from the source both horizontally and vertically downward. Bhardwaj et al.[26] estimated that the soil water distribution at 0–15 and 15–20 cm depth was uniform under drip irrigation and decreased as the soil depth and distance from the dripper increased. Root growth and distribution were not influenced significantly. The water application efficiency of the drip irrigation system was 44% greater than that of flood irrigation.

Warrick[185] examined the effect of limiting flow from subsurface emitters on irrigation uniformity by using soil data from a field in the Arava Valley, Israel. He observed that soil variability can affect the flow rate of water from subsurface trickle emitters. The averages of the calculated ratio of actual discharge to the designed discharge were found as 0.905, 0.825, and 0.704 for designed discharges of 1, 2, and 4 L/h, respectively. Corresponding coefficients of variability were 0.072, 0.124, and 0.193, respectively. Christiansen's uniformity's were 0.95, 0.91, and 0.85.

Hassan[76] found that the emission uniformity is a sound indicator of the efficiency of micro irrigation system. The emission uniformity values for systems operating in one or more than one seasons are excellent if the value is greater than 90%, good—80–90%, fair—70–80%, and poor—less than 70%. The study revealed that poor emission uniformity would lead to over irrigation, resulting in low efficiency and excessive energy consumption at the pump, resulting in contaminating ground water and leaching of fertilizers below the root zone. High emission uniformity is a prerequisite for efficient irrigation. Capra and Scicolone[32] have mentioned that a sample of 16 emitters in a drip system is sufficient to test uniformity. To be a true representative of the total population of emitters in the system, the sample emitters must be chosen in different positions on the lateral with respect to the water inlet.

Satish and Patil[147] stated that the pattern of wetting front will be different for different soils due to variation in soil texture, permeability, the quantity of water applied per irrigation, the discharge rate of the emitter, and the initial moisture content of the soil. They also indicated that the soil moisture content was higher in different depths of soil as well as at different horizontal distances with an increase in the quantity of water application. Bharambe[25] reported that the salt content was low near the dripper and increased with water front laterally. More salt spreading was observed from the trickle source with a higher application rate of water and maximum salt accumulation was at the periphery of the wetted zone of soil. However, more salt accumulation was observed in the root zone under surface irrigation in check basin as compared to irrigation applied through drip.

The wetting patterns during application generally consist of two zones: (i) a saturated zone close to the drippers and (ii) a zone where the water content decreases toward the wetting front. Increasing the discharge rate generally results in an increase in the wetted soil diameter and a decrease in the wetted depth.[17]

12.2.1.2 WATER USE EFFICIENCY UNDER DRIP IRRIGATION

WUE should be estimated to assess the water used by the plants, water saved throughout the growing period of the crop and productivity per unit of water. Pawar et al.[131] stated that application of 100% recommended dose of liquid fertilizer through drip irrigation recorded the highest WUE (250 kg/ha-cm) in garlic.

Micro irrigation techniques can be used to improve the irrigation efficiency in vegetable gardens by reducing soil evaporation and drainage losses and by creating and maintaining soil moisture conditions that are favorable to crop growth. Water balance experiments in Zimbabwe showed that over 50% of the water applied as surface irrigation on traditional irrigated gardens can be lost as soil evaporation. Micro irrigation techniques that were evaluated included low-head drip irrigation, pitcher irrigation, and subsurface irrigation using clay pipes was found to be particularly effective in improving yield, crop quality, and WUE.[35]

Higher WUE (2.34 kg m^{-3}) was recorded at an application of 100% recommended dose of water-soluble fertilizer (WSF) through drip irrigation in capsicum, which was on par with 75% recommended dose.[115] At CAZRI, irrigation studies on gourds and melons showed that trickle irrigation resulted in higher yield and WUE. Drip irrigation increased the yield of gourds by 13.5% as compared to furrow irrigation.[133] Shirahatti et al.[161] reported that by applying same quantity and 50% of water through drip as of surface irrigation (control), the cotton yield increased by 28% and 10%, respectively. When water applied through drip irrigation of only 25% of the control, the yield reduced by just 0.5% but, highest WUE was observed. The soil moisture content along the vertical direction increased and laterally decreased.

The amount of applied IW with drip system was lower than that of flood irrigation. Agronomic WUE and FUE for drip irrigation were always higher than those for flood irrigation.[33] Singandhupe et al.[165] reported that 3.7–12.5% of higher tomato fruit yield with 31–37% saving of water was obtained in the drip irrigation. WUE in drip irrigation on an average over nitrogen level was 68–77% higher over surface irrigation. Webber et al.[186] conducted an experiment to check the WUE of two water saving irrigation technologies for two legumes grown as a second crop, conventional and alternate furrow irrigation, and three irrigation schedules were used. The WUE for root biomass in bean (0.15 kg m^{-3}) was slightly higher than in green gram (0.13 kg m^{-3}). WUE increased in green gram when deficit irrigation or alternate furrow irrigation was practiced, whereas it remained constant in bean for all treatment combinations. Alternate furrow irrigation and deficit irrigation are appropriate methods to increase WUE.

Wan-Shu et al.[184] conducted an experiment to investigate the effects of saline water on cucumber yield, irrigation quantity, and IW use efficiency (IWUE) under drip irrigation. The results indicated that when IW salinity was greater than 1.1 dS/m, cucumber yield decreased. The treatment with a higher SMP (−10 kPa) resulted due to higher irrigation quantity and lower IWUE; whereas the treatment with a lower SMP (−50 kPa) resulted due to lower irrigation quantity and higher IWUE. In the semihumid regions with annual precipitation of approximately 600 mm, where no enough fresh water is available for irrigation, 2.2–4.9 dS/m saline water can be used by drip.

Kirnak et al.[95] studied the effect of preharvest water stress on fruit yield, quality, and some other physiological parameters of watermelon at southeastern Turkey. The results revealed that the fruit yield was significantly lowered by reduced water rates. But the deficit irrigation in the ripening stage significantly increased the WUE. Ali et al.[6] evaluated the effects of irrigation method and water quality on sugar beet yield, percentage of sugar content and IWUE. The irrigation methods investigated were subsurface drip, surface drip, and furrow irrigation. The highest root yield (79.7 Mg ha^{-1}) and highest IWUE (9 kg m^{-3}) were obtained by using surface drip irrigation.

12.2.2 RESEARCH ON FERTIGATION

Fertigation is the application of fertilizers through IW. When fertilizers were applied along with water through drip irrigation system, there was a considerable saving of fertilizers besides increased yield and water saving compared to the surface method of irrigation. In sugarcane and orange, there was a saving of 25–30% of nitrogen when fertigation through drip was done compared to surface banding fertilizer application. Solomon[168] reported that when using subsurface drip irrigation, IW and injected chemicals, like fertilizers are supplied directly to the roots. This is a special advantage for nutrients that have low mobility into the soil. Kaminwar and Rajagopal[87] reported highest dry pod yield of 1773 kg/ha in chili cv. Sindhur at 100 kg N/ha. Sontakke et al.[169] also reported a significant increase in red dry pod yield of chili with increasing levels of nitrogen. Highest fruit yield of 1782 kg/ha was obtained at 120 kg N/ha.

Shashank et al.[156] stated that the increase in plant height and number of leaves might be due to the fact that nitrogen with synthesized carbohydrates was metabolized into amino acids and proteins, which allowed the plants to grow faster. Application of 75% and 100% recommended dose of fertigation (RDF) produced longer roots than other levels. Sharma[154] observed linear response on yield and yield components of chili up to an application of 125 kg N, 75 kg P_2O_5, and 110 kg K_2O/ha. At Jabalpur, the best treatment to promote yield and profitable in chili was 120 kg N + 30 kg P_2O_5/ha. Srivastava[164] reported that in capsicum cv. Hybrid Bharat, first and 50% flowering was delayed by 4 and 4–6 days, respectively in plants receiving the highest rate of fertilizers. The highest number of fruits and yield were obtained in plants treated with 250:200:200 kg NPK/ha. Deshmukh et al.[44] reported that 30% of N, P, and K fertilizers can be saved by the use of liquid fertilizers through drip irrigation system in comparison with recommended fertilizer levels applied conventionally under flood irrigation.

Applying fertilizer according to growth stage and plant needs will increase the efficiency in terms of uptake and reduction of losses. Growers could also decrease the amount of fertilizer by applying fertilizer directly to the root zone. Lamm et al.[97] reported saving nitrogen applications for corn (*Zea mays*) using a subsurface drip system. Tumbare and Bhoite,[178] who conducted an experiment at Rahuri, reported the application of 100% recommended dose of fertilizer (RDSF) through fertigation and recorded significantly higher green chili yield during summer. However, it was on par with the application of 70% N, 80% P, and K through fertigation, indicating a saving of 30% N while P and K of 20% and saving of water use is nearly 51%. Patil et al.[129] reported that application of 10% of RDF as basal and 90% of RDF through fertigation in 19 equal splits at 5-day interval from 30 to 120 days after sowing (DAS) recorded 35% higher yield than soil application of RDF.

Patel and Rajput[128] reported that fertigation of nutrients significantly increased saving of fertilizer nutrients up to 40% without affecting the yield of crops compared to the conventional method of nutrient application. They also reported that fertilizer application at 100% of recommended dose through fertigation recorded an increase in yield of 25.21% from 23.0 t/ha to 28.8 t/ha, in the year 2000 and 16.5% from 23.56 t/ha to 27.47 t/ha, in the year 2001 as compared to broadcasting method of fertilizer application. Ardell[11] reported that application of N and P fertilizer will frequently increase crop yields, thus increasing crop WUE. Adequate levels of essential plant nutrients are needed to optimize crop yields and WUE. Sathya et al.[146] found that right combination of water and nutrients is the key to increasing the yield and quality of the produce. The availability of N, P, and K nutrient was found to be higher in root zone area of drip fertigated plot. Nitrogen and potassium moved laterally from point source up to 15 cm and vertically up to 15–25 cm and P moved 5 cm both laterally and vertically and thereafter dwindled. Fertigation frequency enables to reduce the concentration of immobile elements such as P, K, and trace elements in IW. Fertigation of nutrients significantly increased saving of fertilizer nutrients up to 40% without affecting the yield of crops compared to the conventional method of nutrient application.

Ahmed et al.[5] conducted experiment on irrigation and fertigation levels on capsicum and they revealed that there was a significant improvement in yield, quality, water, and fertilizer use efficiencies of capsicum under drip irrigation and fertigation. However, the combined effect of drip irrigation and fertigation was found superior than their individual effects. The treatment combination of 80% ET through drip and 80% recommended NPK through fertigation registered maximum fruit yield (36,648 kg/ha). The highest WUE (294 kg/ha-cm) was observed with the treatment combination of 60% ET through drip + 80% recommended NPK through fertigation. However, the FUE was found the maximum [nitrogen use efficiency (NUE)—489 kg/kg N, phosphorous use efficiency (PUE)—653 kg/kg P, and potassium use efficiency (KUE)—979 kg/kg K] with the treatment combination of 80% ET through drip + 60% recommended NPK through fertigation.

Paramaguru et al.[124] reported that in big onion var. Agri Found Dark Red, the fertigation with 75% RDSFs (i.e., 75:112.5:56.25 kg of NPK ha^{-1}) registered higher bulb yield (10.30 and 12.70 t/ha). In small onion var. COOn5, the fertigation with 75% RDSFs (i.e., 45:45:22.5 kg of NPK ha^{-1}) registered higher bulb yield (8.34 and 11.05 t/ha) compared to soil application of fertilizer. The nutrient uptake pattern also increased with application of 75% RDSF as fertigation.

Basavarajappa et al.[23] reported that in drip irrigation system with 100% RDF was more profitable as compared to furrow irrigation due to the increase in yield. Anitta and Muthukrishnan[10] conducted an experiment on fertigation levels in crop maize and higher maize grain yield of 7300 kg/ha was recorded under drip fertigation of 100% RDF with 50% P and K through WSF followed by application of 150% RDF through drip (7050 kg/ha). The yield increase over drip irrigation with soil application of fertilizer was 39%.

12.2.2.1   EFFECTS OF FERTIGATION ON THE YIELD OF TOMATO

Cook and Sanders[39] examined the effect of fertigation frequency on tomato yields in a loamy sand soil. Daily or weekly fertigation significantly increased yield compared to less frequent fertigation, but there was no advantage of daily over weekly fertigation. Locascio and Smajstria[102] observed that the marketable yield of large fruits of tomato and total marketable yield were 30% and 10% higher respectively with 60% of N and K applied with drip irrigation than with all fertilizers applied preplant. Yields for the daily and weekly fertigation treatments were similar.

Cultivation experiment for practical use and compared with the same system, which used organic liquid fertigation and the conventional manner (basal dressing and surface application). The experiment showed that using corn steeper liquor as the only one macronutrient fertilizer, tomatoes grow well and about the same yield was obtained as the ordinal cultivation using inorganic fertilizer.[116]

A 2-year study was conducted during 2002–04 at Ludhiana to investigate the effect of irrigation and fertigation on greenhouse tomato. Drip irrigation at 0.5 Epan along with fertigation of 100% recommended nitrogen resulted in an increase in fruit yield by 59.5% over control (recommended practices) inside the greenhouse and by 116.2% over control (recommended practices) outside the greenhouse, respectively. The drip irrigation at 0.5 Epan irrespective of fertigation treatments gave a saving of 48.1% of IW and resulted in 51.7% higher fruit yield as compared to recommended practices inside the greenhouse.[64]

The higher yield of fruits per plant under liquid fertilizer treatments could be due to a continuous supply of NPK from the liquid fertilizers as reported by Kadam and Karthikeyan[86] in tomato. Kavitha et al.[91] conducted an experiment to elucidate the effect of shade and fertigation on yield and quality of tomato under open and shade (35%) as the main plot and three levels of (50%, 75%, and 100% RDF) each of water-soluble and straight fertilizers as subplot treatments. The results revealed that the application of 100% WSF under shade improved the growth parameters namely plant height, primary branches per plant, leaf area index, and dry matter production at different stages of growth. The nitrate reductase activate was higher at flowering stage, which declined towards maturity. Early flowering was noticed with the application of 100% WSF under the open condition, whereas number of flowers per cluster, flowers per plant was the highest at 100% WSF under shade. The economics of shade and fertigation showed that the treatment with 100% straight fertilizers under shade registered the highest benefit–cost ratio (B:C ratio) of 2.90, 3.13, and 3.18 during different seasons.

Essam et al.[49] revealed that tomato yields, water and fertilizer-use efficiencies had been enhanced with about 25.6%, 49.3%, and 20.3% under surface drip compared with solid set sprinkler irrigation systems, respectively. On the other hand, data indicated that there was a positive proportional trend with the applied nutrient amounts and the NPK residues in the fruits under the investigated irrigation systems. Sanchita et al.[144] conducted a study to find out the effect of fertigation level of N and K through drip irrigation on growth, marketable yield, fruit quality, and economics in semi-determinate tomato cultivar Arka Abha. Results indicated that plant height, branch number, fruit setting percentage, fruit number per plant, individual fruit weight, and marketable yield were maximum with 100% fertigation of the recommended dose of N and K at the rate of 75 and 60 kg/ha. Uday[180] revealed that that drip fertigation with 150% recommended fertilizer doses and farm yard manure (FYM) application to soil resulted in maximum shoot growth. The fruit parameters were found to be influenced positively by flood irrigation and soil application of fertilizers. But, the ultimate objective of yield maximization was achieved with drip fertigation taking 100% recommended doses of fertilizers and adding FYM to the soil.

12.2.3   RESEARCH ON PLASTIC MULCHING

A significant increase was observed in strawberry runners and fruits with the use of BPM as compared to clear plastic, white plastic, and bare ground treatments.[77] Mulching with black or clear plastic increased total plant growth and led to an increased rate of branching and early flowering in tomato.[187] Hanada[71] reported that mulching with appropriate materials has a number of effects: it increases the soil temperature, conserves soil moisture, texture, and fertility; controls weeds, pests and diseases. Soil temperature and soil moisture were highest under polyethylene.[176] Salau et al.[143] reported that mulching significantly enhanced vegetative growth and increased plantain bunch yield in both first- and second-year crops. Increase in total yield (first- and second-year crops) on an average was about 41% higher with mulched treatments than with the control. Gutal et al.[67] reported that black polyethylene film increased yield by 55% yield, reduced weed growth intensity by 90%, and saved IW by about 28% over control.

Mulches create a microenvironment by retaining soil moisture and changing root-zone temperatures and the quantity and quality of light reflected back to the plants which alter plant growth and development.[40] Organic or inorganic soil mulches influence the crop in a number of ways. Plastic mulches can offer a barrier against weeds, moisture loss, nutrient loss, erosion, insect, and disease injury while encouraging plant establishment and an earlier crop of potentially higher quality.[114] In the winter season, the conservation of soil moisture may help in preventing the loss of water through evaporation from the soil facilitating maximum utilization of moisture by the plants. Mulching with plastic is a method by which soil moisture can be conserved.[145] The polythene mulch helps to improve soil structure and soil microflora, reduces fertilizer leaching, evaporation, and weed problem. Therefore, polythene mulch has a positive effect on growth, yield, and quality of maize.[96] Shinde et al.[160] studied the effects of six micro irrigation systems (MIS) and three mulches on microclimate growth and yield of summer chili. Soil temperature was highest in the control and lowest under sugarcane trash mulch. The average humidity was greatest with microtubing at 08.30 h and with the rotary microsprinkler at 14.30 h. Plant height and number of branches were greatest with sugarcane trash mulch. The yield of green chili was highest (12.2 t/ha) with sugarcane trash mulch. The weekly crop coefficient values were in the ranges of 0.47–0.95, 0.42–0.86, 0.40–0.84, and 0.38–0.8 for summer chili treated with no mulch, transparent plastic, black plastic, and sugarcane trash, respectively.

Plastic mulches affect plant microclimate by modifying the soil energy balance and restricting soil water evaporation, thereby affecting plant growth and its yield.[173] Anil[9] reported that the pod yield of vegetable pea was 42.5%, 33.8%, 18.0%, 15.7%, and 8.1% more with black polyethylene, white polyethylene, pine needle, ridge-sowing, and deep-sowing respectively over control. It is also observed that growth of weeds was less under polyethylene sheet as compared to transparent sheet, while other plots were badly infested with weeds. Jain et al.[85] studied drip and surface irrigation with and without mulch on potato using three levels of moisture regimes in a sandy loam soil. The potato yield for treatments irrigated with drip system at 80% irrigation moisture regime in combination with plastic mulch was found to be maximum as 30.45 t/ha and minimum it being 18.44 t/ha for the control, that is, surface irrigation at 100% moisture level without mulch. The yield for other treatments varied from 19.58–20.41 t/ha. The highest WUE was 3.24 t/ha-cm for the treatment irrigated with drip system at 80% level with mulch as compared to 2.17 t/ha-cm for the control treatment.

Brown et al.[29] reported that bell peppers grown on BPM alone or in combination with drip irrigation increased pepper yields by 18 and 16 metric t/ha, respectively when compared with bare soil. Mulching with plant residues and synthetic material is a well-established technique for increasing the profitability of many horticultural crops.[57] Shinde et al.[159] studied the effects of micro irrigation, in combination with mulching, on the production of chili (*Capsicum annuum*) in Dapoli, Maharashtra, India. The treatments comprised 50% or 70% microjet irrigation with or without mulching, and 40%, 50%, and 60% drip irrigation with or without mulching. Microjet irrigation (50%) with mulching resulted in the highest plant spread (39.93), average number of fruits per hill (248.60), average weight of fruits per hill (538.93 g), average weight of fruits (2.19 g), yield (2034 kg/ha), and WUE was highest in 25% drip irrigation with mulching (447.18 kg/ha-cm) followed by 50% microjet irrigation with mulching (312.92 kg/ha-cm).

Aniekwe et al.[8] concluded that leaf area and fresh root tuber yield of cassava varieties were significantly improved by the application of BPM with 100% weed control as compared to bare soil. Luis et al.[106] found that total yield of bell pepper was increased by BPM alone or combined with row covers by around 10 t/ha compared to control. The same treatments had a positive effect relative to control in leaf area, specific leaf area, and net assimilation rate. Ibarra et al.[83] reported that dry weight of cucumber plants grown under plastic mulch or mulch combined with row covers (at 50 and 110 days after seeding) were significantly different from bare soil plants. An early yield with BPM was 2.1 times greater when compared with control (10 t/ha). A partial covering of mulch residue on the soil will strongly affect the runoff dynamics and reduce the amount of runoff.[52,138] Zhang et al.[188] reported that mulching increased the leaf area index, reduced the soil evaporation rate by 40–50 mm, and also saved plant water usage for transpiration during dry periods. Patil and Patil[130] reported that the maximum growth in terms of bush height (29.7 cm), number of branches/ plant (7.0), and maximum yield of capsicum (960 kg/ha) was observed in the treatment consisting of drip irrigation with BPM, which was significantly superior over all other treatments. WUE in this treatment was 2.5 times more as compared to conventional irrigation method with no mulch.

Mishra and Paul[113] conducted a study to evaluate effects of three irrigation levels namely, V, 0.8 V and 0.6 V with drip and black linear low-density polyethylene (LLDPE) mulch on biometric and yield response of eggplant. The results of surface irrigation either alone with black LLDPE mulch were compared with drip irrigation. The study indicated better yield under drip irrigation with plastic mulch. Abdul et al.[1] evaluated the effect of colored plastic mulch on growth and yield of chili. They concluded that transparent and blue plastic mulches encouraged weed population which was suppressed under black plastic. Plant height, number of primary branches, stem base diameter, number of leaves, and yield were better for the plants on plastic. At the mature green stage, fruits had the highest vitamin C content on the black plastic. Mulching produced the fruits with the highest chlorophyll-a, chlorophyll-b, and total chlorophyll contents and also increased the number of fruits per plant and yield. However, mulching did not affect the length and diameter of the fruits and number of seeds per fruit. Plants on BPM had the maximum number of fruits and highest yield.

12.2.3.1 EFFECTS OF PLASTIC MULCHES ON PLANT GROWTH AND YIELD

A general increase in plant growth and fruit size in hot peppers was observed by the use of plastic mulch while clear plastic mulch increased the early and total yield by 39% and 19%, respectively.[125] Plastic mulches improved stand establishment[38] and fruit yields relative to unmulched control. Vegetable crops grown under plastic mulches have shown earlier 7–14 days and increased yields 2–3 times over vegetable crops grown on bare soil.[98] The combined effects of soil temperature, soil moisture, and weed suppression not only work to improve crop growth but they also facilitate hand picking and lead to higher yield and increased fruit size.[148] Chili plants grown on plastic mulch had significantly higher N and K contents in leaf tissues at an early fruiting stage when compared with bare soil.[76] Plastic mulches increased crop growth (3.2–4.0 cm), dry root mass (12.2–50.1%), nitrogen fixing activity (3.3–12.8%), leaf chlorophyll content (41–78%), more reproductive buds (63.3–94.1%), and starts flowering 9 days earlier in groundnut than unmulched control.[79]

Farias-Larios et al.[50] found increased fruit weight (2.94 kg) and yield 25.5 t/ha in watermelon by the application of clear plastic mulch as compared to unmulched soil. However, no change was observed in total soluble solids of watermelon fruits by different types of plastic mulches but both clear and white plastic mulch increased fruit length. Total plant and leaf fresh weights in plots with BPM were higher as compared to bare soil.[126] Hallidri[69] stated that plant height and number of leaves were higher in black and transparent polythene mulch than control (bare soil) while no significant difference was observed in case of stem diameter in cucumber. However, cucumber plants grown on transparent polythene mulch gave the highest number of fruits per plant and yield.

Ibarra et al.[82] concluded that watermelon plants grown under plastic mulch and row cover showed greater plant biomass, specific leaf area, relative growth rate, and net assimilation rate than bare soil plants. Similarly, time to anthesis (appearance of perfect flowers) was 45 and 55 DAS for BPM and control plants respectively. Niu et al.[121] concluded that improved productivity was related to increased root dry weight under mulches and larger rooting systems resulted in greater ability to take up water and nutrients that led to higher grain yield with mulched wheat.[181] Color affects the surface temperature of the mulch and underlying soil temperature. BPM, the predominant color used in crop production, is an opaque black body absorber and radiator.[8,98,99] The number of leaves per plant or dry weight per plant better explains the changes in watermelon yield than net photosynthesis rate.[83] Similarly, plant height, number, and length of main roots, fresh and dry weights of roots as well as number of flowers were significantly higher in plants grown on mulch as compared to bare soil.[75] Karp et al.[89] reported that mulching treatment significantly influenced nutrient content of leaves and chlorophyll contents (381 SPAD units) were significantly lower in control plants compared with plants grown on different mulches (498 and 542 SPAD units). Balakrishnan et al.[21] concluded that under plastic mulch soil properties like soil temperature, moisture content, bulk density, aggregate stability, and nutrient availability have been improved. Plant growth and yield also positively influenced by the plastic mulch due to the modification of soil microclimate.

12.2.3.2 EFFECTS OF PLASTIC MULCHING ON YIELD OF TOMATO

BPM doubled the yield of tomatoes as well as increasing the amount of early production for some cultivars when compared with unmulched control.[1] Kaniszewski[88] reported that the mulching generally increased the total and marketable yield of tomato in all the years and for the early yield in 1 year. Only black polyethylene mulch gave better yields than the other two materials. The combined treatments of irrigation and mulching with black polyethylene showed the highest increase in total yield and also in the marketable yield. Shrivastava[163] studied effects of drip irrigation and mulching on tomato using three moisture regimes at 0.4, 0.6, and 0.8 of pan evaporation combined with no mulch, black plastic, and sugarcane trash mulch. Highest yield of 51 t/ha and 44% saving in IW were obtained by using the combination of trickle irrigation at 0.4 level of pan evaporation and polyethylene mulch. This treatment also gave the maximum yield of 163 kg/ha/mm of water applied. The treatment combining drip irrigation at 0.4 pan evaporation and polyethylene with BPM reduced weed infestation by 95% increased the yield by 53%, resulting in 44% saving in IW as compared to surface flooding without mulch.

Anderson et al.[7] reported that tomatoes grown in commercially available black paper mulch show similar yields and earliness to tomatoes grown in black polyethylene mulch, even though the latter results in slightly higher soil temperatures. Asokaraja[16] indicated that drip irrigation had twin benefits of yield increase and water saving in tomato. Drip irrigation at 75% of surface irrigation had registered 46% and 50% increase in yield and 35% and 28% water saving as compared to surface irrigation at 0.8 IW/CPE ratios with 5 cm depth.

Mulches ameliorated soil hydrothermal regime, improved vegetative growth, advanced flowering, and fruit yield of tomato plants when compared with bare soil.[3] Higher tomato yields were reported when BPM and row covers were used together is partially due to increasing air and soil temperatures around the plant growing environment.[188] Meena et al.[110] concluded that use of black polyethylene mulch plus drip irrigation further raised tomato yield to 57.87 t/ha. Plant height, leaf area index, dry matter production, fruit weight, and yield increased significantly with the use of drip irrigation alone and in conjunction with polyethylene mulch compared to surface irrigation alone or with mulch. Agrawal et al.[4] indicated that drip irrigation with red plastic mulch showed superior yield and yield attributing characters as compared to other mulched treatments. The yield of tomato in red plastic mulch, BPM, white plastic mulch, and control plots were 33575, 32462, 31218, and 23072 kg/ha, respectively. These results showed that the red, black, and white plastic mulch increased the yield of tomato by 45.52%, 40.06%, and 35.30%, respectively over the control. WUE and water savings were found to be highest under red plastic mulch and lowest under nonmulch condition. The net income was recorded higher under red plastic mulch (Rs. 85,800) and lowest in without plastic mulch (Rs. 38,020).

12.2.3.3   EFFECTS OF PLASTIC MULCHING ON SOIL TEMPERATURE

Different forms of plastic mulch are available varying from woven plastic to smooth plastic and embossed plastic films. Nowadays, 100% compostable and biodegradable mulches are also available in advanced countries and these are more eco-friendly. In addition to the surface structure, the color and thickness of the mulch create a lot of variations which have an effect on the plant microclimate and in particular the soil temperature. Soil temperatures can be increased in the field by applying plastic mulches. Haddadin[68] found that in tomato field, the average soil temperature was highest under clear plastic mulch followed by the BPM and bare soil treatments. But the differences in soil temperature became narrow as mulches were partially shaded by the plant.

Clear plastic mulch is often used for soil sterilization (solarization): the plastic film is fixed over the wet soil to trap solar heat which kills weeds and soil pathogens. Clear plastic is believed to achieve higher soil temperatures than black plastic. This happens because much of the incident radiation is absorbed by colored films[12] and does not pass through to the soil. Changes in root zone temperature can affect the uptake and translocation of essential nutrients, therefore influencing root and shoot growth.[174] This increase in soil temperature consistently improves root development in vegetables grown under mulches. Gupta and Acharya[66] observed increased root mass under black polyethylene mulch was attributed to the resultant increase in soil temperature and nutrient uptake.

Himelrick et al.[77] found that soil temperatures were warmest with clear plastic mulch followed in order of decreasing temperatures by black-on-white, black, white-on-black, and bare ground. The black plastic film is the most common form of mulch and has been shown to cause a significant temperature rise in soils.[187] Ham et al.[70] showed that the placement of the mulch was important to raise the temperature. Results indicated that clear plastic heated soil less than black plastic if it was placed tightly across the soil with good contact between the soil surface and the mulch. They also suggested that if clear plastic mulches placed loosely over the soil, an insulating air layer develops which results in the soil heat storage and reducing heat loss. Soil temperature is increased by 5–10°C by the application of plastic mulches when compared to bare soil.[47] This well-documented temperature rise is often used as an explanation for the increased production of crops grown on plastic mulch.[43,63] Hu et al.[79] reported that mulches are known to increase the soil temperature since the sun's energy passes through the mulch and heats the air and soil beneath the mulch directly and then the heat is trapped by the "greenhouse effect."

Hasan et al.[75] reported that mulching is practically beneficial in chili production. They concluded that increased plant growth for mulched plants may be related to soil moisture content because plant dry weight was positively correlated with soil temperature and moisture content. Most suitable soil temperature distribution was observed by the application of clear plastic mulch and it was more effective on first blossoming and harvesting time, leaf area and total yield in squash, while lowest plant growth and yield values were observed in bare soil.[177] Hummel et al.[81] indicated that the type of ground cover significantly affected the temperature in the upper 12 cm of the soil. The highest soil temperatures were observed under BPM followed by bare ground. Locher et al.[103] revealed that use of dark-colored mulch is the safest solution because even in case of high air temperature and solar radiation, the soil does not warm to a harmful degree. They observed that in case of light-colored mulches (clear, violet, and light green), the soil temperature increased 2.5–2.9°C higher than in case of the unmulched control. They also mentioned in their studies that dark colored mulches (black, dark green, and red) increased soil temperature 1.4–2.1°C compared to the unmulched (control). Overall studies indicated that higher yields of sweet peppers were achieved from mulched treatments due to higher soil temperatures than the unmulched treatment.

12.2.3.4   EFFECTS OF PLASTIC MULCHING ON SOIL MOISTURE

The use of plastic mulch can be improved if under-mulch irrigation is used in combination with soil moisture monitoring. The influence of rainfall events is not as great when plastic mulch is used, necessitating active irrigation management. Plastic mulch conserved 47.08% of water and increased yield by 47.67% in tomato when compared to unmulched control.[53] Palada et al.[126] concluded that plastic mulching resulted in 33–52% more efficient use of IW in bell pepper compared to bare soil.

12.2.3.5   EFFECTS OF PLASTIC MULCHING ON WEED CONTROL

Mulching as weed control method is used in agriculture throughout the world.[65] Since weed seed germination is affected by soil moisture and temperature, mulch not only suppresses weeds, but also maintains soil moisture at higher levels compared with nonmulched soil.[46] Very little weed growth occurs under the mulch as the mulches prevent penetration of light or exclude certain wavelengths of light that are needed for the weed seedlings to grow.[123]

Mulching for weed control can take a number of forms: inorganic or organic mulches can be applied and left in situ to control the weeds; living mulches can be grown to choke out weeds before planting the mulches are either killed with chemicals or complete their life cycle before the growing season of the herb. Solarization uses an inorganic mulch and solar energy to disinfect the soil, the mulch being removed prior to planting. Similarly, 100% weed control was observed in cassava peel with BPM as compared to bare soil.[8] BPM is both effective at warming the soil and reducing weed competition. Clear plastic mulch provides greater soil warming, but it does not reduce the weed competition.[98] Dark colored mulches lay across the soil and around the crop reduce the amount of light reaching the soil and thus inhibit weed germination and smother emerging weeds. Weed control in crops is a difficult, time consuming, and expensive task. Plastic mulches have the potential to alter soil temperature, crop water use, improve crop quality, and in some cases reduce weed competition, thereby improving crop development and increasing yields.[98,119]

12.2.4   PLASTIC MULCHING AND FERTIGATION

Mulched treatments generally showed significantly greater total uptake of N, P, and K than corresponding nonmulched ones.[2] Aruna et al.[13] observed increased plant height (127.20 cm) by mulching with black polythene mulch along with the application of 100% of the recommended dose in the form of urea + phosphoric acid + potassium sulfate. Earlier flowering (29 days) was observed when mulched with black polythene mulch with 100% of the recommended dose in the form of urea + phosphoric acid + potassium sulfate. Increased number of fruits per plant (32.7), single fruit weight (65.25 g), and yield per plant (6.40 kg) were also observed when mulched with black polythene mulch along with the application of 100% of the recommended dose in the form of urea + phosphoric acid + potassium sulfate.

Lincoln et al.[101] conducted a study to evaluate the interaction between fertilization rates and irrigation treatments, and to quantify nitrate leaching in a plastic mulched/drip irrigated zucchini squash[51] production systems. They reported that application of N above the standard recommended rate of 145 kg/ha did not increase yield, although yields were reduced at the lowest N-rate. Ganga[54] reported minimum fruit drop and fruit cracking in treatment MId3F2 (black polyethylene mulch + drip irrigation at 100% of estimated IW requirement + 100% RDF) and the maximum in control (conventional irrigation + no fertilizer) in both the years. The maximum yield (40.4 kg/plant) was recorded in treatment MId3F2 (black polyethylene mulch + drip irrigation at 100% of estimated IW requirement + 100% RDF) with highest B:C ratio of 6.52.

12.2.5 RESEARCH ON COST ECONOMICS

Jadhav et al.[84] showed that yields of tomato were 48 t/ha for drip irrigation system with pressure compensating emitters and 32 t/ha when furrow irrigation was used. The B:C ratio was 5.15 and 2.96, respectively for drip and furrow methods. The drip system showed a 31% saving in IW. Hapase et al.[74] conducted a significant work in the economic analysis of drip irrigation for sugarcane crop. They indicated that for one crop season, daily drip irrigation registered 50–55% water saving, 12–37% increase in yield, and 2–7 times higher WUE compared to conventional furrow irrigation.

Aswani and Manoj[18] worked out the B:C ratio for various fruit and vegetable crops and reported that for vegetable crops the B:C ratio was 2.35 excluding the water saving and 3.09 including water saving. According to Shrestha et al.,[162] an economics analysis of the factors that affect the choice of drip irrigation revealed that the yield increase of about 1.7 t of sugar per acre or a net gain of US$ 578 per acre per crop and considerable saving of water up to 12% are major contributing factors to the rapid adoption of drip irrigation.

Sivanappan[167] reported that an extra income of Rs 49,280/ha could be obtained through drip irrigation in tomato over surface irrigation and the payback period for the cost of drip system was only 6 months. Khan et al.[93] reported that drip fertigation with 100% WSF applied to potato recorded higher net profit of Rs. 38,720/ha when compared to drip fertigation with 100% normal fertilizers (Rs. 33,604/ha) and furrow irrigation with 100% normal fertilizers (Rs. 32,583 /ha). Dalvi et al.[41] studied that the cost economics of micro irrigation system and optimization was performed to assess minimum input cost of tomato, considering the advent of mechanically moved portable drip sets, with every second day irrigation approximately 50% saving on initial investment of drip set can be achieved as the same set will irrigate double the area.

A study has been conducted on an economic comparison of drip and furrow irrigation methods at New Mexico.[55] The results indicated that yield was estimated 25% greater when employing drip irrigation and with increased fixed and capital expenditures, drip irrigation would produce a greater net operating profit (approximately 12%) than the furrow irrigation method. Manjunatha et al.[108] studied the economic feasibility of micro irrigation system for various vegetables and reported that the gross B:C ratio of 2.56, 3.24, 3.19, and 2.49 were achieved for drip emitter, drip micro tubes, microsprinklers, and surface irrigation, respectively. The net profit achieved per mm application of water used for potato was highest for drip emitter (Rs. 377) followed drip micro tube (Rs. 299), microsprinkler (Rs. 203), and lowest in case of surface irrigation (Rs. 151). Similar results were also reported for eggplant, chili, and cauliflower.

Narayanamoorthy and Deshpande[118] showed that investment in drip irrigation is economically viable even for the farmers who own one hectare of land. They further showed that farmers can regenerate the capital cost of drip set from the profit of the very first year even without availing the subsidy from government schemes. Shinde et al.[159] studied the effects of micro irrigation, in combination with mulching, on the production of chili in Dapoli, Maharashtra, India. The treatments comprised 50% or 70% microjet irrigation with or without mulching, and 40%, 50%, and 60% drip irrigation with or without mulching as well as the highest gross income (Rs. 244,080/ha), net returns (Rs. 100,956.24/ha), B:C ratio (1.70), and net extra income over the control (Rs. 51,628.05/ha). WUE was highest in 25% drip irrigation with mulching (447.18 kg/ha-cm) followed by 50% microjet irrigation with mulching (312.92 kg/ha-cm).

12.3   MATERIALS AND METHODS

12.3.1   LOCATION OF THE STUDY AREA

The study area was selected at the PFDC research and demonstration plot (NA5) situated in the eastern section of the farm of Tamil Nadu

Agricultural University (TNAU), Coimbatore. The field is located at 11°N latitude and 77°E longitude with mean altitude of 426 m above the mean sea level. The topography of the experimental plot was uniform and leveled. The mean annual rainfall of the study area is 612 mm. About 55% of annual rainfall is received during northeast monsoon season and 30% during southwest monsoon. The annual maximum mean and annual minimum mean temperatures were 32.5°C and 20.1°C, respectively. The average relative humidity of the area is 56.8% and sunshine hours range from 3 to 10 h. The mean daily evaporation ranges from 3.5 to 7.6 mm.

12.3.2 CROP AND VARIETY

The Hybrid Tomato variety COTH 2 from TNAU horticulture seed center was chosen for the study since it has a vibrant market potential in the domestic market. The duration of the crop is 120 days.

12.3.2.1 SOIL

The experimental field is having soils with clay loam texture at a pH of 7.7 and a good electrical conductivity of 0.78 dSm^{-1}. Physical and chemical properties of the soil are given in Table 12.1.

TABLE 12.1 Initial Physical and Chemical Properties of Soil.

Soil characteristics		Composition
Physical characteristics	Bulk density, g/cc	1.4
	Particle density, g/cc	2.41
	Porosity, %	42
Chemical characteristics	Available N, kg/ha	131.2
	Available P, kg/ha	10.8
	Available K, kg/ha	352.3
	pH	7.7
	EC, dSm^{-1}	0.78

12.3.2.2   IW QUALITY

Quality of IW is moderately saline. The water was analyzed for pH, EC, total alkalinity, Cl_2, SO_4, Ca, Mg, Na, K, SAR, and total soluble salts (Table 12.2).

TABLE 12.2   Quality of IW.

Water quality parameters	Value
Bicarbonate (HCO_3), meq/L	2.43
Calcium (Ca), meq/L	13.20
Chloride (Cl_2), meq/L	24.00
EC, dS/m	1.93
Magnesium (Mg), meq/L	12.10
pH	7.56
Potassium (K), meq/L	0.41
Sodium adsorption ratio (SAR)	4.23
Sodium (Na), meq/L	11.65
Soluble sodium percentage (SSP)	36.35

12.3.3   EXPERIMENTAL LAYOUT

The experiment was carried out in the open field of PFDC farm. The field layout plan for the experiment is depicted in Figure 12.1. The length and width of the field were 35 m and 15 m, respectively. The total area is divided into various strips of 4.5 m × 1.2 m according to the treatments. The experiment and treatment details are given in Table 12.3. The factorial randomized block design (FRBD) was used for irrigation, mulching, and fertigation treatments (Table 12.4). Each treatment combination was replicated thrice.

12.3.4   FIELD PREPARATION, LAYOUT OF THE PLOTS, AND CULTURAL OPERATIONS

The experimental plot was thoroughly plowed with disc plow and repeatedly tilled with a cultivator to bring optimum soil tilth. Then the layout was taken up forming 27 raised beds of 1.2 m × 4.5 m size and drip system was

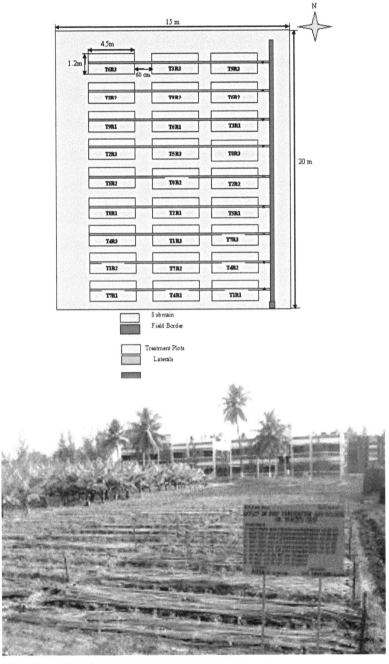

FIGURE 12.1   Field layout.

TABLE 12.3 Details of Experiment.

Parameter	Description
Crop	Tomato
Variety	CO-TH2
Date of sowing	December 19, 2011
Date of first harvesting	February 25, 2012
Duration, days	120
Replication	3
Plant to plant spacing, cm	30
Treatments	9
No. of plants under each treatment in one replication	30

TABLE 12.4 Different Treatments Showing Mulching, Fertigation Levels, and Irrigation Levels.

Treatment	Description
T1	Black plastic mulch of 25 μm thickness with 80% RDF
T2	Black plastic mulch of 25 μm thickness with 100% RDF
T3	Black plastic mulch of 25 μm thickness with 120% RDF
T4	Black plastic mulch of 50 μm thickness with 80% RDF
T5	Black plastic mulch of 50 μm thickness with 100% RDF
T6	Black plastic mulch of 50 μm thickness with 120% RDF
T7	No mulch with 80% RDF drip fertigation
T8	No mulch with 100% RDF drip fertigation
T9	No mulch with 120% RDF drip fertigation
M1	Black plastic mulch of 25 μm thickness
M2	Black plastic mulch of 50 μm thickness
M3	No mulch (control)
F1	80% of recommended dose of N, P, and K
F2	100% of recommended dose of N, P, and K
F3	120% of recommended dose of N, P, and K

installed. Paired row geometry was adopted for the treatments with 30 cm between rows in a pair and 60 cm spacing in between two pairs of rows. The laterals were laid in each bed. Online drippers were used at a spacing of 60 cm. Over the drip line according to the treatment, mulching sheets were spread in each plot and holes were punched where seedlings were to be established. Both ends of the plastic sheet were buried into the soil up to a depth of 10 cm. The field preparation details are shown in Figure 12.2.

FIGURE 12.2   Field preparation.

Tomato seedlings of 25–30 days old from the nursery were transplanted in double rows with a spacing of 30 cm × 60 cm on both sides of the bed as per the treatment schedule in the evening followed by irrigation. Total 30 plants were used in each treatment plot in single replication having an effective area of 0.18 m². Gap filling was done 1 week after transplanting with the reserved plants of the same variety. Crop protection measures were taken when needed. Quinalphos was sprayed against semilooper eating leaves and fruit borer at 2.5 mL/L dosage, $ZnSO_4$ at the flowering stage for easy fruit set and $CaCl_2$ sprayed against blossom end rot of tomato due to Ca deficiency at 5% dosage level. Weeding was done at monthly interval in the control plots.

12.3.5   STATISTICAL DESIGN FOR THE STUDY

The statistical design selected for the study was FRBD with nine treatments and three replications. The treatments include three mulching levels (50 µm, 25 µm, and control) and three levels of fertigation (80%, 100%, and 120% of RDF).

12.3.6   IRRIGATION AND FERTIGATION DETAILS

IW source was from a nearby bore well from which water was pumped using a 7.5 HP pump and conveyed through screen filters to the PVC main line pipes of 63 mm diameter. PVC submain of 50 mm diameter was connected to the main line to which, LDPE laterals of 16 mm diameter were connected. Each lateral was provided with individual taps for controlling irrigation. Along the laterals, online drippers of 4 L/h were installed at a spacing of 60 cm. Submains and laterals were closed at the end with end caps. After installation, the trial run was conducted to assess mean dripper discharge and uniformity coefficient. This was taken into account for fixing the IW application time. During the irrigation period, an average uniformity coefficient of 90–95% was observed. Design data are given below:

Length of each lateral from submain (16 mm diameter LLDPE): 15 m
Number of laterals from submain: 9

Number of emitters per lateral: 25
Emitter-type pressure compensation: online dripper
Emitter discharge rate: 4 L/h

12.3.7   IRRIGATION SCHEDULING

Irrigations were scheduled on the basis of climatological approach in mulch and control plots. First irrigation was given immediately after sowing up to 10 days and subsequent irrigations were scheduled once in 3 days based on the following formula and applied each time as per the treatment schedule. The Kc values for tomato for different stages are given in Table 12.5.

$$\text{WRc} = \text{CPE} \times Kp \times Kc \times Wp \times A \qquad (12.1)$$

where WRc = computed water requirement (liters per plant); CPE = cumulative pan evaporation for 2 days (mm); Kp = pan factor (0.8); Kc = crop factor; Wp = wetted area (80%); and A = area per plant.

TABLE 12.5   Crop Factor (Kc) Values for Tomato.

Crop stage	Days	Kc
Initial stage	25	0.45
Flowering stage	35	0.75
Fruiting stage	35	1.15
Late season stage	25	0.8

Source: FAO Irrigation and Drainage Paper 56.

Time of operation of drip system to deliver the required volume of water per plot was computed as below:

$$\text{Time of operation} = [\text{Volume of water required}]/$$
$$\text{Emitter discharge} \times \text{No. of emitters}] \qquad (12.2)$$

Details of irrigation scheduling for tomato are presented in Table 12.6.

TABLE 12.6 Total Water Requirements for Tomato.

Crop date and stage	Quantity applied per plant	Duration of irrigation each day	Total quantity applied
	Lpd	min/day	L/plant/stage
Initial stage (December 19–January 12) 1–25 days	0.47	33.08	2.4
Vegetative stage (January 13–February 16) 25–60 days	1.14	63.94	5.68
Fruit setting stage (February 17–March 22) 60–95 days	1.55	87.27	7.76
Final stage (March 23–April 16) 95–120 days	0.91	50.89	2.71

12.3.8 FERTIGATION

The recommended soluble fertilizers were applied simultaneously in a combined form to the plant root zone. Urea and muriate of potash[104,105] were applied through fertigation system with fertilizer tank and venturi. The fertilizers were dissolved in water in the ratio of 1:5 and the solution was diluted in fertigation tank. With venturi injectors, water is extracted from the main line, and a pressure differential is created by a valve in the main line forcing water through the injector at high velocity. The high-velocity water passing through the throat of the venturi creates a vacuum or negative pressure, generating suction to draw chemicals into the injector from the chemical tank. The 120%, 100%, and 80% recommended NPK was regulated by operating the tap connected at the starting end of each lateral (Table 12.4). Phosphorus was applied manually as a basal dose.

During vegetative stage, the fertilizer was applied at weekly intervals. During flowering stage, the fertilizer was applied at three days intervals and during fruiting stage, it was applied again at weekly intervals. The total quantity of fertilizers required is shown in Table 12.7. The 100% total RDF was given as 50:300:50 kg/ha for basal dose and N and K each 150 kg/ha in equal splits at various crop growth stages after transplanting. The quantity of fertilizer applied in the plot area is given in Table 12.8.

TABLE 12.7 Fertilization: Method, Types, and Quantity.

Stage	Type of fertilizer	Amount (kg/ha)
Basal dose	Urea	108.69
	Single super phosphate	1875
	Muriate of potash	83.33
Top dressing	Urea	326.08
	Muriate of potash	250

TABLE 12.8 Details of Quantity of Fertilizer Applied Per Plot Area (kg).

Stage	Type of fertilizer	% of recommended dose of fertigation		
		80%	100%	120%
Basal dose	Urea	1.217	1.512	1.825
	Single super phosphate	0.933	1.167	1.41
	Muriate of potash	21	26.25	31.5
Transplanting to plant establishment (1–10 days)	Urea	0.304	0.380	0.456
	Muriate of potash	0.235	0.290	0.349
Vegetative stage (10–30 days)	Urea	0.304	0.380	0.456
	Muriate of potash	0.235	0.290	0.349
Flower initiation to first picking (30–60 days)	Urea	0.304	0.380	0.456
	Muriate of potash	0.235	0.290	0.349
Harvesting stage (60–105 days)	Urea	0.304	0.380	0.456
	Muriate of potash	0.235	0.290	0.349

12.3.9 ASSESSMENT OF DISCHARGE UNIFORMITY

The discharge rate of the emitters at selected points in selected laterals was measured by collecting the water for a known time directly under the emitter with the help of a measuring jar and stopwatch at 0.5 kg/cm² operating pressure, which was maintained throughout the experiment. The efficiency of drip irrigation system depends on the uniform distribution of water through the system. Cv, SU, and CU are used to determine the efficiency of drip irrigation system. These parameters were evaluated for performing discharge measurements in the field.

12.3.9.1   COEFFICIENT OF VARIATION (CV)

The coefficient of manufacturing variation was determined for the drip irrigation system from flow rate measurements of several identical emission devices and was computed with the following equation:

$$Cv = \frac{\left[q_1^2 + q_2^2 + q_3^2 + \cdots + q_n^2 - n\bar{q}^2 \right]^{1/2}}{\bar{q}[n-1]^{1/2}} \tag{12.3}$$

where q_1, q_2, q_3, and q_n = discharges from different segments; q = average discharge for the total segments; and n = no. of segments.

12.3.9.2   STATISTICAL UNIFORMITY

$$SU = 100\,(1-Cv) \tag{12.4}$$

where SU = statistical uniformity and Cv = coefficient of variation.

12.3.9.3   COEFFICIENT OF UNIFORMITY

The discharge rate of drippers was recorded at randomly selected emitter points on 1st, 5th, 10th, 15th, 20th, and last one on each lateral to work out the uniformity of drip system. The uniformity coefficient was computed by the following formula:

$$E_u = 100\left[1 - \frac{1.27}{\sqrt{Ne}}Cv \right]\frac{Q_{min}}{Q_{avg}} \tag{12.5}$$

where E_u = emission uniformity in %; Ne = number of point source segments; Cv = the manufacture's coefficient in the system in L/h; Q_{min} = the minimum discharge rate; and Q_{avg} = the average rate in L/h.

12.3.10   SOIL MOISTURE DISTRIBUTION PATTERNS

The wetting pattern of soil under different mulches was analyzed by taking moisture content at different horizontal distances and soil depths. In order to study the soil moisture distribution, samples were collected at a distance of 0, 15, 30, and 45 cm from emitter along the horizontal direction and at a depth of 0, 10, 20, and 30 cm. The samples were collected immediately after irrigation, after 1 day and 2 days of irrigation and just before the next irrigation. Using gravimetric method, the soil moisture was calculated. Soil samples were taken using tube type soil augers and were kept in moisture boxes and covered immediately with lids. The samples were weighed along with the moisture box (W_2) and then placed in an oven at 105°C for 24 h until all moisture was driven off. It was then weighed again and the weight (W_3) was noted. Soil moisture contour maps were plotted by using the computer software package "surfer" of the windows version. The soil moisture content is expressed as percentage by weight on dry basis.

$$\text{Moisture content, } \% = 100 \left[(W_2 - W_1) / (W_3 - W_1) \right] \qquad (12.6)$$

where W_1 = weight of the empty container with a lid (g); W_2 = weight of the container with a lid and moist soil (g); and W_3 = weight of the container with a lid and dry soil (g).

12.3.10.1 WETTED ZONE DIAMETER

Field observations were done to measure the horizontal movement of the wetting front from the emitter. The diameter of the wetting front was measured over different periods of time during emission and the wetting front advance equation was developed.

12.3.11 PHYSICAL AND CHEMICAL ANALYSIS OF SOIL

12.3.11.1 PHYSICAL ANALYSIS

The soil from each treatment before and after harvest was analyzed by standard procedures for physical characters like bulk density, particle density, and porosity. These values were calculated as follows:

$$\text{Bulk density (BD)} = \frac{\text{Weight of dry soil}}{\text{Volume of soil (including pore space)}} \qquad (12.7)$$

$$\text{Particle density (PD)} = [\text{Weight of particle}]/ \\ [\text{Final volume} - \text{Initial volume}] \qquad (12.8)$$

$$\text{Porosity} = [1 - (BD/PD)] \times 100 \qquad (12.9)$$

12.3.11.2   CHEMICAL ANALYSIS

The soil samples were collected from each plot before transplanting the crop and after the final harvest. The samples were dried under shade, powdered, and sieved through a 2-mm sieve and used for analysis of N, P, and K. The various soil samples before transplanting and after harvest were analyzed for the available NPK content (ppm) using the procedure in Table 12.9. Initial and final pH and EC of soil were measured using pH and EC meters and values are presented in Table 12.10.

TABLE 12.9   Analytical Methods for Chemical Analysis of Soil.

Name of the analysis	Methodology	Procedure
Nitrogen	Alkaline permanganate method	Digestion of 20 g soil + 100 mL 0.32% $KMnO_4$ + 100 mL 2.5% NaOH; collected in 2% boric acid + double indicator (bromocresol green + methyl red) and titrated with N/10 H_2SO_4.
Phosphorus	Olsen method	5 g soil + 50 mL M $NaHCO_3$. pH adjusted to 8.5, shaken for 30 min, filtered; 5 mL filtrate + 4 mL reagent B, the volume is made up to 25 mL and the intensity of the blue color is read in spectrophotometer at 660 nm.
Potassium	Neutral normal ammonium acetate method	5 g soil + 25 mL neutral ammonium acetate, shaken for 5 min, filtered, and the sample was fed to a flame photometer.

TABLE 12.10   Initial pH and EC Values of Soil.

Soil	pH	EC (ds/m)
Initial	7.7	0.78

12.3.12   CROP PARAMETERS

12.3.12.1   SOIL TEMPERATURE

Soil temperature at 5 cm depth was measured using soil thermometer at 9 a.m., 12 noon, and 4 p.m. every day in the field.

12.3.12.2   HARVEST

Tomatoes were harvested by hand at ripening stage and continued until no economic yield was attained and yield particulars are revealed.

12.3.12.3   GROWTH PARAMETERS

Five plants from each different treatment were selected at random and tagged for observation on growth and yield characters. The growth parameters were observed from these tagged plants. The height of the plant (cm) from the bottom of the plant to the tip of the plant was measured at 15, 30, 45, 60, and 90 days after transplanting (DAT). The number of days taken for first flower opening and fruit set from the date of transplanting was recorded and expressed in days.

After harvesting, weights (g) of selected fruits from tagged plants were measured and then mean values were worked. Other parameters were: diameter of fruit; the number of fruits in each plant; and weight of fruits (kg) from each plant was determined by picking ripe fruits. Finally, the total weight of fruits (kg) from each plot was recorded.

12.3.12.4   WEED PARAMETERS

Weeding was carried out on monthly basis for each plot. The total number of weeds present in the area, their types, and the dry matter content of weeds were measured and recorded.

12.3.12.5 WATER USE EFFICIENCY

WUE is the ratio of the yield of the crop in kg/ha and total water used in mm.

$$\text{WUE} = \frac{Y}{W \cdot A}$$
(12.10)

where WUE = water use efficiency, kg/ha-mm; Y = yield of the crop, kg/ha; and $W{\cdot}A$ = total water used during the season, mm.

12.3.12.6 FERTILIZER USE EFFICIENCY

FUE was calculated separately for N, P, and K for each treatment, which is the ratio of the yield of the crop in kg/ha and total nitrogen, potassium, and phosphorus applied in kg/ha.

$$\text{FUE} = \frac{Y}{F \cdot A}$$
(12.11)

where FUE = fertilizer use efficiency; Y = yield of the crop, kg/ha; and $F{\cdot}A$ = total fertilizer used.

12.3.13 COST ECONOMICS

Economics of tomato production under plastic mulching was estimated in terms of total expenditure. The total cost of cultivation was calculated, which is the sum of mulching sheet cost, irrigation drip lines cost, land preparation and management, and other input cost like fertilizer, harvesting, planting material cost, etc. Total revenue from the field for one crop season was also calculated based on the yield from the different treatments. Then benefit cost for the different treatments was determined.

$$\text{Benefit–cost ratio, BCR} = [\text{Gross income per ha}]/$$
$$[\text{Total cost of cultivation per ha}]$$
(12.12)

12.3.14 STATISTICAL ANALYSIS

The data were analyzed in AGRESS package for FRBD for the crop grown in various experimental plots. Wherever the treatment differences were found significant ("F" test), critical differences were determined at 5% probability level.

12.4 RESULTS AND DISCUSSION

12.4.1 PHYSICAL AND CHEMICAL ANALYSES OF SOIL

Studies on vegetable crops have demonstrated that mulches provide several benefits to crop production through soil and water conservation, improved soil physical and chemical properties, and enhanced soil biological activity.[174,182]

12.4.1.1 PHYSICAL ANALYSIS

Bulk density, particle density, and porosity of the initial and final soil under various treatments were analyzed (Table 12.11). In order to improve WUE in horticultural plants, we need a better understanding of physical properties and hydraulic properties.[72]

TABLE 12.11 Physical Parameters of Soil Before Transplanting and After Harvest.

Treatment	Bulk density, g/cm³		Particle density, g/cm³		Porosity	
	Initial	After harvest	Initial	After harvest	Initial	After harvest
T1	1.4	1.34	2.41	2.43	42	44.8
T2	1.4	1.33	2.41	2.37	42	43.9
T3	1.4	1.36	2.41	2.47	42	45
T4	1.4	1.32	2.41	2.35	42	43.8
T5	1.4	1.35	2.41	2.42	42	44.2
T6	1.4	1.33	2.41	2.38	42	44
T7	1.4	1.38	2.41	2.42	42	43
T8	1.4	1.39	2.41	2.45	42	43.2
T9	1.4	1.38	2.41	2.46	42	43.9

Bulk density is a soil parameter that is used to quantify soil compactness. It decreases in mulching treatments compared to control. Ghuman et al.[56] concluded that mulching decreases the bulk density of the surface soil. Initially, the bulk density of the soil was 1.4 g/cm³. After the crop harvest, the bulk density in mulched plots decreased to a range of 1.32–1.35 g/cm³ whereas in control plots, there was not that much change in the bulk density of the soil compared to the mulching treatments. The increased porosity and decreased compaction due to decreased soil bulk density in plastic film mulched plots may have enhanced aeration and microbial activities in the soil thus resulting in increased root penetration and cumulative feeding area leading to increased plant growth and yield in line with the observations of Mbah et al.,[110] and Obi and Ebo.[122] Soil particle density depends on the chemical composition and structure of the minerals in the soil. It is in direct proportion to bulk density.

In control plots, porosity was lower compared to the mulching treatments due to increased bulk density. Maragatham and Paul[109] indicated that total porosity is directly related to aeration porosity, whereas water holding capacity is inversely related. The amount of pore space was significantly higher in nonmulched control than polyethylene film mulched treatments.

12.4.1.2   CHEMICAL ANALYSIS OF SOIL

The initial pH of the soil was 7.7 at the time of transplanting. During crop period due to the application of nutrients, there was a slight increase in the pH value. The pH change is different in different treatments. EC also has the same effect as that of pH, but it increased more in 100% and 120% RDF fertigation treatments due to the nutrient content increase. The initial and final pH and EC values are shown in Table 12.12.

The variation of nutrient content N, P, and K of soil before transplanting and after final harvest was analyzed and results are presented in Table 12.13. The recommended fertigation was applied during cropping period according to fertigation schedule. After final harvest, it was observed that nutrients values were less in treatments with 80% RDF. It indicates that the fertigation applied in that treatment was fully taken up by the plants for their growth and yield. Also, we can see that the final nutrient has increased than the initial content in all the cases due to the application of the fertilizer.

TABLE 12.12 pH and EC of Soil Before Transplanting and After Harvest.

Treatment	pH		EC	
	Initial	After harvest	Initial	After harvest
T1	7.7	7.8	0.78	0.83
T2	7.7	7.85	0.78	0.86
T3	7.7	7.9	0.78	0.89
T4	7.7	7.85	0.78	0.84
T5	7.7	7.9	0.78	0.86
T6	7.7	7.94	0.78	0.9
T7	7.7	7.7	0.78	0.8
T8	7.7	7.8	0.78	0.82
T9	7.7	7.95	0.78	0.83

TABLE 12.13 Quantities of Available N, P, and K in Soil Before Transplanting and After Harvest.

Treatment	N		P		K	
	Initial	After harvest	Initial	After harvest	Initial	After harvest
				ppm		
T1	131.2	121.8	10.8	9.21	352.3	324.3
T2	131.2	133.8	10.8	10.94	352.3	354.2
T3	131.2	142	10.8	12.31	352.3	371.7
T4	131.2	126.5	10.8	9.41	352.3	328.6
T5	131.2	134.7	10.8	11.2	352.3	356.6
T6	131.2	145	10.8	12.93	352.3	378.5
T7	131.2	118	10.8	9.05	352.3	321.4
T8	131.2	132	10.8	10.9	352.3	351.9
T9	131.2	137	10.8	11.72	352.3	364.8

The nutrient status in the mulching treatment was higher compared to the control due to a reduction in the leaching and losses of nutrients because of the mulch cover. These results are in close agreement with those by Bhella,[27] who reported that plastic film mulch promoted early yields, did not immobilize N and increased nutrient availability by the way of reduced leaching which is in close correlation with the current results. The highest uptake of N, P, and K was observed in polyethylene-mulched plots. Increased uptake of N, P, and K under mulch may be due to increased growth and yield characters.[141,183]

12.4.2   ASSESSMENT OF DISCHARGE UNIFORMITY

The efficiency of drip irrigation depends on the uniformity of distribution of water throughout the field. The discharge from the drippers at different points of emission was measured for a particular period of time at 0.5 kgf/cm² pressure and parameters such as Cv, SU, and CU were evaluated from the observed discharge. Volumetric method was used for calculating the uniformity coefficients (UC) of drip irrigation system.[135]

The coefficient of manufacturing variation (Cv) for drip irrigation system is calculated for the pressure 0.5 kgf/cm² as 0.0198%. Statistical uniformity of the system was calculated as 98%. The uniformity coefficient of drip irrigation system was found to be 96.87%. The high value of uniformity coefficient indicated the excellent performance of drip irrigation system in supplying water uniformly throughout the laterals.

The diameter of the horizontal wetted zone during different times of emission is graphically represented in Figure 12.3. As the elapsed time increased, the rate of increase of wetted zone diameter decreased. This was

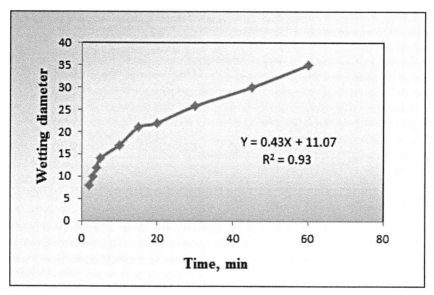

FIGURE 12.3   Diameter of the horizontal wetted zone.

due to the increased area for downward movement of water as the lateral wetting increased. Linear regression equation ($Y = AX + B$) was fitted to the horizontal advancement data. The behavior of horizontal wetted diameter versus time confirmed the findings of Remadevi,[139] Selvaraj,[149,150] and Arunadevi.[14]

12.4.3 SOIL MOISTURE DISTRIBUTION PATTERNS

The soil moisture content at 0, 0–10, 10–20, and 20–30 cm soil depth and a horizontal distance from the emitter were estimated just before irrigation, immediately after irrigation, 1 day after irrigation, and 2 days after irrigation. The mean maximum soil moisture content (38.3%) was observed below the emitter at the depth of 10 cm immediately after irrigation. The soil moisture contents were plotted by using computer software package "surfer" of windows version and are shown in Figure 12.4. Higher moisture content in the lower horizons may be due to water stored in soil pores with minimum evaporation loss. Soil moisture content was lower at the surface layer than at depths and away from the emitter. This might be due to more evaporation from the soil surface compared to lower layers.

Similar results were reported by Philip,[132] who mentioned that the moisture content was gradually decreased with the increase in distance from the emitter. Chakraborty et al.[34] also reported similar findings.

12.4.4 EFFECT OF MULCHING AND FERTIGATION ON SOIL TEMPERATURE

Soil temperature readings were taken with the help of soil thermometer, at 5 cm depth (Table 12.14). Soil temperature varied significantly with the type of mulching and time of the day. Soil temperature was lower in the early morning and gradually increased from 12 noon to 4 p.m. in all treatments and then declined. Temperature under mulching was higher than that of the control plots for all times. The 25 μm thickness BPM produced higher soil temperatures than 50 μm thickness mulch. A difference of 2–5°C was observed between mulched and nonmulched treatments. Data regarding soil temperature under different mulch treatments revealed that plastic mulches increased soil temperature significantly than nonmulched control plots.

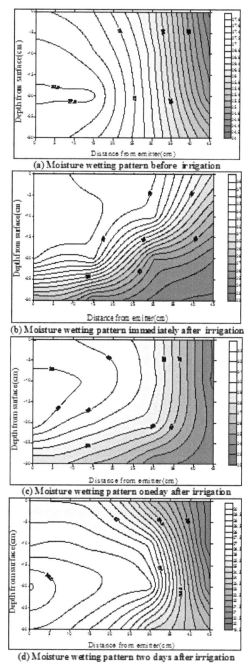

(a) Moisture wetting pattern before irrigation

(b) Moisture wetting pattern immediately after irrigation

(c) Moisture wetting pattern oneday after irrigation

(d) Moisture wetting pattern two days after irrigation

FIGURE 12.4 Moisture wetting pattern with respect to dripper location.

TABLE 12.14   Effects of Mulching and Fertigation on Mean Soil Temperature.

Treatment	Clock		
	10 a.m.	12 a.m.	4 p.m.
T1	29.71	35.22	35.72
T2	28.25	34.04	34.00
T3	27.87	32.25	32.06
T4	28.31	34.32	35.62
T5	27.20	33.17	33.21
T6	27.05	33.02	31.68
T7	25.77	31.68	32.52
T8	25.82	31.07	32.23
T9	24.80	28.81	29.78
Mean	27.19	32.61	32.97
SEd	0.38	0.73	0.65
CD (0.05)	0.81	1.55	1.38
CV	3.21	4.69	4.12

Suwon and Judah[172] reported that soil temperature was increased with the use of plastic mulch. This is because polythene mulches allowed part of the radiation to pass through it but acted as a barrier against outgoing thermal radiation.[127] The temperature increase under plastic mulch is due to high soil moisture content, which leads to more heat flux for thermal conductivity.[36,140] The temperature increase in the BPM condition might be because the black film absorbed incoming solar radiation and radiated much of this energy as sensible heat to the air (above) and soil (below). Soil temperature is increased by 5–10°C by the application of plastic mulches as compared to bare soil.[47] This well-documented temperature rise is often used as an explanation for increased production of crops grown under plastic mulch.[43,63]

12.4.5   EFFECTS OF MULCHING AND FERTIGATION ON GROWTH PARAMETERS

The growth components (height, days to first flowering, fruits set, fruit weight, number of fruits per plant, fruit yield per plant, and total yield) were recorded and analyzed for their significance on the treatments. The growth stages from planting to harvesting are shown in Figure 12.5.

FIGURE 12.5   Plant growth at different growth stages.

12.4.5.1   PLANT HEIGHT

Plant height was measured at 15, 30, 45, 60, and 90 DAT. Mean plant height are presented in Table 12.15 and Figure 12.6 for different treatments. The height of the plant was significantly influenced by mulching and fertigation levels. The height of the crop recorded at 15 DAT showed that the maximum plant height of 35 cm was recorded under 50 μm thickness plastic mulch at 120% RDF (T6). The plant heights of 33 cm and 29 cm were recorded for T5 and T3 cm and lowest height of 22 cm was recorded for the treatment T7. The statistical analysis depicted that there was statistical significance for both mulching and fertilizer levels on the plant height.

At 30 DAT, maximum plant height of 64 cm was recorded in the treatment T5 (100% drip fertigation with BPM of 50 μm thickness), which was significantly superior to rest of the treatments. At 30 DAT, fertigation with 100% RDF gave significantly higher plant height compared to rest two treatments, which were on par. At 45 DAT, maximum plant height of 91 cm was observed in the treatment T6 followed by T5 (88 cm). The minimum plant height was for the control plot T7 (71 cm). It showed

TABLE 12.15   Effect of Mulching and Fertigation on Plant Height (cm) of Tomato.

Treatment	Plant height of tomato, cm					Days to	
	15 DAT	30 DAT	45 DAT	60 DAT	90 DAT	First flowering	First fruit set
T1	26	52	82	92	100	34	41
T2	28	56	84	95	101	33	38
T3	29	58	86	97	106	32	37
T4	28	58	86	98	108	33	37
T5	33	64	88	102	114	31	35
T6	35	63	91	105	115	28	35
T7	22	51	71	87	96	34	44
T8	25	55	80	91	99	35	42
T9	27	58	83	95	104	35	40
Mean	28.11	57.22	83.29	95.85	104.77	33	39
SEd	0.95	1.99	1.40	1.40	1.57	0.52	0.59
CD (0.05)	2.02	4.23	2.98	2.98	3.33	1.11	1.25
CV	5.19	8.33	3.98	3.5	3.09	3.27	3.39

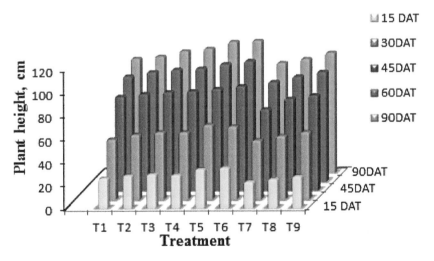

FIGURE 12.6   Effects of mulching and fertigation on plant height (cm) of tomato.

significant effects on mulching and fertigation treatments, but there was no significant effect on the interaction. The plant height at 45, 60, and 90 DAT also showed significant effects on mulching and fertigation treatments, but there was no significant effect on the interaction. Similar results were reported by Hallidri,[69] who indicated that plant height was maximum in plants grown under black and transparent polythene mulch than that in the control (bare soil). Plants attained maximum height under mulches, particularly in 50 µm thickness plastic mulch, which might be due to the increased soil temperature.[177] The increase in plant height is attributed to moisture conservation and weed suppression due to the application of mulches.

12.4.5.2 DAYS TO FIRST FLOWERING

The earliest flowering was recorded in T6 (50 µm thickness plastic mulch at 120% RDF) after 28 days of transplanting followed by T5 after 31 days. The control treatments took more number of days (35 DAT) for first flowering. The results are shown in Table 12.15. The both factors and interaction showed a significant effect on days to flowering. 50 µm thickness plastic mulch and 120% RDF gave better results compared to other two treatments. Devi et al.[45] also observed early flowering (by 5 days) in plastic mulch treated groundnut crop.

TABLE 12.16   Effects of Mulching and Fertigation on Fruit Parameters.

Treatment	Fruit weight (g)	Diameter (cm)	Fruit yield (kg/plant)	Number of fruits/plant	Total yield (t/ha)
T1	64.3	4.7	1.56	31	73.41
T2	67.3	5.2	1.76	35	76.41
T3	79.7	5.9	2	39	81.22
T4	75	5.6	1.93	40	77.40
T5	83.3	6.4	2.41	48	83.16
T6	85.5	7	2.59	52	85.96
T7	44	4	1.38	28	57.98
T8	49	4.9	1.42	28	59.40
T9	53	5.4	1.55	31	62.06
Mean	66.77	5.47	1.85	36	73.20
SEd	0.71	0.19	0.015	0.29	0.23
CD (0.05)	1.51	0.32	0.032	0.51	0.48
CV	2.24	5.44	1.89	1.88	0.71

12.4.5.3   DAYS TO FIRST FRUIT SET

Table 12.15 indicates that the fruit set was earlier in T6 and T5 (35 days) and late in T7 (44 days). The statistical analysis revealed that the best treatment was T6 (50 μm thickness in case of mulching and 120% RDF in case of fertigation). For interaction between fertigation and mulching, there was no significant difference in case of fruit set.

12.4.5.4   FRUIT DIAMETER AND FRUIT WEIGHT

The maximum fruit diameter of 7 cm was recorded in T6 (50 μm thickness plastic mulch at 120% RDF), followed by T5 (6.4 cm). In the control treatment, fruits had a lower diameter of 4 cm. The results are shown in Table 12.16 and Figure 12.7. Fruit weight was significantly different among the mulching and fertigation treatments. The maximum fruit weight was observed in the treatment T6 (85.5 g) followed by T5 (83.3 g) and T3 (79.5 g).

FIGURE 12.7   Effects of mulching and fertigation on fruit size.

12.4.5.5   NUMBER OF FRUITS PER PLANT

For each harvesting, a number of fruits from each plant were counted and weighed. Mulching produced more fruits per plant compared to control, probably due to the conservation of moisture and improved microclimate both beneath and above the soil surface. A maximum of 52 fruits were obtained in the treatment T6 and 28 fruits per plant for the control. These results are given in Table 12.16 and Figure 12.8.

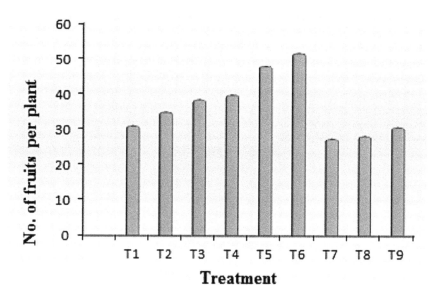

FIGURE 12.8   Effects of mulching and fertigation on number of fruits per plant.

12.4.5.6   FRUIT YIELD PER PLANT

Fruit yield per plant was significantly different among mulching and fertigation treatments. The maximum yield was observed in treatment T6 (2.59 kg) followed by T5 (2.41 kg) and T3 (2 kg). The results are shown in Table 12.16 and Figure 12.9.

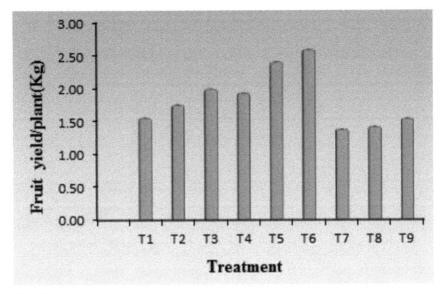

FIGURE 12.9 Effects of mulching and fertigation on fruit yield per plant.

12.4.5.7 TOTAL YIELD

The highest yield was 85.96 t/ha in T6 (50 μm thickness plastic mulching at 120% RDF) followed by T5 (83.16 t/ha) and T3 (81.22 t/ha) as shown in Table 12.16 and Figure 12.10. Lowest yield was 57.98 t/ha in T7 (control). The increased yield in fertigation treatments might be due to better availability of plant nutrients and IW throughout the crop growth period under drip fertigation system. This is in accordance with the findings of Gutal et al.[67]

The results for mulching treatments are in agreement with those by Anderson et al.,[7] who indicated the highest yield of tomato under black paper or black polyethylene mulches. Similarly, Shin et al.[158] obtained much better yield of garlic with polyethylene mulch. Ramakrishna et al.[136] also reported that polythene mulched plots produced highest yields and 94.5% higher than the nonmulched plots in groundnut. Also, higher fruit yield under polyethylene mulch may also be ascribed to reduced nutrient losses due to weed control and improved hydrothermal regimes of soil.[15,27]

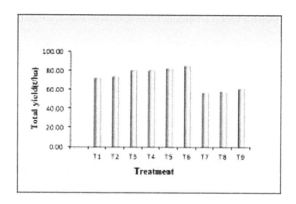

FIGURE 12.10   Effects of mulching and fertigation on total yield of tomato.

12.4.6   EFFECTS OF MULCHING AND FERTIGATION ON WEED POPULATION

Weeding was done four times during the crop period at 15, 30, 60, and 90 DAT. During weeding, weed types, a number of weeds present in the plot, and the dry weight of weeds were recorded. In mulching treatments, weeds were completely absent (Fig. 12.11). Similar results were previously reported by Schonbeck.[149] According to him, BPM produced weeds only through the punch and no weeds were found under the plastic, which might be due to lack of percentage of light through black plastic. BPM blocked the weeds, except a few, which emerged through the planting holes. Zhang et al.[188] reported that black plastic film mulch resulted in 100% control of all weeds in maize crop, thus supporting the results in this chapter.

Control plot

Mulching Pbt

FIGURE 12.11   Effects of mulching on weed control.

12.4.6.1   NUMBER OF WEEDS

The total weed population was significantly reduced by the mulching practices. The general appraisal revealed that plastic mulching decreased the total weed population to zero. Weeds were found only in the control plot and their number was increased with increase in fertilizer application. There was no significant difference in number of weeds between the fertigation treatments. The number of weeds were 12, 9, and 7 in the treatments T9, T8, and T7, respectively.

12.4.6.2   TYPES OF WEEDS

The weed flora observed in the experimental field during the course of study consisted of grasses, sedges, and broad leaved weeds. The major weeds were: *Cynodon dactylon, Dactyloctenium aegyptium, Elytrigia repens* in grasses; *Cyperus rotundus* in sedges, *Digera arvensis, Parthenium hysterophorus, Trianthema portulacastrum, Datura metal, Portulaca oleracea, Chamaesyce maculata, Tribulus terrestris,* and *Acalypha indica* in broad-leaved weeds.

12.4.6.3   DRY WEIGHT OF WEEDS

The dry matter production of weeds was recorded during the crop period. Increased dry matter production was observed with the advancement of crop growth. The mean dry matter production of weeds in the 80% RDF was lower compared to the other two fertigation treatments. The dry weight of weeds in the three control plots at different times of weeding was recorded and total dry weight of weeds in T7, T8, and T9 was 651, 801, and 821 g, respectively.

These results are in agreement with previous reports on the beneficial effects of black polyethylene mulch through its effective weed control, conservation of soil moisture and increasing soil temperature.[37,57,134,183]

12.4.7   WATER USE EFFICIENCY

The Appendix I presents the sample example to calculate the water requirement of the tomato crop. The WUE for all treatments is presented in Table 12.17 and in Figure 12.12. The highest WUE of 386 kg/ha-mm was recorded in treatment T6 (50 μm plastic mulch with 120% RDF). The least WUE of 261 kg/ha-mm was observed in control plot (T7), due to lower yield in the control.

TABLE 12.17   Water Use Efficiency Under Different Treatments.

Treatment	Total yield (kg/ha)	WUE (kg/ha-mm)
T1	73,412	330
T2	74,555	335
T3	81,216	365
T4	81,105	365
T5	83,158	37
T6	85,955	386
T7	57,979	261
T8	59,398	267
T9	62,059	279
Mean	72,998	329
SEd	231	1.32
CD (0.05)	490	2.79
CV	0.71	0.71

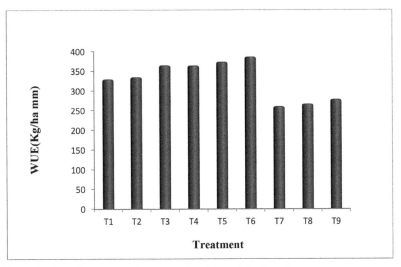

FIGURE 12.12   Water use efficiency under different treatments.

These results are in agreement with the findings of Seyfi et al.[151,152] who indicated that drip irrigation with BPM markedly decreased the amount of water applied, increased WUE and increased crop yield due to increase in number of fruits per plant, fruit weight, and fruit thickness.

12.4.8 FERTILIZER USE EFFICIENCY

FUE of nitrogen, potassium, and phosphorus were calculated for each treatment (Fig. 12.13). According to Malik et al.[107] crop response to fertilizer application through drip irrigation has been excellent and frequent nutrient applications had improved the FUE. Bar-Yosef and Sagiv[22] also reported fertilizer saving and increase in tomato yield due to fertigation [20,31].

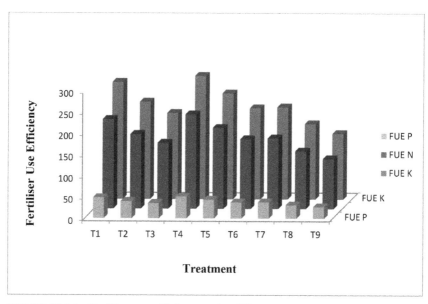

FIGURE 12.13   Fertilizer use efficiency under different treatments.

12.4.8.1 FERTILIZER USE EFFICIENCY OF NITROGEN

The effects of N FUE are shown in Table 12.18. Increased FUE with the decrease level of fertigation dose was observed. The highest N FUE of 223 kg/ha.kg of N was recorded in T4 (50 μm thickness plastic mulch at 80 %

of fertigation) followed by T1 (211 kg/ha.kg of N) and the lowest FUE of N was observed in control plots T9 (119 kg/ha.kg of N). There were statistically significant differences among various treatments and interactions. In the case of fertigation, 80% fertigation was the best.

These results are in accordance with Bharambe et al.,[25] who reported N use efficiency was considerably increased with the N fertigation and it was highest under 75 kg of N/ha applied through drip application in a Parbhani clayey soil at Maharashtra. This can be due to improved distribution of fertilizer with minimum leaching beyond the root zone or runoff. However, Huett and Dettmann[80] reported that tomato is highly responsive to N, but the application of excessive rates of N rarely negatively affects quality.

TABLE 12.18   Fertilizer Use Efficiency of Nitrogen Under Different Treatments.

Treatment	Total yield (kg/ha)	FUE of N (kg/ha.kg of N)
T1	73,412	211
T2	74,555	176
T3	81,216	156
T4	81,105	223
T5	83,158	191
T6	85,955	165
T7	57,979	167
T8	59,398	137
T9	62,059	119
Mean	72,998	171
SEd	231	0.53
CD(0.05)	490	1.32
CV	0.71	0.72

12.4.8.2   FERTILIZER USE EFFICIENCY OF PHOSPHORUS

The FUE of phosphorus was also highest in T4 (52 kg/ha.kg of P) followed by T1 (49 kg/ha.kg of P) and the lowest was noted in T9 (28 kg/ha.kg of P). The variation in FUE of phosphorus is shown in Table 12.19. The highest FUE of phosphorus in T4 is due to the minimum application of fertilizer in that treatment.

TABLE 12.19   Fertilizer Use Efficiency of Phosphorus Under Different Treatments.

Treatment	Total yield (kg/ha)	FUE (kg/ha.kg of P)
T1	73,412	49
T2	74,555	41
T3	81,216	36
T4	81,105	52
T5	83,158	44
T6	85,955	38
T7	57,979	39
T8	59,398	32
T9	62,059	28
Mean	72,998	40
SEd	231	0.12
CD(0.05)	490	0.26
CV	0.71	0.73

12.4.8.3   FERTILIZER USE EFFICIENCY OF POTASSIUM

The effects of fertigation of K on FUE are shown in Table 12.20. FUE was highest in T4 (290 kg/ha.kg of P) followed by T1 (275 kg/ha.kg of P) and the lowest was noted in T9 (155 kg/ha.kg of P) under various mulch and fertigation treatments.

TABLE 12.20   Fertilizer Use Efficiency of Potassium Under Different Treatments.

Treatment	Total yield (kg/ha)	FUE (kg/ha.kg of K)
T1	73,412	275
T2	74,555	229
T3	81,216	203
T4	81,105	290
T5	83,158	249
T6	85,955	215
T7	57,979	217
T8	59,398	178
T9	62,059	155
Mean	72,998	223
SEd	231	0.69
CD (0.05)	490	1.48
CV	0.71	0.72

12.4.9   COST ECONOMICS

Appendix II presents the procedure on cost economics of tomato. In order to study the feasibility of cultivation of tomato under the plastic mulching cost of cultivation, fixed cost, net income, and B:C ratio under different treatments were determined and given in Table 12.21 and Appendix II. The interest rate and repair/maintenance cost of the system were assumed as 10% and 2% per annum of the fixed cost, respectively.[137]

The fixed cost of installation of drip irrigation system and plastic mulching was Rs. 183,753/ha for 25 μm thickness and Rs. 192,553 for 50 μm thickness treatments. The treatment T_6 registered the highest gross income of Rs. 687,642/ha because of high yield due to effective and optimal growing conditions for 50 μm thickness plastic mulch with 120% RDF. The B:C ratio was also higher (1.86) in this treatment (T6) compared to all other treatments. In control plot (T7) with 80% RDF, the B:C ratio was 1.44, which is lower than the other treatments. Similar beneficial effects of black LLDPE mulch in tomato and okra have also been reported by Shrivastava.[163]

12.5   SUMMARY

Land and water are the indispensable resources of crop production. Water is the most limiting natural resource in arid and semiarid areas for the economic development of a country. In most of the regions, the only water available is from the rainfall. Hence, for successful agriculture, proper utilization of water is of paramount importance, which implies to increase the WUE of a crop. The water loss takes place due to evaporation, transpiration, and percolation. In many countries, fresh water is relatively scarce, but there are considerable resources of saline water, which could be utilized for irrigation if proper crops, soil, and water management practices are established. During the last three decades, micro irrigation systems owing to their capability to apply water efficiently, low labor and energy requirement, and increase in quantity and quality of crop yield/produce have made a breakthrough around the globe. Micro irrigation encompasses drip/trickle systems, surface and subsurface drip tapes, microsprinklers, sprayers, microjets, spinners, rotors, and bubblers. In this system, water is applied at low rates frequently near the root zone of the plant. In recent years, fertilizer is

TABLE 12.21 Cost of Cultivation of Tomato (Rs./ha) Under Mulching and Fertigation Treatments.

S. no	Cost economics	M1			M2			M3		
		F1	F2	F3	F1	F2	F3	F1	F2	F3
I	Cost of plastic mulch	25,200	25,200	25,200	44,000	44,000	44,000	0	0	0
	For 1 year	25,200	25,200	25,200	22,000	22,000	22,000	0	0	0
II	Irrigation and fertigation system cost (life period 5 years)	148,550	148,550	148,550	148,550	148,550	148,550	148,550	148,550	148,550
	For 1 year	29,710	29,710	29,710	29,710	29,710	29,710	29,710	29,710	29,710
	Fixed cost, Rs.	54,910	54,910	54,910	51,710	51,710	51,710	29,710	29,710	29,710
III	Repair and maintenance cost at 2 % of fixed cost	1100	1100	1100	1040	1040	1040	600	600	600
IV	Interest on fixed cost at 10%	5490	5490	5490	5170	5170	5170	2970	2970	2970
V	Land preparation and management and other inputs (including harvesting, planting material, and fertilizer)	31,880	36,800	41,720	31,880	36,800	41,720	80,880	75,800	70,720
VI	Total cost of cultivation, Rs./ha	93,380	98,300	103,220	89,790	94,720	99,640	114,150	109,080	104,000
VII	Yield, kg/ha	73,410	74,560	81,220	81,110	83,160	85,960	57,980	59,400	62,060
VIII	Revenue from the field in Rs/ha = yield × 8.00 (considering cost of tomato at Rs. 8/kg)	440,470	447,330	487,300	486,630	498,950	515,730	347,880	356,390	372,360
IX	Net income, Rs.	347,100	349,030	384,080	326,840	388,240	416,100	233,720	247,320	268,360
X	Benefit–cost ratio = IX/VI	3.71	3.55	3.72	3.63	4.09	4.17	2.04	2.26	2.58

also applied along with water through drip irrigation to get higher FUE besides increased yields.

This chapter discusses effects of drip fertigation and mulching on tomato production. Cv, SU, and CU were evaluated from the observed discharges. Cv was obtained as 0.0198, SU as 98%, and CU as 96.87%.

Soil temperature was low in the early morning. Temperature under mulching was higher than that in the control plots. In mulching treatments, weeds were completely absent. The highest yield was 85.96 t/ha in 50 μm thickness plastic mulching at 120% RDF followed by T5 and T3 (83.16 t/ha) and (81.22 t/ha), respectively. Lowest yield was recorded in T7 (57.98 t/ha), that is, in control.

The highest WUE of 386 kg/ha-mm was recorded in 50 μm plastic mulch with 120% RDF. The least WUE of 261 kg/ha-mm was noted in the control. Increased FUE with the decreased level of fertilizer dose through drip was also observed.

The fixed cost of installation of drip irrigation system and mulching sheets was Rs. 54,910/ha for 25 μm thickness treatments and Rs. 51,710 for 50 μm thickness treatments per year. The treatment T6 registered the highest gross income of Rs. 416,100/ha. The B:C ratio was 4.17 in for 50 μm thickness treatments compared to all other treatments. In control plot with 80% RDF, the B:C ratio was 2.04.

KEYWORDS

- benefit–cost ratio
- coefficient of uniformity
- coefficient of variation
- drip irrigation
- fertilizer use efficiency
- soil moisture
- soil temperature
- statistical uniformity
- water scarcity
- water use efficiency

REFERENCES

1. Abdul, M.; Ashrafuzzaman, M.; Mohd Razi, I.; Shahidullah, S. M.; Alamgir, M. H. Effect of Plastic Mulch on Growth and Yield of Chili (*Capsicum annuum* L.). *Braz. Arch. Biol. Technol.* **2011,** *54*(20), 321–330.
2. Acharya, C. L.; Sharma, P. D. 1994. Tillage and Mulch Effects on Soil Physical-environment, Root-growth, Nutrient-uptake and Yield of Maize and Wheat on an Alfisol in North-west India. *Soil Tillage Res.* **1994,** *32*, 291–302.
3. Agele, S. O.; Iremiren, G. O.; Ojeniyi, S. O. Effects of Tillage and Mulching on the Growth, Development and Yield of Late-season Tomato (*Lycopersicon esculentum* L.) in the Humid South of Nigeria. *J. Agric. Sci.* **2000,** *134*, 55–59.
4. Agrawal, R.; Agrawal, N.; Panigrahi, H. K.; Sharma, D. Effect of Different Color Mulches on the Growth and Yield of Tomato Under Chhattisgarh Region. *Indian J. Hort.* **2010,** *67*(Special Issue), 295–300.
5. Ahmad-Bhat, F. N.; Gupta, A. J.; Feza, M. Studies on Yield, Quality, Water and Fertilizer Use Efficiency of Capsicum Under Drip Irrigation and Fertigation. *Indian J. Hort.* **2010,** *67*(2), 213–218.
6. Ali, M. H.; Shahram, A.; Simon, B. Evaluation of the Influence of Irrigation Methods and Water Quality on Sugar Beet Yield and Water Use Efficiency. *Agric. Water Manag.* **2010,** *97*(2), 357–362.
7. Anderson, D. F.; Garisto, M. A.; Bourrut, J. C.; Schonbeck, M. W.; Jaye, R. A.; Wurzberger, K.; DeGregorio, R. Evaluation of a Paper Mulch Made from Recycled Materials as an Alternative to Plastic Film Mulch for Vegetables. *J. Sustain. Agric.* **1995,** *7*, 39–61.
8. Aniekwe, N. L.; Okereke, O. U.; Aniekwe, M. A. N. Modulating the Effect of Black Plastic Mulch on the Environment, Growth and Yield of Cassava in a Derived Savanna Belt of Nigeria. *Tropicultura* **2004,** *22*(4), 185–190.
9. Anil, K. 2001. Effect of Mulch and Sowing Methods on Pod Yield of Vegetable Pea in Garhwal Himalaya. *Indian J. Soil Conserv.* **2001,** *29*(1), 84–85.
10. Anitta-Fanish, S.; Muthukrishnan, P. Influence of Drip Fertigation on Yield, Water Saving and Water Use Efficiency in Maize (*Zea mays* L.) Based Intercropping System. *Madras Agric. J.* **2010,** *98*(7–9), 243–247.
11. Ardell, D. H. Water Use Efficiency Under Different Cropping Situation. *Ann. Agric. Res.* **2006,** *27*, 115–118.
12. Argall, J. F.; Stewart, K. A. The Effect of Year, Planting Date, Mulches and Tunnels on the Productivity of Field Cucumbers in Southern Quebec. *Can. J. Plant Sci.* **1990,** *70*(4), 1207–1213.
13. Aruna, P.; Sudagar, I. P.; Manivannan, M. I.; Rajangam, J.; Natarajan, S. Effect of Fertigation and Mulching for Yield and Quality in Tomato. *Asian J. Hortic.* **2009,** *2*(2), 50–54.
14. Arunadevi, K. Performance Evaluation of Drip Irrigation and Fertigation on the Yield and Water Use Efficiency of Mulberry. Unpublished Ph.D. Thesis, Submitted to Tamil Nadu Agricultural University, Coimbatore, 2005, p 114.
15. Ashworth, S.; Harrison, H. Evaluation of Mulches for Use in the Home Garden. *HortScience* **1983,** *18*, 180–182.

16. Asokaraja, N. In *Drip Irrigation and Mulching in Tomato*, Proceeding Workshop on Micro Irrigation and Sprinkler Irrigation Systems, New Delhi, April 28–30, Organized by Central Board of Irrigation and Power, Vol. II, 1998, pp 49–54.

17. Assouline, S. The Effects of Micro Drip and Conventional Drip Irrigation on Water Distribution and Uptake. *Soil Sci. Soc. Am. J.* **2002**, *66*, 1630–1636.

18. Aswani, K.; Manoj, D. In *Scope of Drip and Sprinkler Irrigation Systems in India*, Proceeding of Workshop on Sprinkler and Drip Irrigation Systems, Jalgaon, India, 1993, pp 1–10.

19. Aujla, M. S.; Thind, H. S.; Buttar, G. S. Cotton Yield and Water Use Efficiency at Various Levels of Water and Nitrogen Through Drip Irrigation Under Two Methods of Planting. *J. Agric. Water Manag.* **2005**, *71*, 167–179.

20. Bafna, A. M.; Daftardar, S. Y.; Khade, K. K.; Patel, P. Y.; Dhotre, R. S. Utilization of Nitrogen and Water by Tomato Under Drip Irrigation System. *J. Agri. Water Manag.* **1993**, *1*(1), 1–5.

21. Balakrishnan, N. M.; Lalitha, V.; Thilagam, K.; Mansour, M. Effect of Plastic Mulch on Soil Properties and Crop Growth—A Review. *Agric. Rev.* **2010**, *31*(2), 145–149.

22. Bar-Yosef, B.; Sagiv, B. Response of Tomatoes to N and Water Applied via a Trickle Irrigation System. *Agron. J.* **1982**, *74*, 637–639.

23. Basavarajappa, H.; Bhogi, M. G.; Patil, B.; Polisgowdar, S. Effectiveness and Cost Economics of Fertigation in Eggplant (*Solanum melongena* L.) Under Drip and Furrow Irrigation. *Karnataka J. Agric. Sci.* **2011**, *24*(3), 417–419.

24. Batchler, C.; Lovell, C.; Murata, S. Simple Micro Irrigation Technique for Improving Irrigation Efficiency on Vegetable Gardens. *J. Agric. Water Manag.* **1996**, *32*, 37–48.

25. Bharambe, P. R.; Mungal, M. S.; Shelke, D. K.; Oza, S. R.; Vaishnava, V. G.; Sondge, V. D. Effect of Soil Moisture Regime with Drip on Spatial Distribution of Moisture, Salts, Nutrient Availability and Water Use Efficiency of Banana. *J. Indian Soc. Soil Sci.* **2001**, *49*(4), 658–665.

26. Bhardwaj, S. K.; Sharma, I. P.; Bhadari, A. R.; Sharma, J. C.; Tripathi, D. Soil Water Distribution and Growth of Apple Plants Under Drip Irrigation. *J. Indian Soc. Soil Sci.* **1995**, *43*(3), 323–327.

27. Bhella, H. S. Tomato Response of Trickle Irrigation and Black Polyethylene Mulch. *J. Am. Soc. Hort. Sci.* **1998**, *113*(4), 543–546.

28. Bralts, V. F.; Kesner, C. D. Drip Irrigation Field Uniformity Estimation. *Trans. ASAE* **1983**, *26*(5), 1369–1374.

29. Brown, J. E. Black Plastic Mulch and Drip Irrigation Affect Growth and Performance of Bell Pepper. *J. Veg. Prod.* **2001**, *7*(2), 109–112.

30. Byary, S. H.; Sayer, A. R. A. The Influence of Different Irrigation Regimes on Five Greenhouse Tomato. *Egypt. J. Hort.* **1999**, *26*(2), 127–146.

31. Candido, V. V.; Miccolis, M.; Perniola, A. Effects of Irrigation Regime on Yield and Quality of Processing Tomato Cultivars. III International Symposium on Irrigation of Horticultural Crops. *Acta Hortic.* **1999**, *537*, 779–788.

32. Capra, A.; Scicolone, B. Water Quality and Distribution Uniformity in Drip/Trickle Irrigation Systems. *J. Ag. Eng. Res.* **1998**, *70*(4), 355–365.

33. Cassal, S.; Harmasarkar, F. S.; Miller, S. D.; Vance, G. F.; Zhang, R. Assessment of Drip and Flood Irrigation on Water and Fertilizer Use Efficiencies for Sugarbeet. *J. Agric. Water Manag.* **2001**, *46*, 241–251.

34. Chakraborty, D.; Singh, A. K.; Kumar, A.; Uppal, K. S.; Khanna, M. In *Effect of Fertigation on Nitrogen Dynamics in Broccoli*, Proceedings on Micro Irrigation and Sprinkler Irrigation Systems, Central Board of Irrigation and Power, Malcha Marg, New Delhi, April 28–30, 1998, pp 23–30.

35. Charles, B.; Chistopher, L.; Murata, M. Simple Micro Irrigation Technique for Improving Irrigation Efficiency on Vegetable Gardens. *J. Agric. Water Manag.* **1996,** *32,* 37–48.

36. Chen, F.; Dudhia, J. Coupling an Advanced Land Surface Hydrology Model with the Penn. State—NCAR MM5 Modeling System. Part I: Model Implementation and Sensitivity. *Mon. Wea. Rev.* **2001,** *129,* 569–585.

37. Chhangani, S. Effect of Mulches (Synthetic and Non-synthetic) on Water Conservation and Bulb Yield of Irrigated Onion (*Allium cepa* L.) Cultivated in Semi-arid Zone of Borno State. *Niger. J. Ecophysiol.* **2000,** *3,* 5–9.

38. Clough, G. H.; Locasio, S. J.; Olson, S. M. Yield of Successively Cropped Polyethylene-mulched Vegetables as Affected by Irrigation Method and Fertilization Management. *J. Am. Soc. Hort. Sci.* **1990,** *115,* 884–887.

39. Cook, W. P.; Sanders, D. C. Nitrogen Application Frequency for Drip-irrigated Tomatoes. *HortScience* **1991,** *26,* 250–252.

40. Csizinsky, A. A. Evaluation of Color Mulches and Oil Sprays for Yield and Silverleaf Whitefly Control on Tomatoes. *HortScience* **1995,** *30*(4), 755.

41. Dalvi, V. B.; Tiwari, K. N.; Pawade, M. N.; Phire, P. S. Response Surface Analysis of Tomato Production Under Micro Irrigation. *J. Water Manag.* **1999,** *41,* 11–19.

42. Dani, O. Stochastic Analysis of Soil Water Monitoring for Drip Irrigation Management in Heterogeneous Soils. *Soil Sci. Soc. Am. J.* **1995,** *29,* 1222–1233.

43. Davis, J. M. Comparison of Mulches for Fresh-market Basil Production. *HortScience* **1994,** *29,* 267–268.

44. Deshmukh, A. S.; Shinde P. P.; Jadhav, S. B. In *Fertigation Under Drip Irrigation for Sugarcane*, Proceedings of All India Seminar on Modern Irrigation Techniques, June 1996, 217–219.

45. Devi, D.; Naik, P. R.; Dongre, B. N. Effect of Mulching on Soil Temperature and Groundnut Yield During Rabi-summer Season. *Groundn. News* **1991,** *3,* 4–5.

46. Edwards, L.; Burney, J. R.; Richter, G.; MacRae, A. H. Evaluation of Compost and Straw Mulching on Soil Less Characteristics in Erosion Plots of Potatoes in Prince Edward Island, Canada. *Agric. Eco. Environ.* **2000,** *81,* 217–222.

47. Elmer, W. H.; Ferrandino, F. J. Effect of Black Plastic Mulch and Nitrogen Side-dressing on Verticillium Wilt of Eggplant. *Plant Dis.* **1991,** *75*(11), 1164–1167.

48. Enda, A.; Singandhupe, R. B. Impact of Drip and Surface Irrigation on Growth, Yield and WUE of Capsicum (*Capsicum annum* L.). *J. Agric. Water Manage.* **2004,** *65,* 121–132.

49. Essam, A.; Wasif, M.; El-Tantawy, T.; Arafa, Y. E. Impact of Fertigation Scheduling on Tomato Yield Under Arid Ecosystem Conditions. *Res. J. Agric. Biol. Sci.* **2009,** *5*(3), 280–286.

50. Farias-Larios, J.; Orozoc-Santos, M. Effect of Polyethylene Mulch Color on Aphid Populations, Soil Temperature, Fruit Quality and Yield of Watermelon Under Tropical Conditions. *N. Z. J. Crop Hortic. Sci.* **1997,** *25,* 369–374.

51. Feibert, E.; Shock, C.; Stieber, T.; Saunders, M. *Groundcover Options and Irrigation Methods for the Production of Kabocha Squash*; Malheur County Alternative Crops and Marketing Research. Special Report 900: Oregon State University, 1992, pp 26–32.

52. Findeling, A.; Ruy, S.; Scopel, E. Modeling the Effects of Partial Residue Mulch on Runoff Using a Physically Based Approach. *J. Hydrol.* **2003,** *275*, 49–66.

53. Friake, N. N.; Bangal, G. B.; Kenghe, R. N.; More, G. M. Plastic Tunnel and Mulches for Water Conservation. *Agric. Eng. Today* **1990,** *14*(3–4), 35–39.

54. Ganga, J.; Singh, P. K.; Singh, S. K.; Srivastava, P. C. Effect of Drip Fertigation and Mulching on Water Requirement, Yield and Economics of High Density Litchi. *Progress. Hortic.* **2011,** *43*(2), 237–242.

55. Garcia-Villanova, R. J.; Cordon, C.; Paramas, A. M. G.; Aparicio, P.; Rosales, M. E. G. *Economic Comparison of Drip and Furrow Irrigation Methods*; College of Agriculture and Home Economics, Cooperative Extension Service, Circular 573: New Mexico State University, 2000, p 12.

56. Ghuman, B. S.; Sur, H. S. Tillage and Residue Management Effects on Soil Properties and Yields of Rain-fed Maize and Wheat in a Sub-humid Subtropical Climate. *Soil Till. Res.* **2001,** *58*, 1–10.

57. Gimenez, C.; Otto, R. F.; Castilla, N. Productivity of Leaf and Root Vegetable Crops Under Direct Cover. *Sci. Hort.* **2002,** *94*, 1–11.

58. Goel, A. K.; Gupta, R. K.; Kumar, R. Effect of Drip Discharge Rate on Soil Moisture Pattern. *J. Water Manag.* **1993,** *1*(1), 50–51.Goyal, M. R. In *Management, Performance, and Applications of Micro Irrigation Systems;* Book Series: Research Advances in Sustainable Micro Irrigation; Apple Academic Press Inc.: Oakville, ON, Canada, 2014; Vol. 4, p 374.

59. Goyal, M. R. In *Applications of Furrow and Micro Irrigation in Arid and Semi-Arid Regions;* Book Series: Research Advances in Sustainable Micro Irrigation; Apple Academic Press Inc.: Oakville, ON, Canada, 2015; Vol. 5, p 325.

60. Goyal, M. R. In *Water and Fertigation Management in Micro Irrigation;* Book Series: Research Advances in Sustainable Micro Irrigation; Apple Academic Press Inc.: Oakville, ON, Canada, 2015; Vol. 9, p 356.

61. Goyal, M. R.; Chavan, V. K.; Tripathi, V. K. In *Innovations in Micro Irrigation Technology;* Book Series: Research Advances in Sustainable Micro Irrigation. Apple Academic Press Inc.: Oakville, ON, Canada, 2016; Vol. 10, p 410.

62. Grubinger, V. P.; Minotti, P. L.; Wien, H. C.; Turner, A. D. Tomato Response to Starter Fertilizer, Polyethylene Mulch and Level of Soil Phosphorus. *J. Am. Soc. Hort. Sci.* **1993,** *118*(2), 212–216.

63. Mahajan, G.; Singh, K. G. Response of Greenhouse Tomato to Irrigation and Fertigation. *Agric. Water Manag.* **2006,** *84*, 202–206.

64. Gupta, G. N. Effects of Mulching and Fertilizer Application on Initial Development of Some Tree Species. *For. Ecol. Manag.* **1991,** *44*, 211–221.

65. Gupta, R.; Acharya, C. L. Effect of Mulch Induced Soil Hydrothermal Regime on Root Growth, Water Use Efficiency, Yield and Quality of Strawberry. *J. Indian Soc. Soil Sci.* **1993,** *41*(1), 17–25.

66. Gutal, G. B.; Jadhav, S. S.; Takte, R. L. In *Effective Surface Covered Cultivation in Fruit Vegetable Crop Tomato*, Proc. of the Internat. Agri. Eng. Conf., Bangkok, Thailand, Dec 7–10, 1992, pp 853–856.

67. Haddadin, S. H. Effect of Plastic Mulches on Soil Water Conservation, Soil Temperature, and Yield of Tomato in the Jordan Valley. M.Sc. Thesis, University of Jordan,
1982, p 168.

68. Hallidri, M. Comparison of Different Mulching Materials on Growth, Yield And
Quality of Cucumber (*Cucumis sativus* L.). *Acta Hortic.* **2001**, *559*, 49–53.

69. Ham, M.; Kluitenberg, G. J.; Lamont, W. J. Optical Properties of Plastic Mulches
Affect the Field Temperature Regime. *J. Am. Soc. Hort. Sci.* **1993**, *118*(3), 188–193.

70. Hanada, T. *The Effect of Mulching and Row Covers on Vegetable Production*; Publication of Chugoku Agr. Exp. Stn., Japan, 1991; p 23.

71. Hanson, B. R.; Williams, D. Influence of Subsurface Irrigation on Cotton Yield and
Water Use. *J. ASAE* **1985**, *12*, 68–70.

72. Hanson, B. R.; May, D. M.; Bendixen, W. E. Wetting Pattern Under Surface and
Subsurface Drip Irrigation. *Trans. ASAE* **1997**, *23*, 1–10.

73. Hapase, D. G.; Mankar, A. N.; Saikhe, V. In *Techno-economic Evaluation of Drip
Irrigation for Sugarcane Crop*, Proceedings of an International Agricultural Engineering Conference, Bangkok, Thailand, Dec 7–10, 1992; Vol. 3, pp 897–904.

74. Hasan, M. F.; Ahmad, B.; Rehman, M. A.; Alam, M. M.; Khan, M. M. H. Environmental Effect on Growth and Yield of Tomato. *J. Biol. Sci.* **2005**, *5*(6), 759–767.

75. Hassan, A. F. Evaluating Emission Uniformity for Efficient Micro Irrigation. *Irrg. J.*
1997, *47*(4), 24–27.

76. Himelrick, D. G.; Dozier, W. A.; Akridge, J. R. Effect of Mulch Type in Annual Hill
Strawberry Plasticulture Systems. *Acta Hortic.* **1993**, *348*, 207–209.

77. Howell, T. A.; Bucks, D. A.; Chesness, J. L. In *Advances in Trickle Irrigation*,
Proceedings National Irrigation Symposium, 1981, pp 62–64.

78. Hu, W.; Duan, S.; Sui, Q. High Yield Technology for Groundnut. In *International
Arachis Newsletter*; ICRISAT, 1995, Vol. 15, pp 1–22.

79. Huett, D. O.; Dettmann, E. B. Effect of Nitrogen on Growth, Fruit Quality and
Nutrient Uptake of Tomatoes Grown in Sand. *Aust. J. Exp. Agric.* **1988**, *28*, 391–399.

80. Hummel, R. L.; Walgenbach, J. F.; Barbercheck, M. E.; Kennedy, G. G.; Hoyt, G.
D.; Arellano, C. Effects of Production Practices on Soil-borne *Entomopathogens* in
Western North Carolina Vegetable Systems. *Environ. Entomol.* **2002**, *31*(1), 84–91.

81. Ibarra, L.; Flores, J.; Diaz-Perez, J. C. Growth and Yield of Muskmelon in Response
to Plastic Mulch and Row Covers. *Sci. Hort.* **2001**, *87*, 139–145.

82. Ibarra, L.; Muguia-Lopez, J.; Lozano-del Rio, A. J.; Zermeno-Gonzalez, A. Effect of
Plastic Mulch and Row Covers on Photosynthesis and Yield of Watermelon. *Aust. J.
Exp. Agric.* **2005**, *44*(1), 91–94.

83. Jadhav, S. S.; Gutal, G. B.; Chougule, A. A. In *Cost Economics of Drip Irrigation
Systems for Tomato Crop*, Proceedings of the 11th International Congress on the Use
of Plastics in Agriculture, New Delhi, India, 1990, 171–176.

84. Jain, V. K.; Shukla, K. N.; Singh, P. K. In *Response of Potato Under Drip Irrigation
and Plastic Mulching*, Proceedings of International Conference on Micro and Sprinkler Irrigation, Organized by Central Board of Irrigation and Power Held at Jalgaon,
India, 2001; Singh, H. P., Kaushish, S. P., Kumar, A., Murthy, T. S., Eds.; pp 413–417.

85. Kadam, J. R.; Karthikeyan, S. Effect of Soluble NPK Fertilizers on the Nutrient
Balance, Water Use Efficiency, Fertilizer Use Efficiency of Drip System for Tomato.
Int. J. Plant. Sci. **2006**, *1*, 92–94.

86. Kaminwar, S. P.; Rajagopal, V. Fertilizer Response and Nutrient Requirement of Rain Fed Chilies in Andhra Pradesh. *Fertil. News* **1993,** *36*(7), 21–26.

87. Kaniszewski, S. Response of Tomatoes of Drip Irrigation and Mulching with Polyethylene and Non-woven Poly Polypropylene. *Biuletyn Warzywniczy* **1994,** *41,* 29–38.

88. Karp, K.; Noormets, M.; Paal, T.; Starast, M. The Influence of Mulching on Nutrition and Yield of North Blue Blueberry. *Acta Hortic.* **2006,** *715,* 301–305.

89. Kataria, D. P.; Michael, A. M. In *Comparative Study of Drip and Furrow Irrigation Methods,* Proc. XI International Congress on the Use of Plastic in Agriculture, New Delhi, 1990, pp 19–27.

90. Kavitha, M.; Natarajan, S.; Sasikala, S.; Tamilselvi, C. Influence of Shade and Fertigation on Growth, Yield and Economics of Tomato (*Lycopersicon esculentum* Mill.). *Int. J. Agric. Sci.* **2007,** *3*(1), 99–101.

91. Keller, J.; Karmeli, D. Trickle Irrigation Design Parameters. *Trans. ASAE* **1974,** *17*(4), 678–684.

92. Khan, M. M.; Shivashankar, K.; Krishna-Manohar, R.; Sree-Rama, R.; Kasiyanna, P. In *Fertigation in Horticultural Crops, Proc. Advances in Micro Irrigation and Fertigation,* Dharwad, India, June 21–30, 1999, pp 181–197.

93. Khepar, S. D.; Neog, P. K; Kaushal, M. P. In *Moisture and Salt Distribution Pattern Under Drip Irrigation,* Proceedings 2nd National Seminar on Drip Irrigation, Coimbatore, India, March 1983, pp 76–85.

94. Kirnak, H.; Dogan, E.; Bilgel, L.; Berakatoglu, K. Effect of Pre-harvest Deficit Irrigation on Second Crop Watermelon Grown in an Extremely Hot Climate. *J. Irrig. Drain. Eng.* **2009,** *135*(2), 141–148.

95. Kulkarni, G. N.; Kalaghatagi, S. B.; Mutanal, S. M. Effect of Various Mulches and Scheduling of Irrigation on Growth and Yield of Summer Maize. *J. Maharashtra Agric. Univ.* **1998,** *13*(2), 223–224.

96. Lamm, F. R.; Schlegel, A. J.; Clark, G. A. In *Optimum Nitrogen Fertigation for Corn Using SDI,* Proceedings of Irrigation Assn. International Technical Conference, Nashville, TN, Nov 2–4, 1997, pp 251–258.

97. Lamont, W. J. Plastics: Modifying Microclimate for the Production of Vegetable Crops. *HortTechnology* **2005,** *15*(3), 477–481.

98. Lamont, W. J. Plastic Mulches for the Production of Vegetable Crops. *HortTechnology* **1993,** *3*(1), 35–39.

99. Lightfoot, D. R. Morphology and Ecology of Lithic-mulch Agriculture. *Geogr. Rev.* **1994,** 172–185.

100. Lincoln, Z.; Dukes, M. D.; Scholberg, J. M.; Hanselman, T.; Le Femminella, K. A.; Munoz-Carpena, R. Nitrogen and Water Use Efficiency of Zucchini Squash for a Plastic Mulch Bed System on a Sandy Soil. *Sci. Hortic.* **2008,** *116,* 8–16.

101. Locascio, S. J.; Smajstria, A. G. In *Water and N-K Application Timing for Drip Irrigation Tomato,* Proceedings of XXXVII Annual Meeting of the Inter-American Society for Tropical Horticulture, 1992, *36,* pp 140–142.

102. Locher, J.; Ombodi, A.; Kassai, T.; Dimeny, J. Influence of Colored Mulches on Soil Temperature and Yield of Sweet Pepper. *Eur. J. Hort. Sci.* **2005,** *70,* 135–141.

103. Ludwick, A. E. Don't Forget Soil-test Calibration. News and Views. Regional Newsletter Published by Potash (PPI) and Phosphorus Institute, Canada, 1998; pp 50–54.

104. Luis, I. J.; Cedeno-Ruvalcaba, B.; Hernandez-Castillo, F.; Flores, V. J. Effects of Soil Mulch and Row Covers on Growth and Yield of Bell Pepper. *Phyton* **2002**, *1*, 101–106.

105. Malik, R. S.; Kumar, K.; Bhandari, A. R. Effect of Urea Application Through Drip Irrigation System on Nitrate Distribution in Loamy Sand Soils and Pea Yield. *J. Indian Soc. Soil Sci.* **1994**, *42*(1), 6–10.

106. Manjunatha, M. V.; Shukla, K. N.; Chauhan, H. S.; Singh, P. K.; Singh, R. In *Economic Feasibility of Micro Irrigation Systems for Various Vegetables*, Proceedings of International Conference on Micro and Sprinkler Systems, Jalgaon, Maharashtra, India, July 2001, pp 360–364.

107. Maragatham, J. D.; Paul, S. Physical Characteristics as a Function of Its Particle Size to be Used as Soil-less Medium. *Am. Eurasian J. Agric. Sci.* **2010**, *8*(4), 431–437.

108. Mbah, C. N.; Mbagwu, J. S. C.; Onyia, V. N.; Anikwe, M. Effects of Application of Bio-fertilizers on Soil Densification, Total Porosity, Aggregate Stability and Maize Grain Yield at Abakaliki. *Niger. J. Sci. Technol.* **2004**, 10, 74–85.

109. Meena, M. S.; Singh, R.; Kumar, S.; Nangare, D. D. Drip Irrigation and Black Polyethylene Mulch Influence on Growth, Yield and Water-use Efficiency of Tomato. *Afr. J. Agric. Res.* **2009**, *4*(12), 1427–1430.

110. Mishra, K. K.; Pyasi, S. K. Moisture Distribution Pattern in Drip Irrigation. *J. Agric. Res.* **1993**, *17*, 30–39.

111. Mishra, J. N.; Paul, J. C. Impact of Drip Irrigation with Plastic Mulch on Yield and Return of Eggplant Crop. *J. Water Manag.* **2009**, *17*(1), 1–7.

112. Mugalla, C. I.; Jolly, C. M.; Martin, N. R. Profitability of Black Plastic Mulch for Limited Resource Farmers. *J. Prod. Agric.* **1996**, *9*(2), 175–176.

113. Muralidhar, A. P.; Shivashankar, H.; Shivakumar R.; Kumar, V. In Fertilizer and Irrigation Use Efficiency as Influenced by Furrow and Ferti-drip Irrigation in Capsicum-maize-sunflower Cropping Sequence, Proceedings of National Seminar on Problems and Prospects of Micro Irrigation—A Critical Appraisal, New Delhi, November 19–20, 1999, pp 100–105.

114. Nakano, A.; Uehara, Y.; Watanabe, I. Qualities Organic Certified Fruits/Vegetables. *Jpn. J. Soil Sci. Plant Nutr.* **2002**, *73*, 307–309.

115. Nakayama, F. S.; Bucks, D. A. Water Quality in Drip/Trickle Irrigation: A Review. *Irrig. Sci.* **1991**, *12*, 187–192.

116. Narayanamoorthy, A.; Deshpande, R. S. In *Economic Evaluation of Drip Irrigation—A Study of Maharashtra*, Proceedings of International Conference on Micro and Sprinkler System, Jalgaon, Maharashtra, India, 2001, pp 730–740.

117. Ngouaajio, M.; Ernest, J. Changes in Physical, Optical and Thermal Properties of Polythene Mulches During Double Cropping. *HortScience* **2005**, *40*(1), 94–97.

118. Ngouajio, M.; McGiffen, M. E. Going Organic Changes Weed Population Dynamics. *HortTechnology* **2002**, *12*, 590–596.

119. Niu, J. Y.; Gan, Y. T.; Huang, G. B. Dynamics of Root Growth in Spring Wheat Mulched with Plastic Film. *Crop Sci.* **2004**, *44*, 1682–1688.

120. Obi, M. E.; Ebo, P. O. The Effects of Different Application Rates of Organic and Inorganic Fertilizers on Soil Physical Properties and Maize Production in a Severely Degraded Ultisol in Southern Nigeria. *Biores. Technol.* **1995**, *51*(2–3), 117–123.

121. Ossom, E. M.; Pace, P. F.; Rhykerd, R. L.; Rhykerd, C. L. Effect of Mulch on Weed Infestation, Soil Temperature, Nutrient Concentration, and Tuber Yield in *Ipomoea batatas* (L.) in Papua New Guinea. *Trop. Agric.* **2001**, *78*, 144–151.

122. Pakamaguru, P.; Pugalendhi, L.; Savitha, B. K. Effect of Drip Fertigation on Growth and Yield of Onion. *Indian J. Hort.* **2010**, *67*(Special Issue), 334–336.

123. Pakyurek, A. Y.; Abak, K.; Sari, N.; Goler, Y. H. Influence of Mulching on Earliness and Yield of Some Vegetables Grown Under High Tunnel. *Acta Hortic.* **1993**, *366*, 155–166.

124. Palada, M. C.; Crossman, S. M. A.; Kowalski, J. A.; Collingwood, C. D. Yield and Irrigation Water Use of Vegetables Grown with Plastic and Straw Mulch in the U.S. Virgin Islands. *Int. Water Irrig.* **2003**, *23*(1), 21–25.

125. Park, S. U.; Park, K. Y.; Kang, Y. K.; Jong, S. K. Effect of Polythene Mulching and Tunnel on the Growth and Yield of Early Produced Sweet Corn. *Res. Rep. Rural Dev. Adm. Crops* **1987**, *29*, 245–250.

126. Patel, N.; Rajput, T. B. S. Effect of Fertigation Frequency on Onion (*Allium cepa*) Yield and Soil Nitrate-nitrogen. *Indian J. Agric. Sci.* **2005**, *75*(11), 725–730.

127. Patil, V. C. *Response of Hybrid Cotton to Split Application of NPK Through Fertigation*; Annual Progress Report: University of Agricultural Sciences, Dharwad, India, 2002, pp 20–28.

128. Patil, P. P.; Patil, V. S. Effect of Mulching with Drip Irrigation on Summer Capsicum Yield and Economics. *Indian J. Soil Conserv.* **2009**, *37*(1), 50–54.

129. Pawar, V. S.; Maher, D. P.; Dahiwalker, S. D.; Bhoite, S. Liquid Fertilizer Effect Through Drip Irrigation on Yield and Water Use Efficiency of Garlic. *J. Water Manag.* **1993**, *1*(1), 10–12.

130. Philip, J. R. General Theorem on Steady Infiltration from Surface Source with Application to Point and Line Source. *Soil Sci. Soc. Am. Proc.* **1971**, *35*, 68–71.

131. Prabhakar, M. In *Micro Irrigation and Fertigation in Fruit Crops*, Training Manual on Use of Plastics in Horticulture with Special Emphasis on Drip Irrigation/Fertigation and Plastic Mulching, ICAR, New Delhi, September 11–20, 2000, pp 43–53.

132. Rahman, M. S.; Khan, M. A. H. Mulching—Induced Alteration of Microclimatic Parameters on the Morpho-physiological Attributes in Onion (*Allium cepa* L.). *Plant Prod. Sci.* **2001**, *4*, 241–248.

133. Raina, J. N.; Thakur, B. C.; Verma, M. L. Effect of Drip Irrigation and Polyethylene Mulch on Yield, Quality and Water-use Efficiency of Tomato. *Indian J. Agric. Sci.* **1999**, *69*, 430–433.

134. Ramakrishna, A.; Tam, H. M.; Wani, S. P.; Long, T. D. Effect of Mulch on Soil Temperature, Moisture, Weed Infestation and Yield of Groundnut in Northern Vietnam. *Field Crops Res.* **2006**, *95*, 115–125.

135. Rao, A. S., Ed. In *Drip Irrigation in India,* Indian National Committee on Irrigation and Drainage, Ministry of Water Resources, Government of India, New Delhi, 1994, pp 178–180.

136. Rees, H. W.; Chow, T. L.; Walker, D. F.; Smith, O. M. Potential Use of Under Seeded Barley to Increase Carbon Inputs to a Loam Soil in New Brunswick Potato Belt, Canada. *Can. J. Soil Sci.* **1999**, *79*, 211–216.

137. Remadevi, A. N. In *Soil Moisture Distribution Patterns Under Drip Irrigation System*, Proceedings of 2nd National Seminar on Drip Irrigation, Coimbatore, India, 1983, pp 86–93.

138. Robock, A.; Vinnikov, K. Y.; Srinivasan, G.; Entin, J. K.; Hollinger, S. E.; Speran-skaya, N. A.; Liu, S.; Namkhai, A. The Global Soil Moisture Data Bank. *Bull. Am. Meteorol. Soc.* **2000,** *81*(6), 1281–1299.

139. Romic, D.; Romic, M.; Borosic, J.; Poljak, M. Mulch Decreases Nitrate Leaching in Bell Pepper (*Capsicum annuum* L.) Cultivation. *Agric. Water Manag.* **2003,** *60*, 87–97.

140. Rowe-Dutton, P. In *The Mulching of Vegetables*, Commonwealth Bureau of Horticulture and Plantation Crops, Bucks, UK, 1957, p 51.

141. Salau, O. A.; Opara-Nadi, O. A.; Swennen, R. Effects of Mulching on Soil Properties, Growth and Yield of Plantain on a Tropical Ultisol in Southeastern Nigeria. *Soil Tillage Res.* **1999,** *23*, 73–93.

142. Sanchita, B.; Phookan, D. B.; Barua, P.; Saikia, L. Effect of Drip Fertigation on Performance of Tomato Under Assam Conditions. *Indian J. Hortic.* **2010,** *67*(1), 56–60.

143. Sandal, S. K.; Acharya, C. L. Effect of Tillage on Moisture Conservation, Soil Physical Conditions, Seedling Emergence and Grain Yield of Rain-fed Maize (*Zea mays* L.) and Wheat (*Triticum aestivum*). *Indian J. Agric. Sci.* **1997,** *67*, 227–231.

144. Sathya, S. G.; Pitchai, J.; Indirani, R.; Kannathasan, M. Effect of Fertigation on Availability of Nutrients (NPK) in Soil—A Review. *Agric. Rev.* **2008,** *29*(3), 214–219.

145. Satish, K.; Patil, S. In *Micro Irrigation in Plantation Crops*, Proc. Advances in Micro Irrigation and Fertigation, Dharward, Karnataka, June 21–30, 1999, pp 109–118.

146. Scheerens, J. C.; Brenneman, G. L. In *Effects of Cultural Systems on the Horticultural Performance and Fruit Quality of Strawberries*, Research Circulation Ohio Agriculture, Research and Development Center, 1994, *298*, pp 81–98.

147. Schonbeck, M. W.; Evanylo, G. K. Effects of Mulches on Soil Properties and Tomato Production, I: Soil Temperature, Soil Moisture and Marketable Yield. *Sustain. Agric.* **1998,** *13*(1), 55–81.

148. Selvaraj, P. K. Optimization of Irrigation Scheduling and Nitrogen Fertigation for Maximizing Water Use Efficiency of Turmeric in Drip Irrigation. Unpublished Ph.D. Thesis, Submitted to Tamil Nadu Agricultural University, Coimbatore, 1997, p 151.

149. Selvaraj, P. K.; Krishnamoorthy, V. V.; Manickasundram, P. G.; Martin, J.; Ayyasamy, M. Effect of Irrigation Schedules and Nitrogen Levels on the Yield of Turmeric Through Drip Irrigation. *Madras Agric. J.* **1997,** *84*(6), 344–348.

150. Seyfi, K.; Rashidi, M. Effect of Drip Irrigation and Plastic Mulch on Crop Yield and Yield Components of Cantaloupe. *Int. J. Agric. Biol.* **2007,** *9*(2), 101–103.

151. Shankar, V.; Lawande, K. E. Influence of Micro-irrigation Methods on Growth, Yield and Storage of Rabi Onion. *Indian J. Hortic.* **2010,** *67*(1), 56–59.

152. Sharma, S. K. Effect of Phosphorus and Potassium on Capsicum Seed Production. *Indian J. Hortic.* **1995,** *52*(2), 141–145.

153. Sharmasarkar, F. C.; Sharmasarkar, S.; Miller, S. D.; Vance, G. F.; Zhang, R. Assessment of Drip and Flood Irrigation on Water and Fertilizers Use Efficiencies for Sugarbeet. *Agric. Water Manag.* **2001,** *46*, 241–251.

154. Shashank, S.; Singh, S. K.; Singh, S. B. Effect of Spacing and Various Levels of Nitrogen on Seed Crops of Kharif Onion. *Veg. Sci.* **1994,** *21*, 1–6.

155. Shelke, D. K.; Vaishnava, V. G.; Jadhav, G. S.; Oza, S. R. Optimization of Irrigation Water and Nitrogen to Cotton (*Gossypium sp.*) Through Drip Irrigation System. *Indian J. Agron.* **1999,** *44*(3), 624–633.

156. Shin, K. H.; Park, J. C.; Lee, K. S.; Han, K. Y.; Lee, Y. S. Effects of Planting Dates and Bulb Size on the Growth and Yield of Garlic. *Res. Rep. Rural Dev. Admin. Hort.* **1998,** *30,* 41–52

157. Shinde, P. P.; Ramteke, J. R.; More, V. G.; Chavan, S. A. Evaluation of Micro Irrigation Systems and Mulch for Summer Chili Production. *J. Maharashtra Agric. Univ.* **2002,** *27*(1), 51–54.

158. Shinde, U. R.; Firake, N. N.; Dhotrey, R. S.; Bankar, M. C. Effect of Micro-irrigation Systems and Mulches on Micro-climatic Factors and Development of Crop Coefficient Models for Summer Chili. *J. Maharashtra Agric. Univ.* **1999,** *24*(1), 72–75.

159. Shirahatti, M. S.; Itnal, C. J.; Gouda, D. S. In *Comparison of Drip and Furrow Irrigated Cotton on a Red Soil,* Proceedings of International Conference on Micro and Sprinkler System, Jalgaon, Maharashtra, India, 2001, pp 651–656.

160. Shock, C. C.; Zalewaski, J. C.; Stieber, T. D.; Burnett., D. S. Impact of Early Season Water Deficits on Russet Burbank Plant Development, Tuber Yield and Quality. *Am. Potato J.* **1992,** *69,* 327–340.

161. Shrestha, R. B.; Gopalakrishnan, C. In *Adoption and Diffusion of Drip Irrigation Technology,* Presented at 2nd International Strawberry Symposium, Beltsville, 1993, p 10.

162. Shrivastava, A. K. Effect of Fertilizer Levels and Spacings of Flowering, Fruit Set and Yield of Sweet Pepper (*Capsicum annuum*). *Adv. Plant Sci.* **1996,** *9*(2), 171–175.

163. Singandhupe, R. B.; Rao, G. G. S. N.; Patil, M. T.; Brahmanad, P. S. Fertigation Studies and Irrigation Scheduling in Drip Irrigation System in Tomato Crop (*Lycopersicon esculentum*). *Eur. J. Agron.* **2002,** *19,* 327–340.

164. Sivanappan, R. K.; Ranghaswami, M. V. Technology to Take 100 Tons Per Acre in Sugarcane. *Kisan World* **2005,** *32*(10), 35–38.

165. Sivanappan, R. K. Status of Micro Irrigation System in Crop Production. In *Micro Irrigation and Fertigation*; Ravi Publications: Bangalore, 1996; pp 16–25.

166. Solomon, K. Subsurface Drip Irrigation: Product Selection and Performance. In *Subsurface Drip Irrigation: Theory, Practices and Application*; Jorgensen, G. S., Norum, K. N., Eds.; CATI Publication no. 921001: California State University, Fresno, California, 1993; p 20.

167. Sontakke, N. A.; Singh, N. Optimization of the Rain Gauges for a Representative All-India and Subdivisional Rainfall Series. *Theor. Appl. Climatol.* **1993,** *47,* 159–173.

168. Srivastava, P. K.; Parikh, M. M.; Sawani, N. G.; Raman, S. Effect of Drip Irrigation and Mulching on Tomato Yield. *Agric. Water Manag.* **1994,** *25,* 179–184.

169. Suryawanshi, S. K. *Success of Drip Irrigation in India: An Example to the Third World,* Proceedings of 5th International Micro Irrigation Congress, Orlando, Florida, April 2–6, 1995, pp 347–352.

170. Suwon. M. A; Judah, O. M. Influence of Plastic Mulching on Growth and Yield and Soil Moisture Conservation in Plastic House Tomatoes. *Dirasat* **1985,** *12,* 21–23.

171. Tarara, J. M. Microclimate Modification with Plastic Mulch. *HortScience* **2000,** *35,* 169–180.

172. Tindall, J. A.; Beverly, R. B.; Radcliffe, D. E. Mulch Effect on Soil Properties and Tomato Growth Using Micro-irrigation. *Agron. J.* **1990,** *83,* 1028–1034.

173. Tingwu, L.; Jaun, X.; Jianping, W.; Zhizhong, L.; Jianguo, Z. Effect of Drip Irrigation with Saline Water on Water Use Efficiency and Quality of Watermelon. *Water Res. Manag.* **2003,** *17*(6), 395–408.

174. Tiwari, K. N.; Mal, P. K.; Singh, R. M.; Chattopadhyay, A. Response of Okra (*Abelmoschus esculentus*) to Drip Irrigation Under Mulch and Non-mulch Condition. *Agric. Water Manag.* **1998,** *38,* 91–102.

175. Tuli, A.; Yesilsoy, M. S. Effect of Soil Temperature on Growth and Yield of Squash Under Different Mulch Applications in Plastic Tunnel and Open-air. *Turk. J. Agric. For.* **1997,** *21*(2), 101–108.

176. Tumbare, A. D.; Bhoite, S. U. Effect of Solid Soluble Fertilizers Through Fertigation on the Growth and Yield of Chili (*Capsicum annuum* L.). *Indian J. Agric. Sci.* **2002,** *72*(2), 109–111.

177. Tyson, A. W.; Harrison, K. A. Commercial Production and Management of Pumpkin and Gourds. *Bulletin 1180,* Univ. Georgia, Atlanta, GA, 2009, p 15. https://secure. caes.uga.edu/extension/publications/files/pdf/B%201180_4.PDF

178. Uday, S. Response of Tomato to Irrigation and Fertilizer Management Practices. *J. Hill Agric.* **2011,** *2*(1), 122–125.

179. Ulah, R. A. R.; Khan, A.; Ghulam, S. Effect of Different Mulching Materials on the Growth and Production of Wheat Crop. *Sarhad J. Agric.* 1998, *14*(1), 21–25.

180. Unger, P. W. Role of Mulches in Dryland Agriculture. In *Crop Physiology*; Gupta, U. S., Ed.; Oxford and IBH: New Delhi, 1975; pp 237–260.

181. Vethamoni, P. I.; Balakrishnan, R. Studies on the Influence of Herbicide, Nitrogen and Mulching on the Nutrient Uptake of Okra (*Abelmoschus esculentus*). *Ind. J. Hort.* **1990,** *47*(2), 233–238.

182. Wan-Shu, Q.; Kang, Y.; Wang, D.; Liu, S.; Feng, L. Effects of Saline Water on Cucumber Yields and Irrigation Water Use Efficiency Under Drip Irrigation. *J. Agric. Water Manag.* **2007,** *23*(3), 30–35.

183. Warrick, A. W.; Shani, U. Soil-limiting Flow from Subsurface Emitters. *J. Irrig. Drain. Eng.* **1996,** *122*(5), 296–300.

184. Webber, H. A.; Madramootoo, C. A.; Bourgault, M.; Horst, M. G.; Stulina, G.; Smith, D. L. Water Use Efficiency of Common Bean and Green Gram Grown Using Alternate Furrow and Deficit Irrigation. *J. Agric. Water Manag.* **2006,** *86*(3), 259–268.

185. Wein, H. C.; Minotti, P. L.; Grubinger, V. P. Polythene Mulch Stimulates Early Root Growth and Nutrient Uptake of Transplanted Tomatoes. *J. Am. Soc. Hortic. Sci.* **1993,** *118*(2), 207–211.

186. Zhang, X.; Chen, S.; Liu, M.; Pei, D.; Sun, H. Improved Water Use Efficiency Associated with Cultivates and Agronomic Management in the North China Plain. *Agron. J.* **2005,** *24,* 783–789.

187. Znidarcic, D.; Trdan, S.; Oswald, J. Effect of Row Covers and Black Plastic Mulch on the Yield of Determinate Tomatoes. M.Sc. Thesis, Agronomy Department, Biotechnical Faculty, University of Ljubljana, Jamnikarjeva 101, 1000 Ljubljana, 2004, p 110.

APPENDIX I

Sample example to calculate water requirement of the tomato crop.

The daily water requirement of tomato crop for drip irrigation was calculated by using the following equation:

$$WRc = CPE \times Kp \times Kc \times Wp$$

where WRc = water requirement (mm); CPE = cumulative pan evaporation for 3 days (mm); Kp = pan factor (0.8); Kc = crop factor; Wp = wetted area (80%).

Sample calculation

For the first day, CPE = 9 mm; Kc = 0.45; and WRc = 9 × 0.8 × 0.45 × 0.8 = 2.59 mm/day.

APPENDIX II

Cost economics of tomato crop under mulching and fertigation.

a) Initial cost of drip irrigation system, Rs./ha for a life span of 5 years

S. no.	Particulars		Cost Rs.
1	Submersible pump		14,950
2	Screen filter		5200
3	Main line PVC 63 mm		3200
4	Submain PVC (50 mm)		9583
5	Laterals 16 mm		64,350
7	Control valves		2500
8	Emitters		35,970
9	Gromate take off (16 mm)		4337
10	End caps		2300
11	PVC fitting and accessories		4663
12	Installation charges		1500
		Total cost for 6 months	**148,550**
		For 1 year, the system cost	**297,100**

In this chapter: US $ 1.00 = Rs. 60.00

b) Initial cost of mulching, Rs./ha

50 μm thickness sheet = Rs. 25,200
25 μm thickness sheet = Rs. 44,000
Total fixed cost for 1 year = Rs. 54,910
Repair and maintenance cost at 2% = Rs. 1100
Interest on initial cost at 12% = Rs. 5490

c) Total operating cost, Rs./ha

Operations	Treatments								
	T1	T2	T3	T4	T5	T6	T7	T8	T9
Ploughing	400	400	400	400	400	400	400	400	400
Bed preparation	500	500	500	500	500	500	500	500	500
Seed	100	100	100	100	100	100	100	100	100
Nursery preparation and transplantation	200	200	200	200	200	200	200	200	200
Pesticide application	1000	1000	1000	1000	1000	1000	3000	3000	3000
Fertilizer cost (Rs.)									
1. Urea	3200	4200	5100	3200	4200	5100	3200	4200	5100
2. SSP	14,280	17,850	21,420	14,280	17,850	21,420	14,280	17,850	21,420
3. MOP	1920	2400	2880	1920	2400	2880	1920	2400	2880
Irrigation	4000	4000	4000	4000	4000	4000	4000	4000	4000
Weeding operation	0	0	0	0	0	0	7000	7000	7000
Harvesting	5000	5000	5000	5000	5000	5000	5000	5000	5000
Total operating cost	31,880	36,800	41,720	31,880	36,800	41,720	80,880	75,800	70,720

Total cost of cultivation = Rs. 93,380

d) Gross income (kg/ha)

Total yield = 73,410 kg/ha
Selling cost of 1 kg tomato = Rs. 8/kg
Total revenue = Total yield × Cost = 73,410 × 8 = Rs. 440,470
Net income = Total revenue – Total cost of cultivation = Rs. 347,100

e) Benefit–cost ratio

Benefit–Cost ratio = Gross income/Cost of cultivation = [347,100]/[73,410] = 3.71

CHAPTER 13

PERFORMANCE OF DRIP-IRRIGATED EGGPLANT UNDER BEST MANAGEMENT PRACTICES

M. B. VINUTA[1], S. V. KOTTISWARAN[1],
BASAMMA K. ALADAKATTI[2,*], and TRUPTIMAYEE SUNA[1]

[1]*Department of Soil and Water Conservation Engineering, Agricultural Engineering College and Research Institute (AEC & RI), Tamil Nadu Agricultural University (TNAU), Coimbatore 641003, Tamil Nadu, India*

[2]*Senior Research Fellow (Hydrology), Sujala-III Project, Department of Soil Science and Agril. Chemistry, College of Agriculture, University of Agricultural Sciences, Dharwad- 580005, India;*

[*]Corresponding author. E-mail: basammaka@gmail.com

CONTENTS

ABSTRACT

Field studies were conducted at Precision Farming Development Center farm of Tamil Nadu Agricultural University, Coimbatore to evaluate the crop water and fertigation requirements for grafted brinjal under drip irrigation system. The experiments were laid in strip-plot design with 27 treatments. The daily water requirement of grafted brinjal at different stages was found to be 1.9 L/day (initial stage), 3.6 L/day (vegetative stage), 4.2 L/day (flowering stage), and 5.2 L/day (harvesting stage). The total water requirement for the growth period under drip irrigation was 396.00 mm with mulch at 80% ET_0 level and 495.00 mm without mulch at 100% ET_0 level. It can be concluded that drip irrigation can save 16.17% of water with mulch as compared without mulch condition. The highest yield of 83.3 t ha^{-1}, highest water use efficiency of 62.87 kg m^{-3}, and fertilizer use efficiency of 416.5 kg ha^{-1} kg of N, 555.3 kg ha^{-1} kg of P, and 833.00 kg ha^{-1} kg of K) were recorded in treatment under 25 μ thickness plastic mulching at 80% ET_0 level and 100% RDF followed by T_6 (81.1 t ha^{-1}) and T_{14} (79.2 t ha^{-1}) and lowest yield of 18.1 t ha^{-1}, least water use efficiency (19.38 kg m^{-3}), and fertilizer use efficiency (113.13 kg ha^{-1} kg of N, 150.83 kg ha^{-1} kg of P, and 226.25 kg ha^{-1} kg of K) were found in control plot treatment with 60% ET_0 and 80% RDF. Increased fertilizer use efficiency with the decreased level of fertilizer dose through drip was observed.

13.1   INTRODUCTION

Water is considered as liquid gold and land is one of the platforms for survival of many living things for performing several activities. Conservation and management of these resources is top most priority to overcome the problems of water scarcity, as they both go hand in hand. Agriculture is the source for the Indian belly and also to fulfill the needs of human beings; its value is dependent on the health of land/soil and timely availability of water resources, which are declining day by day at rapid rate and demand is growing every moment. Therefore, by considering the growth of demand for the resources, the future security and safety, it is necessary to go for the adaptation of technologies, which put hands in conserving and managing scarce resources in agriculture by giving more importance to production quality as well as quantity. To achieve this with the available scarce quantity of water, it is necessary to increase the water use efficiency (WUE), which can be achieved through the adaptation of micro irrigation system.

Micro irrigation technology is rapidly expanding throughout the world, especially in the water scarce areas of developing countries. It is very popular in the United States, Israel, and some parts of Europe. North America and Europe have the highest area under micro irrigation utilization while Asia is in the development stage. Due to the decreasing availability of water resources and increasing competition for water between different users, improving agricultural WUE is vitally important in many parts of the world that have limited water resources.

It has been estimated that the irrigated area in the world is 253 million hectare. The gross irrigated area of India in 2005–06 had increased to 82.6 million hectare from 22.6 million hectare in 1951–52 and increase being more than 250% during the last five decades. Efficient use of water through scientific irrigation management is of utmost importance in providing the best insurance against weather-induced fluctuations in food production. The application of irrigation water by traditional method causes 27–42% loss of water through deep percolation depending on the soil type.[1] Due to depletion of water sources and high labor costs, micro irrigation has a significant adaptability all over the world. Drip irrigation is an effective tool for conserving water resources and studies have revealed significant water saving ranging from 40% to 70% under drip irrigation compared with surface irrigation. Drip irrigation helps to increase WUE by reducing soil evaporation and drainage losses, maintaining soil moisture conditions that are favorable to crop growth and helps to sustain the productivity of the land. Productivity can be increased by adopting an improved package of practices, particularly in situ moisture conservation by mulching.

The plastic mulch was first adopted in the United States. Even with the rapid growth in production and use of plastics in India, the per capita consumption of plastics is only 2.2 kg which is very low compared to 60 kg in developed countries such as the United States, Germany, and Japan. World average of per capita consumption of plastic is 16.2 kg.[12] Sweet corn, tomatoes, cucumber, strawberry, lettuce, watermelon, okra, and grapes are primary crops that have been evaluated under plastic mulch.

The notable advantage of the use of plastic mulch is its impermeability, which prevents direct evaporation of moisture from the soil and thus cuts down the water losses.[3] Plastic such as high-density polyethylene (HDPE), low-density polyethylene (LDPE), and linear low-density polyethylene (LLDPE) has been used as plastic mulch. Among these types of plastics, LDPE mulch is most commonly used. Recently, LLDPE has been scoring over LDPE as a mulch material due to its two associated characteristics of better down gauging and puncture resistance, and checks weeds growth.

The use of plastic mulch to achieve early and higher yield of vegetable crops is increasing. This is especially true for warm-season species such as peppers, corn, tomatoes, and vine crops. Water that evaporates from the soil under the plastic film condenses on the lower surface of the film and falls back to the soil droplets. Therefore, soil moisture is preserved and is available for the crop. The mulch film prevents the direct impact of rain on the soil and maintains a porous soil structure. Thus, better moisture movement and gaseous exchange take place in mulched soils. This process increases the concentration of carbon dioxide around the mulch film and improves photosynthesis.

Fertigation offers the best solution for intensive and economical crop production where both water and fertilizers are delivered to crop through drip irrigation. It provides essential elements directly to active root zone, thus minimizing fertilizer losses, and ensuring high-quality yield along with saving time, labor, and energy resulting in reduced cost of cultivation. Experiments have indicated that fertigation can save 40–50% of nutrients. Soluble fertilizers can be injected into drip irrigation system by selecting appropriate application methods, such as available pumps, bypass through tank, valves, venturi, and aspirators. Among these, venturi is the cheapest, simpler, and economic one, though it creates high pressure loss.

Solanum melongena L. is a staple vegetable and is also known as eggplant. India is the second largest producer of eggplant after China with the production of 11.89 million tons production from an area of 0.68 m ha. In Tamil Nadu, it is grown over an area of 12,400 ha with a production of 0.2 million tons in 2010–11 (www.nhb.gov.in). Furthermore, continuous use of the same field for the cultivation of eggplant or related hosts susceptible to a number of pathogens leads to an increase in the soil inoculum. Therefore, to overcome the problems such as difficulties in chemical control measures and absence of crop rotation, one of the short-term practical solutions is to graft susceptible eggplant cultivars onto rootstocks possessing biotic and abiotic stress resistances.[6,14]

The technology of vegetable production with grafts was originated in Japan and Korea to avoid serious loss caused by soil borne diseases aggravated by successive cropping. Grafting is also effective in ameliorating crop losses caused by adverse environmental conditions.[8] Cultivation of vegetable grafts is widely recognized and has advantages of disease tolerance and high crop yields. This practice is now rapidly spreading and expanding over the world. The number and size of commercial vegetable seedling and grafted plant producers have increased among the farmers.[17]

The use of grafting as an integrated pest management tool to manage biotic stress is most successful when complemented with sustainable farming system practices.[9,15] Grafting of eggplant cultivars on perennial and wild species increased the yield and availability period of the fruits, while standardization of cultural practices, irrigation, and nutritional requirements under different soils and climatic conditions helps in better crop stand.[18]

By keeping these facts in mind, authors made an attempt to study the effects of different levels irrigation and fertigation under mulch with the following objectives:

- To evaluate the crop water and fertigation requirements for grafted eggplant under drip irrigation system.
- To find out the optimum water and fertilizer requirements under drip irrigation for grafted eggplant.
- To study the effect of plastic mulch on yield and growth parameters of grafted eggplant.

13.2 MATERIALS AND METHODS

13.2.1 LOCATION

The experiment was conducted at Precision Farming Development Center (PFDC) research farm in the Eastern Block of Tamil Nadu Agricultural University, Coimbatore at 11.0183°N latitude and 76.9725°E longitude with mean altitude of 426 m above the mean sea level and topography of experimental plot was uniform.

The mean annual rainfall is 720 mm. About 55% of annual rainfall is received during northeast monsoon season and 30% during southwest monsoon. The annual maximum and minimum mean temperatures were 32.50°C and 20.10°C, respectively, and the average relative humidity of the area is 56.8%, and the mean daily evaporation ranged from 3.14 to 7.05 mm/day. Table 13.1 indicates monthly average climatic data for maximum and minimum temperatures, maximum and minimum relative humidity, pan evaporation (E_{pan}), and rainfall of last 22 years from 1991 to 2013; these were collected from Agro Climate Research Centre, Tamil Nadu Agricultural University, Coimbatore.

TABLE 13.1   Monthly Average Values for Climatic Data at Coimbatore (1991–2013).

Month	Evaporation (mm)	Relative humidity (%)		Rainfall (mm)	Temperature (°C)	
		Max	Min		Max	Min
January	4.3	86.82	42.68	6.02	30.25	18.67
February	5.2	83.72	37.25	18.02	32.21	19.34
March	6.1	81.64	34.65	19.49	34.41	21.63
April	6.2	82.41	40.69	43.33	35.25	23.82
May	6.1	81.79	46.70	36.62	34.27	23.79
June	6.3	78.15	50.96	49.84	32.02	23.37
July	5.9	78.39	52.75	43.10	30.89	23.03
August	5.7	81.89	52.69	49.62	31.32	22.54
September	5.5	84.61	50.05	55.97	31.29	22.32
October	4.3	87.82	56.24	147.87	30.55	22.02
November	3.4	89.54	58.27	127.32	28.89	20.81
December	3.6	88.25	51.01	46.01	28.81	19.06

This table was prepared from the raw data from Agro Climate Research Centre, Tamil Nadu Agricultural University, Coimbatore. All data generated by this center is for public domain.

13.2.2   CROP

The grafted eggplant which was developed by Department of Vegetable Crops at TNAU, Coimbatore by using two *Solanum* species *S. torvum*, known as Turkey berry, a wild species used as rootstock and the scion was COBH2 and Ravaiya which exhibit the tolerance to shoot and fruit borer incidence and cultivated in all types of soils under water stress conditions in semiarid regions; this gives better yield performances and has a vibrant market potential in domestic market. The harvested crop can be used as ratoon crop for the next season.

13.2.3   SOIL

The texture of soil is sandy clay loam soil with clay 30.8%, silt 28.7%, fine sand 19.5%, and coarse sand 20.5%; with pH of 8.07, EC of 0.78 dS m^{-1}, available N of 185.6 kg ha^{-1}, available P of 9.0 kg ha^{-1}, and available K of 356.7 kg ha^{-1}; with water holding capacity of 39.41%, pore space of 42.73%, hydraulic conductivity 0.38 cm h^{-1} and infiltration rate of 0.73 cm h^{-1}.

13.2.4   LAND PREPARATION

Farm yard and poultry manure at 25 t ha^{-1} was applied to the experimental plot and it was thoroughly plowed with disc plow and repeatedly tilled with a cultivator to bring optimum soil tilth.

13.2.5   DRIP IRRIGATION SYSTEM

Bore well water was used for irrigating the crop, which was moderately saline with pH of 7.56 and EC of 1.93 dS m^{-1}. The layout was taken up forming 81 strips of 6 m × 1.2 m size and drip system was installed. The drip system was laid out with 75 mm diameter polyvinyl chloride (PVC) main pipe line and 63 mm diameter PVC submain with fertigation tank and venturi. LLDPE laterals of 16 mm diameter were connected to submain. Each lateral was provided with individual taps for controlling irrigation and fertigation. Along the laterals, online drippers of 4 L/h were fixed at the spacing of 1.2 m. Submains and laterals were plugged at the end with end caps. After installation, the trial run was conducted to assess mean dripper discharge and uniformity coefficient. Morning time was preferred for irrigation since evaporation was less at that time.

13.2.6   MULCHING

Black polythene mulch (BPM) of 25 μm thickness LLDPE and 50 μm thickness LLDPE were used for the study. Over the drip line, according to the treatment, mulching sheets were spread in each plot and both ends of the plastic sheet were buried into the soil up to a depth of 10 cm and holes were punched.

13.2.7 WATER REQUIREMENT

The daily water requirement was calculated as follows:

$$\text{Daily Water Requirement (DWR)} = Ep \times Kp \times Kc \times Wp \times A, \quad (13.1)$$

where DWR = computed daily water requirement (lit plant^{-1}); Ep = average pan evaporation of the day (mm); Kp = pan factor (0.8); Kc = crop factor (the crop factor values were 0.6 at initial stage up to 1–10 days, 1.05 for vegetative stage and flowering stage for 11–70 days, and 0.9 at harvesting stage from 71–180 days); Wp = wetted percentage (80%); and A = area per plant.

The water was supplied to the plant daily in each treatment. Time of operation of drip system to deliver the required volume of water per treatment was calculated using the equation:

$$\text{Time of operation (min)} = [\text{Volume of water required}]/$$
$$[\text{Emitter discharge} \times \text{No. of emitters}] \quad (13.2)$$

13.2.8 TRANSPLANTING

The healthy seedlings of 25 days old were transplanted followed by irrigation at 1.2 m ×1.2 m geometry in the main field. Total five plants were used in each treatment having an effective area of 1.44 m^2.

13.2.9 GAP FILLING

Gap filling was done 7 days after transplanting (DAT) to ensure optimum plant population.

13.2.10 PLANT PROTECTION MEASURES

The different plant protection measures were taken against pest and diseases during the period of investigation is as follows: The crop was regularly sprayed with organic insecticides such as neem oil (25 ml L^{-1}), fluradon granules 1 g/plant, and methalaxyl 8% + mancozeb 64% Wp to control thrips, mites, and root/shoot borers fruit borers.

13.2.11   FERTILIZER REQUIREMENT OF GRAFTED EGGPLANT

The quantity of fertilizer required for the study area of 584 m² was calculated based on quantity recommended for per hectare. The recommended dose of fertilizers (RDF) was 200:150:100 kg ha⁻¹ and basal dose of 75% (112.5 kg phosphorus) of phosphorus was applied as single super phosphate at 703 kg ha⁻¹ with 25% phosphorus as water soluble fertilizer (urea of 340 kg ha⁻¹), 19:19:19 of 79.00 kg ha⁻¹, murate of potash at 189 kg ha⁻¹, and potassium nitrate at 37 kg ha⁻¹ were applied simultaneously in a combined form to the plant root zone. Water-soluble fertilizers were used in this experiment. The fertilizers were dissolved in water in the ratio of 1:5 and the solution was diluted in fertigation tank. The fertilizer was applied at weekly intervals during the vegetative stage, flowering stage, and fruiting stage.

13.2.12   EXPERIMENT DETAILS

The field area was 584 m² with length of 32.40 m and width of 18.00 m. The total area was divided into various strips of 6 m × 1.2 m according to the treatment. The field layout plan for the experiment is shown in Figure 13.1.

13.2.13   TREATMENTS DETAILS

Crop	Grafted eggplant
Spacing	1.2 m ×1.2 m
Treatments	27
Replication	3
Design	*Strip Plot Design (SPD)* with three factors namely: (1) plastic mulch, (2) irrigation, and (3) fertigation with three levels each.

T_1 Black polythene mulch of 25 μm thickness, irrigation level 60% ET_0, and fertigation with 80% RDF.

T_2 Black polythene mulch of 25 μm thickness, irrigation level 60% ET_0, and fertigation with 100% RDF.

T_3 Black polythene mulch of 25 μm thickness, irrigation level 60% ET_0, and fertigation with 120% RDF.

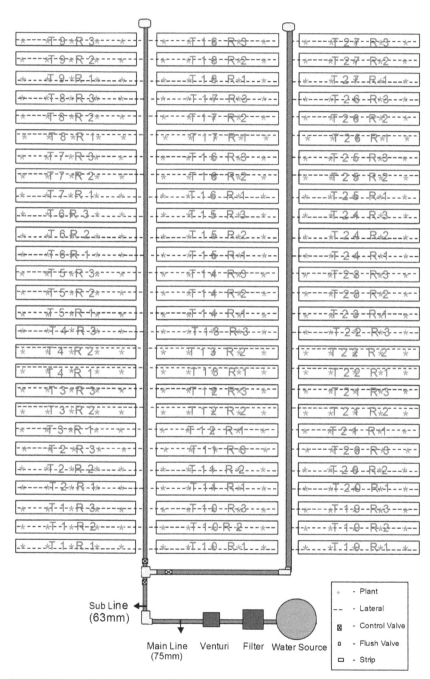

FIGURE 13.1 Field layout plan for the experiment.

T_4 Black polythene mulch of 25 µm thickness, irrigation level 80% ET_0, and fertigation with 80% RDF.

T_5 Black polythene mulch of 25 µm thickness, irrigation level 80% ET_0, and fertigation with 100% RDF.

T_6 Black polythene mulch of 25 µm thickness, irrigation level 80% ET_0, and fertigation with 120% RDF.

T_7 Black polythene mulch of 25 µm thickness, irrigation level 100% ET_0, and fertigation with 80% RDF.

T_8 Black polythene mulch of 25 µm thickness, irrigation level 100% ET_0, and fertigation with 100% RDF.

T_9 Black polythene mulch of 25 µm thickness, irrigation level 100% ET_0, and fertigation with 120% RDF.

T_{10} Black polythene mulch of 50 µm thickness, irrigation level 60% ET_0, and fertigation with 80% RDF.

T_{11} Black polythene mulch of 50 µm thickness, irrigation level 60% ET_0, and fertigation with 100% RDF.

T_{12} Black polythene mulch of 50 µm thickness, irrigation level 60% ET_0, and fertigation with 120% RDF.

T_{13} Black polythene mulch of 50 µm thickness, irrigation level 80% ET_0, and fertigation with 80% RDF.

T_{14} Black polythene mulch of 50 µm thickness, irrigation level 80% ET_0, and fertigation with 100% RDF.

T_{15} Black polythene mulch of 50 µm thickness, irrigation level 80% ET_0, and fertigation with 120% RDF.

T_{16} Black polythene mulch of 50 µm thickness, irrigation level 100% ET_0, and fertigation with 80% RDF.

T_{17} Black polythene mulch of 50 µm thickness, irrigation level 100% ET_0, and fertigation with 100% RDF.

T_{18} Black polythene mulch of 50 µm thickness, irrigation level 100% ET_0, and fertigation with 120% RDF

T_{19} No mulch at irrigation level 60% ET_0 with 80% RDF drip fertigation.

T_{20} No mulch at irrigation level 60% ET_0 with 100% RDF drip fertigation.

T_{21} No mulch at irrigation level 60% ET_0 with 120% RDF drip fertigation.

T_{22} No mulch at irrigation level 80% ET_0 with 80% RDF drip fertigation.

T_{23} No mulch at irrigation level 80% ET_0 with 100% RDF drip fertigation

T_{24} No mulch at irrigation level 80% ET_0 with 120% RDF drip fertigation.

T_{25} No mulch at irrigation level 100% ET_0 with 80% RDF drip fertigation.

T_{26} No mulch at irrigation level 100% ET_0 with 100% RDF drip fertigation.

T_{27} No mulch at irrigation level 100% ET_0 with 120% RDF drip fertigation.

13.2.14   OBSERVATIONS

The parameters to be observed were: soil moisture distribution patterns, the effect of plastic mulch on temperature variations of soil, and biometric parameters. Effects of mulch on temperature variations were measured at the soil surface and at 10 cm depth using digital soil thermometer at different times at 8 a.m., 1 p.m., and 4 p.m. in the field and were compared with ambient temperature.

13.2.15   BIOMETRIC PARAMETERS

Three plants in each treatment were selected at random and utilized for recording biometric observations; mean values were subjected to statistical analysis. The height of the plant (cm) from cotyledonary node to the tip of the plant was measured at 15, 30, 60, and 90 days intervals. The number of leaves in each plant was measured at 15, 30, 60, and 90 days and expressed in number of branches per plant. The number of fruits in each plant in all harvests were counted and expressed in number of fruits per plant. The weight of fruits from each plant in all harvests was recorded and expressed in kilograms. The yield per hectare was estimated for the cropped area based on yield per plot and expressed in tons per hectare.

13.2.16   STATISTICAL ANALYSIS

The data were analyzed using AGRESS package for SPD for eggplant. Wherever the treatment differences were found significant (*F*-test), critical differences were determined at 5% probability level. The factor and level details are as follows:

Factor I—Mulching with three levels: M_1 with 25 μm thickness LLDPE; M_2 with 50 μm thickness LLDPE; and M_3 with no mulch.
Factor II—Irrigation with three levels: I_1 at 60% ET_0; I_2 at 80% ET_0; and I_3 at 100% ET_0.
Factor III—Fertigation with three levels: F_1 at 80% RDF; F_2 at 100% RDF; and F_3 at 120% RDF.

13.3   RESULTS AND DISCUSSION

13.3.1   WATER REQUIREMENT OF GRAFTED EGGPLANT

The experiments were conducted under plastic mulch and without mulch with drip irrigation system. The total amount of water applied per plant under different treatments at various crop growth stages of grafted eggplant is presented in Table 13.2.

TABLE 13.2   Amount of Water Application at Various Growth Stages of Grafted Eggplant.

Stage	Amount of water applied at different irrigation levels (L/plant)					
	25 μm LLDPE mulch	**Irrigation levels**				
	50 μm LLDPE mulch	**60% ET**		**80% ET**		**100% ET**
	No mulch	T_1, T_2, T_3		T_4, T_5, T_6		T_7, T_8, T_9
		T_{10}, T_{11}, T_{12}		T_{13}, T_{14}, T_{15}		T_{16}, T_{17}, T_{18}
		T_{19}, T_{20}, T_{21}		T_{22}, T_{23}, T_{24}		T_{25}, T_{26}, T_{27}
		Per day	Per stage	Per day	Per stage	Per day	Per stage
Initial stage (0–10 days)		1.20	12.00	1.6	16.00	1.9	19.00
Vegetative stage (11–40 days)		2.10	63.00	2.9	84.00	3.6	105.00
Flowering stage (41–70 days)		2.50	76.00	3.4	101.00	4.2	126.00
Harvesting stage (71–180 days)		3.10	278.00	4.1	371.00	5.2	462.00
Total		–	429.00		572.00		712.00

The water was applied based on monthly average evaporation for last 22 years data from 1991 to 2013; therefore one has to be careful in applying these results because the contributing factors to the water requirement were location specific. If similar conditions exist, one can use these results with a suitable allowance so that crop growth and yield are not adversely affected.

On the basis of soil moisture and plant growth parameters, the water requirement was evaluated. The rainfall during the crop season was 233.00 mm for 20 rainy days. On considering the effective rainfall which was for 11 days, the maximum was 44.5 mm. Of the total 180 number of days of irrigation, the water supply to the crop was not given for 13 days due to antecedent moisture content. The daily water requirement of grafted eggplant at different stages was 1.9 L/day (initial stage), 3.6 L/day (vegetative stage), 4.2 L/day (flowering stage), and 5.2 L/day (harvesting stage). It was observed that the water requirement was maximum at harvesting stage.

The results showed that the total water requirement under drip irrigation was 396.00 mm for the treatments under mulch at 80% ET_0 level and 495.00 mm for the treatments without mulch at 100% ET_0 level. On comparing with the conventional method of irrigation (which used 600 mm of water for its growth), drip irrigation caused a saving of 33.83% and 17.66% with and without mulch, respectively. It can be concluded that drip irrigation can save 16.17% of water with mulch as compared without mulch condition. Similar results of water saving for eggplant under drip irrigation with different ET levels have been reported by Bhogi et al.[5] Drip irrigation system used less water due to the fact that maximum amount of water will be stored in the root zone and deep percolation losses will be minimum at lower irrigation levels. These results were in agreement with the findings of Tagar et al.[19]

13.3.2 FERTILIZER REQUIREMENT OF GRAFTED EGGPLANT

The quantity of fertilizer required for the study area of 584 m² was calculated based on quantity recommended for per hectare. The RDF was 200:150:100 kg ha⁻¹. Water soluble fertilizers were used in this experiment. The total amount of fertilizer applied under different treatments at various crop growth stages of grafted eggplant was calculated by using recommended dose of fertilizer ratio and is presented in Table 13.3.

The results indicated that the treatment T_5 with 100% RDF under 25 µm thickness at 80% ET_0 level was the best when compared with all other different treatments. This is in agreement with findings by Dalvi et al.,[7] who reported that water and fertilizer management by drip fertigation at 96% of recommended level dose resulted in a maximum yield of tomato.

TABLE 13.3   Details of Quantity of Fertilizer Application for the Study Area (kg).

Stage	Name	80% RDF	100% RDF	120% RDF
Basal dose	Single super phosphate	32.80	41.00	49.20
Transplanting to plant establishment (1–10 days)	NPK 19:19:19	0.616	0.770	0.924
	Murate of potash 13:0:45	0.087	0.108	0.13
	Urea	0.399	0.498	0.598
Vegetative stage (11–40 days)	KNO$_3$ 12:61:0	0.383	0.478	0.574
	Murate of potash 13:0:45	1.384	1.73	2.076
	Urea	2.215	2.768	3.322
Flower initiation to first picking (41–70 days)	NPK 19:19:19	0.616	0.770	0.924
	Murate of potash 13:0:45	0.784	0.98	1.176
	Urea	1.555	1.944	2.333
Harvesting stage (71–180 days)	KNO$_3$ 12:61:0	0.192	0.240	0.288
	Murate of potash 13:0:45	0.692	0.864	1.039
	Urea	1.108	1.384	1.661

13.3.3 EFFECT OF PLASTIC MULCH ON TEMPERATURE VARIATIONS

The soil temperature was measured with the help of digital soil thermometer at 10 cm depth and at different times of 8 a.m., 1 p.m., and 4 p.m. and the data are presented in Table 13.4. Temperature varied significantly with the type of mulching and time of the day. The temperature of the treatments under mulch was higher than that of the control plots for all the times.

The 25 μm thickness black plastic mulch showed higher temperature variation of 1.1–4.35°C than 50 μm thickness mulch and no mulch plots. Treatment T_5 with 25 μm thickness at 80% ET_0 level with 100% RDF resulted in higher temperature than treatment T_{14} with 50 μm thickness at 80% ET_0 level with 100% RDF and treatment T_{23} with no mulch at 80% ET_0 level with 100% RDF, which allowed more radiation to pass through and did not allow to reflect back and hence increase in temperature. Effect of mulch on temperature variations and yield of grafted eggplant was

TABLE 13.4 Studies on Spatial Variation of Temperature Under Different Treatments.

Treatment	Ambient temperature °C at 8 a.m.	8 a.m.	Ambient temperature °C at 1 p.m.	1 p.m.	Ambient temperature °C at 4 p.m.	4 p.m.
T_1	28.5	29.39	30.1	34.2	33.8	37.3
T_2	27.3	28.94	29.4	33.7	33.1	36.86
T_3	26.5	27.54	29.1	33.4	32.9	36.23
T_4	26.9	27.8	30.3	33.6	33.1	37.06
T_5	27.2	28.88	31.2	34.7	34.3	38.54
T_6	26.7	27.5	30.5	34.4	34.1	37.78
T_7	25	26.45	28.7	34	33.7	37.37
T_8	25.8	27.53	28.9	33.9	34.1	37.86
T_9	27.3	28.5	29.0	33.8	33.2	37.46
T_{10}	25.4	27.1	29.6	33.6	31.9	34.8
T_{11}	25.8	27.5	29.2	33.5	31.7	34.95
T_{12}	26.3	28.1	30.3	33.6	31.5	34.88
T_{13}	24.9	26.7	28.6	32.8	31.2	34.65
T_{14}	24.5	26.2	30.1	33.3	31	34.12
T_{15}	25.2	27	29.7	32.7	30.9	34.08
T_{16}	25	26.8	31.2	33.5	30.8	34.01
T_{17}	26.3	28	29.5	32.5	30.5	33.25
T_{18}	26.2	28.3	29.3	32.7	20.1	33.73
T_{19}	24.7	26	28.6	31.6	28.7	32.34
T_{20}	24.2	26.5	28.9	31.3	29.1	32.76
T_{21}	25.6	27	28.6	31.2	29.3	32.43
T_{22}	24	26.4	28.4	31.6	28.5	32.12
T_{23}	24.3	25.9	28.1	31.6	28.1	32.28
T_{24}	24.7	26	28.6	31.6	28.3	32.41
T_{25}	25.7	26.8	28.3	31.9	28.7	32.53
T_{26}	23.9	25.8	29.0	32.2	29.9	32.67
T_{27}	24..9	26.7	29.3	32.6	29.7	32.89
Mean	25.66	27.23	29.35	32.9	30.8	34.72

significantly improved with mulching over control. The highest temperature recorded under mulch treatments was 3.1–6.47°C more than the treatments without mulch. In general, this effect was more evident during the early crop season when grafted eggplant plants shaded less soil surface. It was observed that the temperature recorded in afternoon was higher by 4.53°C and 6.62°C as compared to temperature recorded at 8 a.m. and without mulch condition. Higher temperature under plastic mulch may be due to increased radiation absorption and better thermal conductivity between soil surface and the plastic mulch. These results were in agreement with the findings of Gubbels[10] and Ham et al.[11]

13.3.4 BIOMETRIC PARAMETERS OF GRAFTED EGGPLANT

13.3.4.1 PLANT HEIGHT

The data recorded on plant height under different treatments at 15, 30, 60, and 90 days are presented in Table 13.5. The height of the crop recorded at 15 DAT showed that the maximum plant height of 17 cm was under 25 μm thickness plastic mulch at 80% ET_0 level with 100% RDF (T_5) and 120% RDF (T_6) and lowest height of 13 cm was recorded for the control treatment T_{19}. The statistical analysis indicated statistical significance for mulching, irrigation and fertilizer levels on the plant height and they are significant in their interaction. Mulching levels of 25 and 50 μm were on par with each other. The results indicated that at 30 DAT, the mulch treatment showed the significantly maximum as in case of 15 DAT than the without mulch treatment. The minimum height (19.00 cm) was observed in control treatment at irrigation level of 60% ET_0 with fertigation level of 100% RDF.

TABLE 13.5   Plant Height Under Different Treatments.

Treatments	Plant height (cm)			
	15 DAT	30 DAT	60 DAT	90 DAT
T_1	14.00	26.33	38.67	47.00
T_2	15.00	28.33	45.00	80.33
T_3	14.67	27.33	42.33	76.00
T_4	16.67	32.67	53.00	90.67
T_5	17.00	30.33	51.67	93.33
T_6	17.00	35.00	55.33	93.33
T_7	16.00	31.00	50.00	87.33
T_8	16.00	32.33	51.00	89.33
T_9	16.00	30.67	49.33	86.67

TABLE 13.5　(Continued)

Treatments	Plant height (cm)						
	15 DAT		**30 DAT**		**60 DAT**		**90 DAT**
T_{10}	14.00		25.67		37.00		45.67
T_{11}	15.00		27.33		42.67		78.67
T_{12}	14.33		27.00		41.00		73.67
T_{13}	15.00		29.00		46.67		83.67
T_{14}	16.67		33.67		53.67		91.67
T_{15}	16.33		32.33		52.00		89.33
T_{16}	15.33		29.33		47.67		84.00
T_{17}	16.00		31.33		50.33		88.00
T_{18}	15.67		30.33		48.67		86.00
T_{19}	12.67		18.67		32.00		37.33
T_{20}	13.67		23.67		35.33		40.00
T_{21}	13.00		22.67		33.33		38.00
T_{22}	13.67		24.33		35.67		41.33
T_{23}	14.00		25.33		36.67		43.33
T_{24}	14.00		25.00		36.33		42.33
T_{25}	14.00		26.67		39.00		49.67
T_{26}	15.00		29.00		46.00		81.00
T_{27}	15.00		28.33		45.33		81.00
Mean	29		44		71		71
Effects	S.Ed	CD (0.05)	S.Ed	CD (0.05)	S.Ed	CD (0.05)	S.Ed	CD (0.05)
M	0.15	0.42**	0.26	0.71**	0.41	1.13**	0.52	1.45**
I	0.12	0.36**	0.33	0.92**	0.63	1.75**	0.59	1.63**
M × I	0.17	0.39**	0.42	0.97**	0.54	1.26**	1.00	2.31**
F	0.11	0.22**	0.31	0.63**	0.29	0.60**	0.40	0.81**
M × F	0.19	0.32**	0.53	0.68**	0.51	1.04**	0.69	1.40*
I × F	0.19	0.32**	0.53	0.68**	0.51	1.04**	0.69	1.40**
M × I × F	0.32	0.61[NS]	0.92	1.87**	0.89	1.81**	1.20	2.44**

*Significant; **highly significant.

Similar trends were observed in 60 DAT and at the time of harvest. The treatment with mulch showed the higher height (93.00 cm) than the without mulch (37.00 cm). They were significant in their interactions.

13.3.4.2   NUMBER OF LEAVES

The data pertaining to number of leaves at 15, 30, 60, and 90 DAT are presented in Table 13.6. The results indicated plants in mulch treatment showed the maximum number of leaves. The minimum number of leaves was recorded in without mulch treatments. Among the different treatments, the treatment T_5 receiving water at 80% ET_0 with 100% RDF under 25 μm thickness mulch recorded the maximum number of leaves 10 at 15 DAT, 49 at 30 DAT, 143 at 60 DAT, and 245 at 90 DAT followed by T_6 at 100% ET_0 with 120% RDF. The lowest were observed in control treatment T_{19} with 60% ET_0 and 80% RDF. The interaction effects were significant. Similar trends were observed at 60 DAT and at the time of harvest.

TABLE 13.6   Number of Leaves per Plant Under Different Treatments.

Treatments	Number of leaves			
	15 DAT	30 DAT	60 DAT	90 DAT
T_1	6	33	107	181
T_2	7	35	113	188
T_3	6	34	111	186
T_4	8	43	129	222
T_5	10	49	143	245
T_6	9	46	133	236
T_7	7	39	123	201
T_8	7	41	126	213
T_9	7	38	122	201
T_{10}	6	32	107	180
T_{11}	7	34	112	187
T_{12}	6	33	110	184
T_{13}	7	36	117	193
T_{14}	8	45	130	227
T_{15}	8	42	127	218
T_{16}	7	37	120	194
T_{17}	7	40	124	206
T_{18}	7	37	121	101
T_{19}	5	25	87	143
T_{20}	5	28	97	170

TABLE 13.6 *(Continued)*

Treatments	Number of leaves							
	15 DAT		30 DAT		60 DAT		90 DAT	
T_{21}	5		28		91		167	
T_{22}	5		29		102		173	
T_{23}	6		30		106		178	
T_{24}	5		30		104		175	
T_{25}	6		33		109		183	
T_{26}	7		35		115		191	
T_{27}	7		35		113		189	
Mean	7		36		115		190	
Effects	S.Ed	CD (0.05)	S.Ed	CD (0.05)	S.Ed	CD (0.05)	S.Ed	CD (0.05)
M	0.25	0.70**	1.09	3.02**	1.54	4.28**	1.01	2.82**
I	0.28	0.49**	0.68	1.89**	1.27	3.52**	1.38	3.83**
M × I	0.30	0.68**	0.93	2.14**	1.81	4.18**	1.59	3.67**
F	0.10	0.20**	0.34	0.69**	0.50	1.02**	0.94	1.89**
M × F	0.18	0.36**	0.59	1.20*	0.87	1.77**	1.62	3.29**
I × F	0.18	0.36**	0.59	1.20**	0.87	1.77**	1.62	3.29**
M × I × F	0.31	0.62**	1.03	2.08**	1.51	3.07**	2.81	5.66**

*Significant; **highly significant.

13.3.5 YIELD AND YIELD ATTRIBUTES

13.3.5.1 NUMBER OF FRUITS PER PLANT

Maximum of 263 of fruits were obtained for treatment T_5 (25 μm thickness at 80% ET_0 level with 100% RDF) followed by T_6 (25 μm thickness 120% RDF), followed by 50 μm thickness at 80% ET_0 level with 100% RDF; and minimum number of fruits per plant (57) were recorded for the control at 60% ET_0 with 80% RDF. Treatments under mulch produced more fruits per plant compared to control. The statistical analysis depicted that all the three factors (mulching, irrigation, and fertilizer levels) and their interaction showed highly significant effects on the total number of fruits. These results are presented in Table 13.7.

TABLE 13.7   Number of Fruits per Plant, Yield per Plant, and Total Yield Under Different Treatments.

Treatments	Number of fruits per plant		Fruit yield per plant (kg)		Total yield (t ha^{-1})	
T_1	134		6.11		42.50	
T_2	170		7.76		53.90	
T_3	159		7.25		50.40	
T_4	233		10.63		73.80	
T_5	263		12		83.30	
T_6	256		11.68		81.10	
T_7	206		9.4		65.30	
T_8	219		9.99		69.40	
T_9	202		9.22		64.00	
T_{10}	118		5.38		37.40	
T_{11}	164		7.48		52.00	
T_{12}	153		6.98		48.50	
T_{13}	181		8.26		57.30	
T_{14}	250		11.41		79.20	
T_{15}	224		10.22		71.00	
T_{16}	188		8.58		59.60	
T_{17}	213		9.72		67.50	
T_{18}	193		8.8		61.10	
T_{19}	57		2.6		18.10	
T_{20}	83		3.79		26.30	
T_{21}	69		3.15		21.90	
T_{22}	94		4.29		29.80	
T_{23}	106		4.84		33.60	
T_{24}	101		4.61		32.00	
T_{25}	136		6.2		43.10	
T_{26}	176		8.03		55.80	
T_{27}	172		7.85		54.50	
Mean	167		8		53.05	
Effects	S.Ed	CD (0.05)	S.Ed	CD (0.05)	S.Ed	CD (0.05)
M	6.79	18.86**	0.26	0.73**	1.82	5.07**
I	8.34	23.17**	0.32	0.88**	1.42	3.94**
M × I	11.9	27.47**	0.62	1.42**	2.27	5.24**

TABLE 13.7   *(Continued)*

Treatments	Number of fruits per plant		Fruit yield per plant (kg)		Total yield (t ha^{-1})	
F	4.25	8.64**	0.18	0.36**	1.76	3.58**
M × F	7.37	13.00**	0.31	0.90**	3.05	5.08**
I × F	7.37	13.00**	0.31	0.90**	3.05	5.08**
M × I × F	12.7	36.11**	0.54	1.60**	5.29	9.05**

**Highly significant.

Data showed more number of fruits per plant in mulched treatments than the control. This increase in number of fruits per plant was probably associated with the conservation of moisture, reduced number of weeds and improved microclimate both beneath and above the soil surface. These mulching results were similar with those by Awodoyin et al.,[4] where difference between treatments in total number of fruits per plant were highly significant (131 fruits) in black plastic mulch as compared to the other mulched treatments such as grass (104 fruits), wood-chip (96 fruits), hand weeded (89 fruits), and unweeded (42 fruits) plots.

These results are also in agreement with the findings of Agrawal et al.,[1,2] who found out that the yield attributes (number of fruits per plant, fruits per cluster, diameter of fruits, and weight of fruits) under polythene mulch were highest and same characters were lowest in the control.

13.3.5.2   YIELD PER HECTARE

The maximum yield was observed in the treatment T_5 (12.00 kg) followed by T_6 (11.68 kg) and T_{14} (11.41 kg). Lowest fruit yield of 2.6 kg was observed in T_{19} (control at 60% ET_0 with 80% RDF). Higher yield was found in mulched treatments compared to the control. The three factors and interactions showed highly significant effects on the fruit yield per plant. These results are given in Table 13.7. The highest yield (83.3 t ha^{-1}) was recorded in T_5 (25 μm thickness plastic mulching at 80% ET_0 level with 100% RDF) followed by treatment T_6 and T_{14} (81.10 t ha^{-1}) and (79.2 t ha^{-1}), respectively. These results are given in Table 13.7. Lowest yield (18.1 t ha^{-1}) was recorded in T_{19}, that is, control. The statistical analysis depicted that all the three factors (mulching, irrigation, and fertilizer levels) and their interactions showed highly significant effects on total yield. Among the treatments at different irrigation levels, the 80% ET_0 with 100% RDF gave the maximum yield and the minimum yield was recorded in control plot at irrigation level 60% ET_0 and fertigation level of 80% RDF. The complimentary soil moisture will improve growth of the plant. This leads to the increase in yield. The present results are in line with the findings of Jinhui et al.[13]

The drip fertigation treatments with 100% RDF showed a statistically significant higher yield compared with other drip fertigation treatments. This can be explained by the fact that with drip fertigation the root zone is simultaneously supplied with water and nutrients, and nutrients are abundantly available as compared to other plots. Hagin et al. (2002) reported that in a fertigation system, the timing, amounts, concentrations, and ratios of the nutrients are easily controlled. Due to this improved control, crop yields are higher than those produced by a conventional fertilizer application and irrigation. A number of other investigators reported higher yields in different crops when fertilizers were injected through the drip system in comparison with conventional application of fertilizers.

The yields from plants grown on bare soil were significantly lower than those from plants grown with black plastic mulch. The increase of soil temperature below mulch and efficient utilization of water, fertilizers, and nutrients resulting from the use of the plastic mulch might be an important reason for the highest yield. Similar results have been reported by Mukherjee et al.[16]

13.4   FUTURE PROSPECTS

The research study in this chapter suggested that grafted eggplant crop responded well to drip irrigation with plastic mulch and fertigation levels. In the light of these findings, further studies are required as below:

- Studies on the effect of different types mulches with different irrigation levels by taking first grafted eggplant crop as a ratoon crop (it is a perennial crop, so one can cultivate 2–3 ratoon crops during successive 1–2 years).
- Studies on comparing the drip irrigation with conventional irrigation methods to compare the amount of water saving and increase in the yield.

13.5 SUMMARY

The field studies were conducted at PFDC Farm of Tamil Nadu Agricultural University, Coimbatore to evaluate the effects of plastic mulch and fertigation on grafted eggplant (*Solanum melongena* L.) under drip irrigation. The experiments were laid in SPD with three factors and three levels of each with 27 treatments and three replications, which included three mulching levels such as 25 μm thickness plastic mulch, 50 μm thickness plastic mulch, control, and three irrigation levels, that is, irrigation at 60% ET_0, 80% ET_0, 100% ET_0, and three fertigation levels, that is, 80% RDF, 100% RDF, and 120% RDF.

- The daily water requirement of grafted eggplant at different stages was: 1.9 L/day (initial stage), 3.6 L/day (vegetative stage), 4.2 L/day (flowering stage), and 5.2 L/day (harvesting stage). The total water requirement for growth period in drip irrigation was 396.00 mm with mulch at 80% ET_0 level and 495.00 mm without mulch at 100% ET_0 level. Drip irrigation could save 16.17% of water with mulch as compared to without mulch condition.

- The temperature was higher under mulched plots than unmulched plots thus helping in the plant growth. The 25 μm thickness black plastic mulch produced a higher temperature than 50 μm thickness mulch. A difference of 3.1–6.47°C was observed between mulch and without mulch treatments.

- Treatments under mulch recorded higher yield per plant compared to control and it was associated with the conservation of moisture and improved microclimate both beneath and above the soil surface. A maximum of 263 numbers of fruits were recorded for the treatment T_5 under mulch condition with 25 μm thickness at 80% ET_0 with 100% RDF followed by T_6 with 256 fruits per plant and a minimum number of fruits per plant (57) was recorded under the treatment T_{19} without mulch at 60% ET_0 with 80% RDF. The highest yield of 83.34 t ha^{-1} was recorded for the treatment T_5 under mulch with 25 μm thickness at 80% ET_0 with 100% RDF which was followed by T_6 and T_{14} (81.10 t ha^{-1}) and (79.2 t ha^{-1}), respectively. Lowest yield values were recorded under treatment without mulch at 60% ET_0 and 80% RDF (18.1 t ha^{-1}). The increase of temperature and efficient utilization of water, fertilizers, and nutrients resulting from the use of the plastic mulch are important reasons for the highest yield.

KEYWORDS

- drip irrigation
- eggplant
- ET0
- fertigation
- fertilizer saving
- mulching
- plant height
- RDF
- soil temperature
- spatial variation
- water saving

REFERENCES

1. Agarwal, M. G.; Khanna, S. S. *Efficient Soil and Water Management in Haryana*; Haryana Agricultural University: Hisar, Bulletin, **1983**; p 118.
2. Agrawal, N.; Panigrahi, H. K.; Sharma, D. Effect of Different Color Mulches on the Growth and Yield of Tomato Under Chhattisgarh Region. *Indian J. Hortic.* **2010,** *67*(Special Issue), 295–300.
3. Akbari, M.; Dehghanisanij, H.; Mirlatifi, S. M. Impact of Irrigation Scheduling on Agriculture Water Productivity. *Iran. J. Irrig. Drain.* **2009,** *1*, 69–79.
4. Awodoyin, F. I.; Ogbedide, O. O. Effects of Three Mulch Types on Growth and Yield of Tomato and Weed Suppression in Ilbadan, Rainforest-savanna Transition Zone of Nigeria. *Trop. Agric. Res. Ext.* **2007,** *10*, 53–60.
5. Bhogi, B. H.; Polisgowdar, B. S.; Patil, M. G. Effectiveness and Cost Economics of Fertigation in Eggplant (*Solanum melongena* L.) Under Drip and Furrow Irrigation. *Karnataka J. Agric. Sci.* **2011,** *24*(3), 417–419.
6. Bletsos, F.; Thanassoulopoulos, C.; Roupakias, D. Effect of Grafting on Growth, Yield, and *Verticillium* Wilt of Eggplant. *HortScience* **2003,** *38*, 183–186.
7. Dalvi, V. B.; Tiwari, K. N.; Pawade, M. N.; Phire, P. S. Response Surface Analysis of Tomato Production Under Micro Irrigation. *J. Water Manag.* **1999,** *41*, 11–19.
8. Dimitrios, S.; Colla, G.; Youssef, R.; Dietmar, S. Amelioration of Heavy Metal and Nutrient Stress in Fruit Vegetables by Grafting. *Sci. Hortic.* **2010,** *127*, 156–161.
9. Frank, J. L.; Rivard, C. L.; Kubota, C. Grafting Fruiting Vegetables to Manage Soil Borne Pathogens, Foliar Pathogens, Arthropods and Weeds. *Sci. Hortic.* **2010,** *127*, 127–146.

10. Gubbels, G. H. Response of Cabbage Grown North of Latitude 60°N to Plastic Mulch Ridge and Row Orientation. *Can. J. Plant Sci.* **1971,** *51,* 17–20.

11. Ham, J.; Kluitenberg, G. J.; Lamont, W. J. Optical Properties of Plastic Mulches Affect the Field Temperature Regime. *J. Am. Soc. Hortic. Sci.* **1993,** *118*(2), 188–193.

12. Ilyas, S. M. Present Status of Plastics in Agriculture Lecture Note Delivered in Summer School on Application of Plastics in Agriculture. CIPHET, Punjab Agriculture University: Ludhiana, 2001; p 332.

13. Jinhui, X.; Cardenas, E. S.; Sammis, T. W.; Wall, M. M.; Lindsey, D. L.; Murray, L. W. Effects of Irrigation Method on Chile Pepper Yield and Phytophthora Root Rot Incidence. *Agric. Water Manag.* **1999,** *42,* 127–142.

14. Khah, E. M.; Katsoulas, N.; Tchamichian, M.; Kittas, C. Effect of Grafting on Eggplant Leaf Gas Exchange Under Mediterranean Greenhouse Conditions. *Int. J. Plant Prod.* **2011,** *5*(2), 121–134.

15. Kubota, C. Use of Grafted Seedlings for Vegetable Production in North America. *Acta Hortic.* **2008,** *770,* 21–26.

16. Mukherjee, A.; Kundu, M.; Sarkar, S. Role of Irrigation and Mulch on Yield, Evapotranspiration Rate and Water Use Pattern of Tomato (*Lycopersicon esculentum* L.). *Agric. Water Manag.* **2010,** *98,* 182–189.

17. Rouphael, Y.; Dietman, S.; Angelika, K.; Colla, G. Impact of Grafting on Product Quality of Fruit Vegetables. *Sci. Hortic.* **2010,** *127,* 172–179.

18. Sidhu, A. S.; Dhatt, A. S. Current Status of Eggplant Research in India. *Acta Hortic.* **2007,** *752,* 243–247.

19. Tagar, A.; Chandio, F. A.; Mari, I. A.; Wagan, B. Comparative Study of Drip and Furrow Irrigation Methods at Farmer's Field in Umarkot. *World Acad. Sci. Eng. Technol.* **2012,** *69,* 863–867.

CHAPTER 14

PERFORMANCE OF CABBAGE UNDER BEST MANAGEMENT PRACTICES

VASANTGOUDA ROTI[1,*], B. S. POLISGOWDAR[2], and P. S. KASHYAP[1]

[1]Department of SWCE, College of Technology, G. B. Pant University of Agriculture & Technology (GBPUA&T), Pantnagar, Uttarakhand 263145, India

[2]Department of Soil and Water Engineering, College of Agricultural Engineering, University of Agricultural Sciences, Raichur 584101, India

*Corresponding author. E-mail: vaasu0478@gmail.com

CONTENTS

ABSTRACT

A field experiment was conducted on the sandy loam soil at northern transition zone II of India to compare the yield and water use efficiency of cabbage under different level drip (80%, 100%, and 120% ET) and furrow irrigation with mulch and nonmulch condition. The crop was grown in *rabi* season in Main Agricultural Research Station, University of Agricultural Science, Raichur. The study showed that the water use efficiency was higher in mulch with 80% ET (36.95 kg m^{-3}) followed by 80% ET with nonmulch condition (33.85 kg m^{-3}). The water use efficiency was lowest in furrow irrigation with nonmulch plots (10.91 kg m^{-3}). The study also showed that the highest yield was recorded in mulch (81.24 t/ha) than the nonmulch (74.08 t/ha). In the combined effect, the mulch with 100% ET (92.95 t/ha) recorded the maximum yield than the same level of irrigation under nonmulch treatment (84.19 t/ha). From the study, the results showed the yield and water use efficiency of cabbage were higher in 100% ET with mulch.

14.1 INTRODUCTION

Water is considered as liquid gold. It is a precious commodity and its judicious use is essential for maximizing crop yields. Out of the total geographical area of 329 million ha, the net cultivable area in India is about 186 million ha and the net sown area is about 142 million ha. At present, only about one-third of the area has assured irrigation. It has been estimated that the irrigated area is 253 million ha in the world. The gross irrigated area of India has increased to 86.42 million ha in 2009–10 from 22.6 million ha in 1951–52, increases being more than 250% during the last four decades.[2] Efficient use of water through scientific irrigation management is of utmost importance in providing the best insurance against weather induced fluctuations in food production. The application of irrigation water by traditional method causes 27–42% loss of water depending on the soil type.[1] Micro irrigation (drip and micro sprinkler) helps to conserve irrigation water and increase water use efficiency (WUE) by reducing soil evaporation and drainage losses. It also helps to maintain soil moisture conditions that are favorable to crop growth. Therefore, micro irrigation can help increase the productivity of the land.

In arid or semiarid areas, crop growth is mainly dependent on irrigation. Irrigation methods and management are of importance to soil water status and plant water status. Inappropriate irrigation could result in water stress. Drip irrigation provides more efficient water use for crops than furrow irrigation because drip irrigation applies frequent small amounts of water to the root zone and reduces adverse effects of cyclic over irrigation and water stress commonly caused by furrow irrigation. Centin and Bilgel[3] reported that drip irrigation has higher WUE (4.87 kg ha^{-1} mm^{-1}) than the furrow (3.87 kg ha^{-1} mm^{-1}) and sprinkler irrigation (2.36 kg ha^{-1}mm^{-1}).

Water plays an important role in crop production. Irrigation water is often limited and therefore the techniques which help to conserve water in the field are needed. Mulching is a recommended practice of moisture conservation in arid and semiarid regions. Zotarelli et al.[7] reported that adoption of surface drip irrigation system along with plastic mulch saved irrigation water by 15–51% with 11–80% more tomato fruit yield compared to the conventional irrigation system.

The WUE is the quantity of yield obtained from the each meter cube of water. The WUE increases with increase in the yield and decreases with increase in irrigation level. Tiwari et al.[5] compared the WUE of cabbage with plastic mulch and without plastic mulch under different levels of irrigation using drip irrigation method. The study revealed that 80% evapotranspiration (ET) without mulch has WUE lower than the plastic mulch with the same level of irrigation. The efficiency of mulching in water conservation and the effect of the irrigation WUE have been investigated in India. However, reports in interactive effects of irrigation and mulches on water use pattern of vegetable crops are limited.

Therefore, this chapter evaluates the effects of plastic mulch on drip irrigated cabbage.

14.2 MATERIALS AND METHODS

A field study was conducted at the Main Agricultural Research Station, of University of Agricultural Sciences, Raichur, India. The site is located at 16°15′ N latitude and 77°20′ E longitude at an elevation of 389 m above mean sea level (MSL). The minimum and maximum temperature, minimum and maximum evaporation were recorded as 8.5°C and 35.4°C, 1.2 and 5.8 mm/day, respectively during the study period 2012–13.

Infiltration rate was measured by using double ring infiltrometer and was 1.72 cm/h. The study was carried out in a sandy loam soil. The field capacity and wilting point was measured with pressure plate apparatus, and was 19.16% and 11.55%, respectively.

The experiment was laid out in split plot design with two main treatments [mulch, (M1) and without mulch, (M2)] and four subtreatments [80% (T1), 100% (T2), and 120% (T3) of ET and furrow irrigation (T4)]. Each treatment was replicated thrice. The size of the each plot was 0.8 m × 10 m. Among the various approaches for irrigation scheduling, water requirement of a plant was determined based on the canopy, coefficient of evaporation, and ET of study area. The water applied for both mulch and without mulch was same. The amount of water to be delivered in furrow irrigation method was computed as follows:

$$d = [(M_{fc} - M_{bi})/100] \times [A_s \times d_s]$$ (14.1)

where d = net amount of water to be applied during irrigation, (cm); M_{fc} = moisture content at field capacity, (%); M_{bi} = moisture content before irrigation, (%); A_s = soil bulk density, (g cm^{-3}); and d_s = effective root zone depth, (cm).

The quantity of water per plant is given as follows:

$$Q = d \times A \times B$$ (14.2)

where Q = quantity of water required per plant (L); d = net amount of water to be applied during an irrigation (cm); A = gross area per plant (cm^2); and B = extent of area covered by foliage, fraction.

The bulk density was found as 1.53 kg m^{-3}; effective root zone depth was assumed as 25 cm, and the gross area covered by each plant was 50 cm × 45 cm. The extent of the area covered by foliage was assumed as 1.1 fraction of the given area. The WUE of the crop was calculated as follows:

$$WUE = [Y]/[WR]$$ (14.3)

where WUE = water use efficiency (kg m^{-3}); Y = crop yield (kg); and WR = total amount of water used in the field (m^3).

14.3   RESULTS AND DISCUSSION

The water was applied as desired in each treatment. The amount of water delivered to cabbage under different levels of drip irrigation and furrow irrigation are presented in Table 14.1.

TABLE 14.1   Amount of Water Application in Each Treatment.

Treatment	Amount of water per plant (L/day)	Water applied per plot (L)	Water applied during the crop season (m^3/ha)
T1	43.88	1930.73	2413.41
T2	52.51	2310.23	2887.79
T3	61.13	2689.73	3362.17
T4	84.41	3714.04	4642.55

In furrow irrigation, 4642.55 m^3/ha of water was applied during the entire crop period. Among the different levels of drip irrigation, 120% of ET consumed more water (3362.17 m^3/ha) followed by 100% ET (2887.79 m^3/ha) and 80% ET (2413.41 m^3/ha).

The effects of plastic mulching, irrigation methods, and levels of drip irrigation on total marketable cabbage yield (t/ha) are presented in Table 14.2. The plastic mulch with 100% ET (92.95 t/ha) showed maximum yield followed by 80% ET combined with mulching (89.17 t/ha). The lowest yield was observed in furrow irrigation without mulch (50.64 t/ha). This was due to several advantages of the plastic mulch, such as moisture conservation, reduction in weed population, adequate micro-climate around the plant. Also, drip irrigation supplied water as per the plant requirements and moisture at field capacity was maintained throughout the season. These results are in agreement with the findings of Vijay et al.[6]

The effects of plastic mulch, different drip irrigation levels, and irrigation methods on WUE are presented in Table 14.3. Among the main treatments with different levels of irrigation, the irrigation at 80% ET with plastic mulch showed the maximum WUE (36.95 kg m^{-3}) followed by 80% ET without mulch (33.85 kg m^{-3}). The minimum WUE was in plots without plastic mulch in furrow irrigation (10.91 kg m^{-3}) followed by a combination of mulch with control treatment (11.89 kg m^{-3}). This was due to fact that plastic mulch provided favorable metabolic activities of the plant, which led to the increase in cabbage yield with the same application rate of water than the without plastic mulch. These results are in agreement with the findings of Paul et al.[4]

TABLE 14.2   Effects of Plastic Mulch, Irrigation Methods, and Drip Irrigation Levels on Cabbage Yield.

Treatment	Marketable yield of cabbage (t/ha)				
	T_1	T_2	T_3	T_4	Mean
M_1	89.17	92.95	87.63	55.22	81.24
M_2	81.69	84.19	79.80	50.64	74.08
Mean	85.43	88.57	83.72	52.93	–
		SEM±		CD at 5%	
Main treatment		0.78		4.74	
Sub treatment		0.67		2.08	
T at same M		0.95		2.94	
M at the same or different T	0.99			2.98	

TABLE 14.3   Effects of Plastic Mulch, Irrigation Methods, and Different Levels of Drip Irrigation on Water Use Efficiency (WUE).

Treatment	Water use efficiency (kg m^{-3})				
	T1	T2	T3	T4	Mean
M1	36.95	32.19	26.06	11.89	26.77
M2	33.85	29.15	23.74	10.91	24.41
Mean	35.40	30.67	24.90	11.40	–
		SEM±		CD at 5%	
Main treatment		0.27		1.62	
Sub treatment		0.21		0.63	
T at same M		0.29		0.90	
M at the same or different T	0.31			5.64	

14.4   SUMMARY

The highest WUE of 36.95 kg m^{-3} was recorded in plastic mulch with drip irrigation at 80% ET followed by without mulch with 80% of ET (33.85 kg m^{-3}). Increased WUE with a decreased level of water input through drip was noted. The least WUE was found in nonmulched plots under furrow irrigation (10.91 kg m^{-3}).

KEYWORDS

- cabbage
- evapotranspiration
- field capacity
- marketable yield
- mulching
- water use efficiency

REFERENCES

1. Agarwal, M. C.; Khanna, S. S. *Efficient Soil and Water Management in Haryana;* Bulletin 118: Haryana Agricultural University, Hisar, **1983;** p 51.

2. Anonymous, *Statistical Data 2010;* Ministry of Agriculture, Government of India. www.indiastat.com (accessed July 24, 2016).

3. Centin, C.; Bilgel, I. Effect of Different Method If Irrigation on Shedding and Yield on Cotton. *Agric. Water Manag.* **2002,** *54,* 1–15.

4. Paul, J. C.; Mishra, J. N.; Pradhan, P. L.; Panigrahi, B. Effect of Drip and Surface Irrigation on Yield, Water Use Efficiency and Economics of Capsicum (*Capsicum annum* L.) Grown Under Mulch and Non-mulch Conditions in Eastern Coastal India. *Eur. J. Sustain. Dev.* **2013,** *2*(1), 99–108.

5. Tiwari, K. N.; Singh, Ajai; Mal, P. K. Effect of Drip on a Yield of Cabbage (*Brassica oleracea* var. capitata L.) Crop Under Mulch and Non-mulch Condition. *Agric. Water Manag.* **2003,** *58,* 19–28.

6. Vijay, K.; Mouli, A.; Ramulu, V.; Kumar, K. A. Effect of Drip Irrigation Levels and Mulches on Growth, Yield. *J. Res. ANGRAU* **2012,** *40*(4), 73–74.

7. Zotarelli, L.; Scolberg, J. M.; Dukes, M. D.; Munoz-Carpena, R. Tomato Yield, Biomass Accumulation, Root Distribution and Irrigation Water Use Efficiency on a Sandy Soil, as Affected by Nitrogen Rate and Irrigation Scheduling. *Agric. Water Manag.* **2009,** *96,* 23–34.

CHAPTER 15

PERFORMANCE OF CHILI (*CAPSICUM ANNUUM* L) UNDER BEST MANAGEMENT PRACTICES

A. SELVAPERUMAL* and I. MUTHUCHAMY

Department of Soil and Water Conservation Engineering, Agricultural Engineering College and Research Institute, Kumulur, Trichy, Tamil Nadu, India

*Corresponding author. E-mail: selvabtech.agri@gmail.com

CONTENTS

ABSTRACT

Field experiment was conducted in 2013–14 at Precision Farming Development Centre Research Farm, Tamil Nadu Agricultural University, Coimbatore to study the effect of drip fertigation (DF) and plastic mulching (PM) on the plant growth and yield attributes of chilli (*Capsicum annuum* L). The experiment laid out factorial randomized block design which included three fertigation levels 80%, 100%, and 120% recommended dose of fertilizers (RDF) and three different mulching treatments such as 25, 50 μm black plastic mulch (BPM), and no mulch which were replicated thrice. In chilli, maximum yield of 128 numbers of fruits per plant which is worked out as 12.27 t ha^{-1} was observed for the treatments T_3. The total quantity of water applied uniformly to all the treatments was 75.83 L as per the crop water requirement. Maximum water use efficiency observed in T_3 (66.36 kg ha^{-1} mm^{-1}). The maximum N, P, and K fertilizer use efficiency of 109.95, 164.94, and 164.94 kg ha^{-1}, respectively, was observed in T_1. The highest benefit–cost ratio was recorded under both T_2 (BPM of 25 μm thickness with 100% RDF) and T_3 (BPM of 25 μm thickness with 120% RDF). From economic viability point, the T_2 treatment registered results that were economically viable with highest profit. Increased yield in fertigation treatments might be due to better availability of plant nutrients and irrigation water throughout the crop growth period under drip fertigation system.

15.1 INTRODUCTION

In recent trends in India, the irrigated area consists of about 36% of the net sown area. Presently, the agricultural sector accounts for about 83% of all water use. Increasing competition with the other water users in the future would be limiting the water availability for expanding the irrigated area. Mark et al.[4] reported that 33% of India's population will live under absolute water scarcity condition by the year 2025. The per capita water availability in terms of average utilizable water resources in India was 6008 m^3 in 1947 and is expected to dwindle to 760 m^3 by 2025.[3]

Drip irrigation involves supplying water to the soil very close to the plant roots at low flow rates (0.5–10 L/h) from a plastic pipe fitted with drip emitters or outlets. Drip irrigation results in a very high water application efficiency of about 90–95%. Fertigation has the potential to supply a right mixture of water and nutrients to the root zone, and thus meeting plants water and nutrient requirements in most efficient possible manner.[5] With fertigation, water use efficiency (WUE) of the crops has to be increased in order to reduce the water loss from the fields. With the increase in nutrient use efficiency (NUE), the loss of nutrients to the groundwater is reduced.

Mulching is the process or practice of covering the soil to make more conditions favorable for plant growth, development, and efficient crop production. Black plastic mulch is used most widely because it slows down weed growth, resulting in less chemical usage. Ashrafuzzaman et al.[1] recorded the lowest number of weeds in black plastic mulch. Black plastic mulch raises soil temperatures quickly, so the plants can increase growth resulting in earlier and higher yields (possibly up to 15% or more) compared to bare ground production.[6]

Vegetable production in Indian agriculture has a wider scope for increasing the income of the marginal and small farmers. Vegetables have vast potential in gaining foreign exchange through the export. The vegetable growers are looking for new ways to achieve superior quality produce with higher yields. Among the vegetables grown, chili is a spice cum vegetable crop of commercial importance.

This study discusses effects of drip fertigation (DF) and plastic mulching (PM) on the plant growth and yield attributes of chili (*Capsicum annuum* L).

15.2 MATERIALS AND METHODS

This study was carried out in PFDC research farm, Tamil Nadu Agricultural University (TNAU), Coimbatore, situated at 11°N latitude and 77°E longitude with mean altitude of 426 m above the mean sea level. The proposed research experiment was laid out during 2013–14 under irrigated conditions to study the effects of drip fertigation and plastic mulching on the plant growth, and yield attributes of chili var. Cochi (*Capsicum annuum* L.) on a sandy clay loam soil.

The experimental plot was thoroughly plowed with disc plow and tilled twice with a cultivator to bring optimum soil tilth. The length and width of the field was 15 and 15 m, respectively. The total area was divided into strips of 4.5 m × 1.2 m according to the treatments. The spacing of 60 cm × 60 cm is recommended for chili in the package of practices by TNAU.

15.2.1 EXPERIMENTAL TREATMENTS

Nine treatments included combination of three fertigation treatments and three mulching treatments. The experiment was laid out in Factorial Randomized block design having nine treatment combinations and was replicated thrice. The treatment details are shown in Table 15.1.

TABLE 15.1 Treatment Details for Chili.

Treatments	Description
$T_1 M_1$	Black plastic mulch of 25 µm thickness with 80% RDF
$T_2 M_1$	Black plastic mulch of 25 µm thickness with 100% RDF
$T_3 M_1$	Black plastic mulch of 25 µm thickness with 120% RDF
$T_3 M_2$	Black plastic mulch of 50 µm thickness with 80% RDF
$T_5 M_2$	Black plastic mulch of 50 µm thickness with 100% RDF
$T_6 M_2$	Black plastic mulch of 50 µm thickness with 120% RDF
$T_7 M_3$	No mulch with 80% RDF
$T_8 M_3$	No mulch with 100% RDF
$T_9 M_3$	No mulch with 120% RDF

• Main plots: fertigation levels

    F_1: 80% of recommended dose of fertilizer

    F_2: 100% of recommended dose of fertilizer

    F_3: 120% of recommended dose of fertilizer

• Subplots: mulching treatments

    M_1: black plastic mulch of 25 µm thickness

    M_2: black plastic mulch of 50 µm thickness

    M_3: control (no mulch)

RDF: recommended dose of fertilizer

15.2.2 IRRIGATION SCHEDULING

Irrigations were scheduled on the basis of climatological approach in the mulched and control plots. Lifesaving irrigation was given immediately after transplanting and the field was regularly irrigated continuously for 10 days. After the 10th day, subsequent irrigations were scheduled once in 3 days based on the following formula and applied each time as per the treatment schedule. The discharge rate of single dripper was 4 L/h at a nominal operating pressure of 50.66 kPa. The time required for each irrigation is shown in Table 15.2.

$$WR_c = CPE \times K_p \times K_c \times W_p \times A, \tag{15.1}$$

where WR_c = computed water requirement (L plant^{-1}); CPE = cumulative pan evaporation for 3 days (mm); K_p = pan factor (0.8); K_c = crop factor; W_p = wetted fraction (0.8); and A = area per plant, m^2.

TABLE 15.2   Quantity of Water Applied per Plant of Chili.

Crop date	Quantity applied per plant (L/d)	Duration of irrigation (min) each day	Total quantity (L) applied per plant per stage
Initial stage (September 25–October 14) 1–20 days	0.427	20	1.281
Vegetative stage (October 15–November 09) 21–45 days	0.223	10	0.669
Fruit setting stage (November 10–December 24) 46–90 days	0.583	27	6.996
Final stage (December 25–January 23) 91–120 days	1.078	48	10.78

$$\text{Time of operation} = [\text{Water volume required} \times \text{Irrigation interval}]/ [\text{Emitter discharge}] \tag{15.2}$$

15.2.3   FERTIGATION SCHEDULING

Drip laterals were laid along the length of each raised bed at the center with the spacing of 1.20 m between two adjacent laterals. Fertigation to individual plot in each replication was controlled by a manual regulating valve attached to the lateral line to ensure precise delivery of the required inputs, thus, enabling full control of experimental setup. A dosage of 120:80:80 NPK kg ha^{-1} was taken as 100% recommended dose of fertilizer (RDF) and 75% of RDF was phosphorous applied as basal through super phosphate. Fertilizer requirement of chili is shown in Table 15.3.

TABLE 15.3   Details of Quantity of Fertilizers (kg) in the Plot Area.

Basal dose			
75% of RDF—Phosphorus applied as basal as super phosphate = (60×6.25) kg ha^{-1} = 375 kg ha^{-1}			
Top dressing			
Stage	Name	80% RDF	100% RDF	120% RDF
---	---	---	---	---
Transplanting to plant establishment stage (1–10 days)	NPK 19:19:19	0.081	0.102	0.122
	Urea	0.067	0.084	0.101
	SOP	0.029	0.037	0.044
Flower initiation to flowering (10–40 days)	NPK 19:19:19	0.163	0.204	0.245
	Urea	0.338	0.422	0.507
	SOP	0.179	0.224	0.269
Flowering to fruit set (40–70 days)	NPK 19:19:19	0.081	0.102	0.122
	Urea	0.270	0.338	0.405
	SOP	0.149	0.186	0.224
Alternate day from picking (70–120 days)	NPK 19:19:19	0.081	0.102	0.122
	Urea	0.169	0.221	0.153
	SOP	0.089	0.112	0.134

15.2.4   INSTALLATION OF DRIP SYSTEM AND FERTIGATION UNIT

Irrigation water was pumped through 7.5 hp bore well pump and conveyed through the main line of 75 mm diameter polyvinyl chloride (PVC) pipes after filtering through a screen filter. To the main pipe, submain of 63 mm diameter PVC pipes was connected. From the submain, laterals of 16 mm diameter linear low-density polyethylene (LLDPE) pipes were installed on the ground surface. Each lateral was provided with individual tap control for imposing irrigation. Along the laterals, online drippers were installed at a spacing of 60 cm. The number of laterals installed was based on the number of rows of the crop. The discharge rate of single dripper was 4 L/h. Submains and laterals were closed at the end with end caps. Water-soluble N and K fertilizers were used in this experiment. Phosphorus was applied manually as a basal dose. The recommended soluble fertilizers were applied simultaneously in a combined form to the plant root zone. Urea, NPK 19:19:19, and sulfate of potash were fertigated with fertilizer tank and venturi. The fertilizers were dissolved in water in the ratio of 1:5 and the solution was diluted in the fertigation tank. With venturi injectors, water was extracted from the main line, and a pressure differential was created by a valve in the main line forcing water through the injector at high velocity. The high-velocity water passing through the throat of the venturi creates a vacuum or negative pressure, generating suction to draw chemicals into the injector from the chemical tank. The 80%, 100%, and 120% recommended N and K water-soluble fertilizers were regulated by operating the tap connected at the upstream of each lateral.

15.3 RESULTS AND DISCUSSION

15.3.1 EFFECT OF MULCHING AND FERTIGATION ON BIOMETRIC PARAMETERS

The data on plant height, number of branches, fruit length, fruit girth, and green chili yield were observed at 45, 60, and 90 days after transplanting (DAT) as influenced by mulching and fertigation levels. The results for 90 DAT revealed that maximum plant height of 85.46 cm was recorded under 25 μm thickness plastic mulch at 120% RDF (T_3) and lowest plant height of 72.67 cm was recorded for the treatment T_7 (no mulch with 80% RDF). The maximum number of primary branches was recorded in the treatment of 25 μm thickness plastic mulching with 120% RDF level (T_3), which was 7.83 at 90 DAT. The minimum number of primary branches per plant was observed in the T_7 treatment of 80% RDF level with no mulching.

Mulching produced more fruits per plant compared to control. A maximum of 128 numbers of fruits per plant was recorded in the treatment of 25 μm thickness with 120% RDF (T_3) and the minimum number of fruits was obtained in the treatment T_7. Fruit girth had significant effects due to mulching and fertigation treatments. The maximum fruit girth was observed in the treatment T_3 (3.83 cm) followed by T_6 (3.78 cm) and T_2 (3.72 cm). The lowest fruit girth was observed in the treatment T_7 (3.12 cm). The maximum green chili length of 13.06 cm was recorded in treatment T_3 of 25 μm thickness plastic mulching with 120% RDF level of fertigation followed by treatment T_6 and the lowest fruit length of 12.25 cm is observed in treatment T_7 of no mulch with 80% RDF level of fertigation.

The maximum total green chili yield of 12.27 t ha^{-1} was recorded under 25 μm thickness plastic mulch at 120% RDF (T$_3$). The total yield of 11.99 t ha^{-1} and 11.82 t ha^{-1} were recorded for T$_6$ and T$_2$, respectively, and the lowest total green chili yield of 8.91 t ha^{-1} was recorded for the treatment T$_7$ (no mulch with 80% RDF). The interaction effect of mulching and application of drip fertigation level were significant for the total yield of chili.

The increased yield in fertigation treatments might be due to better availability of plant nutrients and irrigation water throughout the crop growth period under drip fertigation system. This is in accordance with the findings of Gutal et al.[2] Hence, the 25 μm thickness black plastic mulch produced higher soil temperature than 50 μm thickness mulch. The difference of 2–5°C was observed between mulch and nonmulched treatments. Mulching increased the soil temperature, prevented soil water evaporation, and retained the soil moisture. The adoption of plastic mulching in chilies resulted in 20% increase in yield over the control treatments.

Weeds were found only in the nonmulched plots and their numbers were increased with the increase in fertilizer application. There were 12, 9, and 7 numbers of weeds present in the treatments T$_9$, T$_8$, and T$_7$, respectively.

15.3.2   WUE AND FERTILIZER USE EFFICIENCY

The total quantity of water applied uniformly to all the treatments was 75.83 L as per crop water requirement. The highest WUE of 66.36 kg ha^{-1} mm^{-1} was recorded in treatment T$_3$ (25 μm plastic mulch with 120% RDF). The lowest WUE of 48.19 kg ha^{-1} mm^{-1} was recorded in nonmulched treatment with 80% RDF (T$_7$), due to lower yield in the treatment. The highest N, P, and K fertilizer use efficiency (FUE) of 109.95, 164.94, and 164.94 kg ha^{-1}, respectively, were recorded in T$_1$ (25 μm thickness plastic mulch at 80% of fertigation); and the lowest N, P, and K FUE of 75.61, 113.42, and 113.42 kg ha^{-1}, respectively, were recorded in T$_9$ (no mulch at 120% of fertigation levels). Increased FUE with the decreased level fertilizer dose through drip was observed. The data on WUE and FUE of N, P, and K in chili crop are shown in Figures 15.1–15.3.

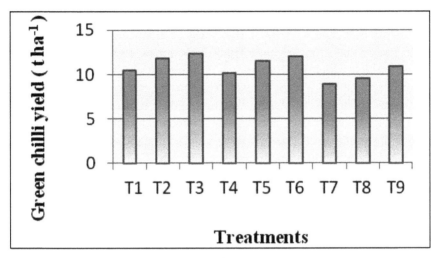

FIGURE 15.1   Effect of DF and PM on chili yield.

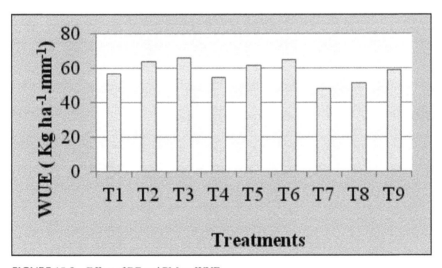

FIGURE 15.2   Effect of DF and PM on WUE.

15.3.3   COST ECONOMICS

Among different levels of fertigation and mulching higher benefit–cost ratio (B:C ratio) was recorded for both 25 μm thickness plastic mulch at 100% RDF (T_2) and 25 μm thickness plastic mulch with 120% RDF (T_3).

Figure 15.4 indicates that among all mulching treatments, only treatment T_2 (25 μm thickness with 100% RDF) was economically viable compared to T_3, since the gross revenue from T_2 treatment exceeded the cost of plastic mulch which gave additional and highest profit in this treatment. All other treatments except T_4 and T_6 gave positive beneficial profits when compared to the cost of plastic mulching in the treatments. In control plot (T_7) with 80% RDF, B:C ratio was 1.82, which is less than the other treatments. From this experiment, it was observed that the crop chili receiving 100% RDF with 25 μm thickness plastic mulch (T_2) registered results that were economically viable than other treatment combinations.

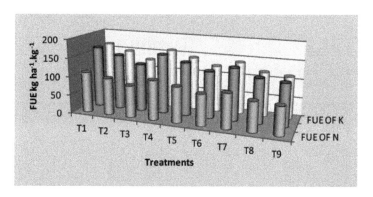

FIGURE 15.3   Effect of DF and PM on FUE of N, P, and K.

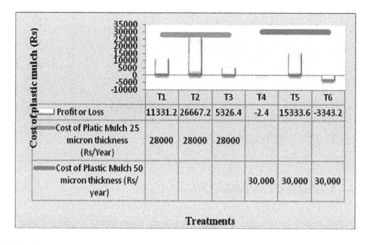

FIGURE 15.4   Economic viability of using plastic mulch.

15.4 SUMMARY

A field experiment was conducted during 2013–14 at Precision Farming Development Center Research Farm, Tamil Nadu Agricultural University, Coimbatore, to study the effect of drip fertigation and plastic mulching on the plant growth and yield attributes of chili (*Capsicum annuum* L). The total quantity of water applied to all the treatments was 75.83 L as per the crop water requirement. Maximum WUE was observed in T_3 (66.36 kg ha^{-1} mm^{-1}). The maximum N, P, and K FUE of 109.95, 164.94, and 164.94 kg ha^{-1} were observed in T_1. The highest B:C ratio was recorded under both T_2. Increased yield in fertigation treatments might be due to better availability of plant nutrients and irrigation water throughout the crop growth period under drip fertigation system. Higher fruit yield under plastic mulch may be due to the reduction of nutrient losses because of weed control and improved hydrothermal regimes of soil.

KEYWORDS

- benefit–cost ratio
- drip irrigation
- fertilizer use efficiency
- green chili
- plastic mulching
- water use efficiency

REFERENCES

1. Ashrafuzzaman, M. M.; Hamid, A. M.; Ismail, M. R.; Sahidullah, S. M. Effect of Plastic Mulch on Growth and Yield of Chili. *Braz. Arch. Biol. Technol.* **2011,** *54*(2), 321–330.

2. Gutal, G. B.; Jadhav, S. S.; Takte, R. L. In *Effective Surface Covered Cultivation in Fruit Vegetable Crop Tomato,* Proceedings of the International Agricultural Engineering Conference, Bangkok, Thailand, Dec 7–10, 1992; pp 853–856.

3. Kumar, A. Fertigation Through Drip Irrigation. In *Micro Irrigation Seminar*; Singh, H. P., Kaushish, S. P., Kumar, A., Murthy, T. S., Jose, C. S., Eds.; Central Board of Irrigation and Power, GOI: New Delhi, 2001; pp 349–356.
4. Mark, W.; Rosegrant, C. X.; Sarah, A. *Global Water Outlook to 2015*; Food Policy Report on Averting on Impending Crises, International Water Management Institute, Colombo, Sri Lanka, 2002; p 3.
5. Patel, N.; Rajput, T. B. S. *Effect of Fertigation on Growth and Yield of Onion*; CBIP Publication 282, Indian Council of Agricultural Research, New Delhi, 2001; p. 451.
6. Wallace, R. W.; French-Monar, R. D.; Porter, P. *Growing Tomatoes Successfully on the Texas High Plains*; Texas AgriLife Extension Bulletin; Texas Cooperative Extension, Texas A&M University: Amarillo, Texas, USA, 1996.

CHAPTER 16

PERFORMANCE OF GARLIC UNDER BEST MANAGEMENT PRACTICES

CHANDRIKA PATTI[1,*], MALLIKAJUNA S. AYYANAGOWDAR[1], UMAPATI SATISHKUMAR[1], B. S. POLISGOWDAR[1], and ASHOK HUGAR[2]

[1]*Department of Soil and Water Engineering, College of Agricultural Engineering, University of Agricultural Sciences (UAS), Lingsugur Road, Raichur 584104, Karnataka, India*

[2]*Department of Horticulture, College of Agriculture, University of Agricultural Sciences (UAS), Lingsugur Road, Raichur 584104, Karnataka, India*

Corresponding author. E-mail: chandrikapatti@gmail.com

CONTENTS

ABSTRACT

This chapter discusses the performance of microirrigated garlic (*Allium sativum* L.). Results in this study indicate that microirrigation system significantly improved growth, quality, yield, and WUE of garlic under Raichur conditions.

16.1 INTRODUCTION

Garlic (*Allium sativum* L.) is a strongly aromatic bulb crop that has been cultivated for thousands of years. It is renowned throughout the world for its distinctive flavor as well as its health-giving properties. Garlic has higher nutritive value than other bulb crops. It is rich in proteins, phosphorous, potassium, calcium, magnesium, and carbohydrates. Ascorbic acid content is very high in green garlic. Though India is the second largest producer of garlic in the world, yet the average marketable bulb yield is low compared to other garlic-producing countries. Among the various reasons, shortage of irrigation water at critical growth stages is an important factor in reducing the yield. In garlic, flood irrigation is widely practiced in India, which results in inefficient use of irrigation water due to losses in deep percolation, distribution, and evaporation.

Microirrigation is the frequent application of water directly on or below the soil surface near the root zone of plants.[3] It delivers required and measured quantity of water in relatively small amounts slowly to the individual or groups of plants. Water is applied as continuous drops, tiny streams, or fine spray through emitters placed along a low-pressure delivery system. Such a system provides water precisely to plant root zone and maintains ideal moisture condition for plant growth. Drip irrigation lends itself readily to establish a nearly constant water regime in the root zone and the fluctuation of the soil water potential (SWP) can be held to a minimum without difficulties, which ensure plants growing under proper soil water for the higher yield. Drip irrigation is the most effective way to convey direct water and nutrients to plants, and not only save water but also increases yields of vegetable crops.[7]

This chapter discusses the performance of microirrigated garlic (*Allium sativum* L.).

16.2 MATERIALS AND METHODS

16.2.1 FIELD LOCATION

A field experiment was conducted during October 2014 to February 2015 at Herbal Garden of College of Agriculture, UAS Raichur of Karnataka State in India. The site is located at 16°15′ N latitude and 77°20′ E longitude at an elevation of 389 m above mean sea level (MSL). The climate is semiarid and average annual rainfall is 612 mm. The soil was sandy clay loam in texture with pH of 7.32.

16.2.2 EXPERIMENTAL SETUP AND SOWING

The field experiment was laid out in a randomized block design with nine treatments and three replications. The treatments comprised drip and microsprinkler irrigation at 60, 80, 100, and 120% evapotranspiration (ET) along with sunken bed irrigation as a control. The cloves were dibbled at 10 cm × 15 cm spacing on beds of 2.4 m width with 10 m length maintaining 20 cm height. Four laterals of 12 mm diameter were used for each bed with a spacing of inline dripper at 60 cm distance having a discharge of 2.9 L per hour (L/h) and microsprinkler with 16 L/h discharge was placed at 2.4 m distance.

16.2.3 DETERMINATION OF CROP WATER REQUIREMENT AND IRRIGATION DURATION

16.2.3.1 DEPTH OF IRRIGATION IN SURFACE IRRIGATION METHOD

Irrigation was provided daily after calculating water requirement based on past 24 h of pan evaporation while in sunken bed irrigation it was scheduled once in 5 days. The amount of water to be delivered in surface method was computed using the following equation

$$d = \frac{M_{fc} - M_{bi}}{100} \times A_s \times d_s \tag{16.1}$$

where d = net amount of water to be applied during irrigation, cm; M_{fc} = moisture content at field capacity, %; M_{bi} = moisture content before irrigation, %; A_s = bulk density of soil, g cc^{-1}; and d_s = effective root zone depth, cm

   In the calculations, the root zone depth was assumed as 15 cm for the first 30 days after sowing (DAS) and for the rest of the period to maturity as 20 and 40 cm. The bed area was 10 m^2 while bulk density collected at 15 and 30 cm averaged depths was 1.4 g/cm^3. The quantity of water (L) required per plant is given below:

$$Q = \frac{d \times A \times B}{1000} \qquad (16.2)$$

where Q = quantity of water required per plant, liters; d = net amount of water to be applied during an irrigation, cm; A = gross area per plant, cm^2; and B = extent of the area covered by foliage, fraction.

16.2.3.2 CROP WATER REQUIREMENT: DRIP AND MICROSPRINKLER IRRIGATION METHODS

The daily water requirements for drip and microsprinkler irrigation were computed using pan evaporation data from USDA Class-A open pan evaporimeter. The water requirement of garlic crop per day under drip and microsprinkler irrigation was computed using the following equation:

$$Q = [A \times B \times C]/[E] \qquad (16.3)$$

where, Q = quantity of water required mm/day; A = daily ET, mm day^{-1} (pan evaporation × pan coefficient); B = amount of area covered with foliage (canopy factor), fraction; C = crop coefficient, fraction; and E = efficiency of irrigation system, %.

16.2.3.3 IRRIGATION SCHEDULING UNDER DRIP AND MICROSPRINKLER

For drip and microsprinkler, the daily water requirement was calculated and water was applied based on the duration of irrigation. The quantity of water to be applied was computed every day as explained above. For the known discharge rate of emitters, the duration of irrigation water application for drip irrigation was calculated using eq 16.4. For the known discharge rate of microsprinkler, the duration of irrigation water application was calculated using eq 16.5.

$$\text{Duration of Irrigation} = \frac{\text{dripper discharge}}{\text{dripper spacing} \times \text{lateral spacing}} \qquad (16.4)$$

$$\text{Duration of Irrigation} = \frac{\text{microsprinkler discharge}}{\text{microsprinkler spacing} \times \text{lateral spacing}} \qquad (16.5)$$

16.2.4   BIOMETRIC PARAMETERS

For periodical field observations, five plants were selected randomly from each treatment and were tagged. Observations on biometric parameters [such as plant height, number of leaves per plant, leaf area index (LAI), the girth of the neck, bulb diameter, average bulb weight, numbers of cloves per bulb, and total soluble solids (TSS)] were taken from selected five plants.

16.2.5   FIELD WATER USE EFFICIENCY

The water use efficiency (WUE) for each treatment was computed using the following formula for surface, drip, and microsprinkler irrigation methods:

$$e_u = \frac{Y}{\text{WR}} \qquad (16.6)$$

where e_u = water use efficiency, kg m^{-3}; Y = crop yield, kg; and WR = total amount of water used in the field, m^3.

16.3   RESULTS AND DISCUSSION

Growth, yield, and yield contributing characters of garlic were influenced significantly by different methods and levels of irrigation (Table 16.1). At 90 DAS, drip irrigation at 100% ET recorded the highest plant height

TABLE 16.1 Effect of Irrigation Methods and Irrigation Levels on Growth Parameters at Different Intervals.

Treatment	Plant height				Number of leaves				Girth of plant (cm)			
	30 DAS	60 DAS	90 DAS	120 DAS	30 DAS	60 DAS	90 DAS	120 DAS	30 DAS	60 DAS	90 DAS	120 DAS
T_1	33.05	59.85	61.49	41.93	5.17	7.93	8.83	6.21	2.15	3.40	3.60	3.07
T_2	35.57	60.95	65.20	47.61	4.77	8.77	11.30	7.27	2.29	3.85	4.47	3.48
T_3	43.15	67.37	70.10	50.83	5.77	10.6	13.47	8.27	2.68	4.47	4.69	3.86
T_4	40.70	64.63	67.90	48.30	4.53	9.07	11.97	7.73	2.53	4.27	4.17	3.77
T_5	31.70	54.68	53.87	37.33	4.30	7.83	8.53	6.00	2.13	3.37	3.56	3.05
T_6	38.27	60.63	61.35	39.33	4.77	8.47	9.27	7.03	2.29	4.31	3.83	3.50
T_7	38.40	62.23	62.13	48.17	5.20	9.00	11.07	8.10	2.41	4.37	4.17	3.68
T_8	41.39	66.15	68.07	50.33	5.37	9.47	12.87	8.30	2.59	4.46	4.63	3.81
T_9	33.27	59.93	60.27	40.89	4.47	8.40	11.30	7.07	2.22	3.63	3.83	3.23
Mean	37.28	61.83	63.37	44.97	44.33	79.53	98.6	65.98	21.29	36.12	36.96	31.46
SEM±	1.03	1.51	1.99	0.39	0.19	0.11	0.19	0.04	0.02	0.12	0.02	0.39
CD (0.05)	3.09	5.45	5.97	1.16	0.57	0.34	0.57	0.13	0.07	0.35	0.07	1.16

(70.1 cm) followed by 120% ET under microsprinkler irrigation (68.07 cm). Plant height is an important yield attribute in garlic and any practice to alter the plant height would influence the bulb yield as reported by Vincent[8] in onion. In the present experiment, the increased plant height in microirrigated plots might be due to better availability of moisture during entire crop growth period which favored the growth attributes.

The number of leaves at 90 DAS was highest in 100% ET under drip irrigation (13.47) followed by 120% ET under microsprinkler irrigation (12.87). The crop would produce a sufficient number of leaves to harness light energy and synthesize adequate photoassimilates for biomass production. The increased growth attributes might be due to adequate availability and supply of water and nutrients in proportion, which ultimately resulted in triggering the production of plant growth hormones namely indole acetic acid (IAA), which helped in maintaining a higher number of leaves throughout the cropping period. In case of the girth of the plant (neck thickness) and LAI, both systems of irrigation (drip irrigation at 100% ET and sprinkler irrigation at 120% ET) recorded the highest girth of plant and LAI (Table 16.2).The present results obtained are in agreement with the findings of Sankar et al.[6]

TABLE 16.2   Effect of Irrigation Methods and Irrigation Levels on Leaf Area Index of Garlic at Different Intervals.

Treatment	LAI			
	30 DAS	60 DAS	90 DAS	120 DAS
T_1	0.376	0.724	1.503	0.852
T_2	0.464	1.061	1.799	1.108
T_3	0.592	1.709	2.184	1.495
T_4	0.517	1.217	1.915	1.216
T_5	0.33	0.685	1.436	0.832
T_6	0.469	0.76	1.657	0.962
T_7	0.516	1.453	1.938	1.303
T_8	0.536	1.573	2.154	1.421
T_9	0.362	0.694	1.59	0.882
Mean	0.462	1.097	1.798	1.119
SEM±	0.037	0.046	0.096	0.037
CD (0.01)	0.149	0.186	0.392	0.152

The increase in plant height, number of leaves, and LAI in drip and microsprinkler irrigation over sunken bed irrigation may be due to frequent application of irrigation water at lower rates, resulting in even distribution of soil moisture in the root zone of the crop; and due to this reason, soil moisture was maintained fairly close to the field capacity throughout the crop season in drip and microsprinkler irrigated plots, which resulted in high level of plant water use, while in case of surface irrigation there existed a soil moisture stress due to scheduling of irrigation once in 5 days. This shows that adequate supply of soil moisture to garlic plant resulted in the development of the required efficient photosynthetic system apart from increased available nutrient in the soil due to mineralization and transformation of nutrients in soil under drip and microsprinkler irrigation.

The garlic crop performed well in terms of yield and yield contributing factors under drip, microsprinkler irrigation as compared to sunken bed irrigation. The better performance of the plant in terms of bulb diameter, average bulb weight, and number of cloves (Table 16.3) may be attributed to the frequent and consistent application of water in the vicinity of the roots, which provides a good soil moisture regime in the crop root zone throughout the life period of the crop. The number of cloves was highest in drip irrigation 100% ET (44.03) followed by 120% ET under

TABLE 16.3 Effect of Irrigation Methods and Irrigation Levels on Number of Cloves, Diameter, Weight of Bulb, TSS, and Pungency.

Treatment	Number of cloves	Diameter of bulb (mm)	Weight of bulb (g)	TSS (°Brix)	Pungency (µmol mL⁻¹)
T_1	27.73	34.7	15.04	42.07	30.63
T_2	30.03	36.24	16.77	41.43	29.76
T_3	44.03	43.27	27.13	40.77	24.16
T_4	31.38	38.81	21.68	40.1	23.87
T_5	24.11	32.29	13.31	42.07	29.73
T_6	24.76	34.95	17.28	41.63	25.88
T_7	31.33	39.9	21.33	41.37	25.1
T_8	37.53	41.99	22.42	40	23.01
T_9	27.93	35.27	16.71	42.83	32.61
Mean	30.98	37.49	19.08	41.36	27.19
SEM±	0.88	0.456	0.37	0.093	0.18
CD (0.01)	3.57	1.857	1.48	0.378	0.52

microsprinkler (37.53). Similarly, in case of bulb weight and diameter, 100% ET under drip and 120% ET under microsprinkler recorded highest values.

The highest marketable bulb yield (7230 kg ha^{-1}) was obtained in drip irrigation at 100% ET followed by microsprinkler irrigation at 120% ET (7081 kg ha^{-1}) (Table 16.4). This confirms the earlier findings of Patel et al.,[5] who also recorded higher marketable bulb yield of garlic under drip irrigation system. The increased yield in drip irrigation system was mostly due to the favorable effect of available soil moisture, uniform distribution of irrigation water during the entire growth period. Another possible reason was the continuous availability of moisture which enhanced the availability and uptake of nutrients throughout the cropping period which resulted in better growth and bulb development.

TABLE 16.4   Effect of Irrigation Methods and Irrigation Levels on Yield and Water Use Efficiency of Garlic.

Treatment	Yield		Total water applied (cm)	Water saving over surface (%)	WUE (kg ha^{-1}cm^{-1})
	kg plot^{-1}	q ha^{-1}			
T$_1$	13.032	54.3	16.01	70.03	339.26
T$_2$	15.03	62.63	21.34	60.04	293.46
T$_3$	17.352	72.3	26.68	50.04	271.04
T$_4$	16.782	69.92	32.01	40.06	218.44
T$_5$	12.36	51.5	18.01	66.28	286.02
T$_6$	14.335	59.73	24.01	55.04	248.79
T$_7$	15.66	65.25	30.01	43.8	217.43
T$_8$	16.993	70.81	36.01	32.56	196.61
T$_9$	13.515	56.31	53.4	–	105.45
Mean	15.006	62.53			
SEM±	0.08	0.35			
CD (0.05)	0.25	1.04			

Any crop production system will be appreciated when it is not only quantitatively superior but also qualitatively found promising. Certain parameters like bulb size were readily influenced by moisture deficit, whereas internal qualities such as TSS and pungency values showed an increasing trend with decreasing irrigation levels (Table 16.3). TSS (°Brix) under control treatment was recorded highest (42.83) followed by 60% ET under drip and microsprinkler (42.07), and lowest was found in 120% ET under drip irrigation (T_4) and microsprinkler (T_8) which was on par with each other (Table 16.3). Pungency under control treatment was recorded highest (32.61) followed by 60% ET under drip (30.63) and microsprinkler (29.73). In order to support the potential gradient required for water absorption in soils under water stress, the plant decreases the osmotic potential by increasing the levels of organic solutes.[1] Higher temperature favors the accumulation of sulfur, pyruvate production, and increased pungency. Similarly, pungency increases in garlic that suffer growing stress. The plants get stressed in all the stages of the crop, but the effect of moisture stress is predominant in January. The increasing TSS and pungency reflected qualitatively in the yield. These results are in agreement with the earlier findings of Daniel and Shinsuke.[2]

The quantity of water applied for garlic crop was worked out as shown in Table 16.4. The minimum water was applied to drip irrigation at 60% ET (16.01 cm) followed by microsprinkler irrigation at 60% ET (18.01 cm), and the maximum water was applied to the sunken bed irrigation (53.40 cm). It is clear from the data that water saving was 70% in drip and 66.28% in microirrigation system at 60% ET. Among the drip and microsprinkler irrigation levels, higher WUE was found in 60% ET under drip treatment (329.03 kg ha cm^{-1}) and 80% ET under drip (284.61 kg ha cm^{-1}) indicating more efficient use of irrigation water, closely followed by 60% ET under microsprinkler level (277.39 kg ha cm^{-1}) and 80% ET under microsprinkler (241.28 kg ha cm^{-1}). The higher WUE at 60% and 80% ET level was mainly due to higher yield and maximum saving in irrigation water. The lowest WUE in sunken bed irrigation (105.45 kg ha cm^{-1}) might be the result of higher irrigation water use with comparatively less yield. The above discussion suggests that higher garlic yields could be achieved by adopting drip and microsprinkler irrigation scheduled at 60% and 80% ET levels. The results fall in line with of the findings of Kumar et al.[4] and Sankar et al.[6]

16.4 SUMMARY

Results in this study indicate that microirrigation system significantly improved growth, quality, yield, and WUE of garlic under Raichur conditions. Among the various methods of irrigation, drip irrigation at 100% ET was superior in terms of improved growth, quality, yield, and WUE followed by 120% ET under microsprinkler irrigation than surface irrigation method.

KEYWORDS

- cabbage
- crop coefficient
- deficit irrigation
- drip irrigation
- dripper
- ET
- garlic
- microirrigation
- water use efficiency
- crop water requirement

REFERENCES

1. Azcon-Bieto, J.; Talon, M. *Fundamental of Vegetable Physiology* (Fundamentos de fisiologia vegetal); McGraw-Hill Interamericana: Madrid, España, 2000; p 522.
2. Daniel, L.; Shinsuke, A. Crop Coefficient-based Deficit Irrigation and Planting Density for Onion: Growth, Yield, and Bulb Quality. *Hortscience* **2012,** *47*(1), 31–37.
3. Goyal, M. R. Performance Evaluation of Micro Irrigation Management Principles and Practices. In *Innovations and Challenges in Micro Irrigation*; Apple Academic Press Inc.: New Jersey, 2016; Vol. 3; p 342 (Book series).
4. Kumar, S.; Imtiyaz, K. A.; Singh, R. Response of Onion (*Allium cepa* L.) to Different Levels of Irrigation Water. *Agric. Water Manag.* **2007,** *89*, 161–166.
5. Patel, B. G.; Khanpara, V. D.; Malavia, D. D.; Kaneria, B. B. Performance of Drip and Surface of Irrigation for Garlic (*Allium sativum* L.) Under Varying Nitrogen Level. *Indian J. Agron.* **1993,** *41*(1), 174–176.
6. Sankar, V.; Lawande, K. E.; Tripathi, P. C. Effect of Micro Irrigation Practices on Growth and Yield of Garlic (*Allium sativum* L.). *J. Spices Aromat. Crops* **2008,** *17*(3), 230–234.

7. Tiwari, K. N.; Singh, A.; Mal, P. K. Effect of Drip Irrigation on Yield of Cabbage (*Brassica oleracea L. var. capitata*) Under Mulch and Non-mulch Conditions. *Agric. Water Manag.* **2003,** *58,* 19–28.

8. Vincent, S. R. Studies on the Effect of Different Levels of N, P and K on Onion. M.Sc. (Agri.) Thesis, Tamil Nadu Agricultural University, Coimbatore, 1980.

CHAPTER 17

MOISTURE DISTRIBUTION PATTERN UNDER BEST MANAGEMENT PRACTICES

BASAMMA K. ALADAKATTI[1,*], K. SHANMUGASUNDARAM[2],
M. B. VINUTA[3], and JAGADEESHA MULAGUND[4]

[1]*Sujala-III Project, Department of Soil Science and Agricultural Chemistry University of Agricultural Sciences, Dharwad- 580005, India*

[2]*Department of Soil and Water Conservation Engineering, Agricultural Engineering College and Research Institute, Tamil Nadu Agricultural University, Kumulur 621712, Tamil Nadu, India*

[3]*Department of Soil and Water Conservation Engineering, Agricultural Engineering College and Research Institute, Tamil Nadu Agricultural University, Coimbatore 641003, Tamil Nadu, India*

[4]*Indian Institute of Horticultural Research, Bengaluru, India (PG outreach Program by IARI, New Delhi)*

Corresponding author. E-mail: basammaka@gmail.com

CONTENTS

This chapter is a modified version and partially taken from "Basamma K. A. Effect of Drip Fertigation and Mulching on Tomato (Solanum lycopersicum L.)—Hybrid Deepthi. M. Tech. Thesis, Department of Soil and Water Conservation Engineering, Agricultural Engineering College & Research Institute, Tamil Nadu Agricultural University, Coimbatore 641003, India, 2014."

ABSTRACT

A field experiment was conducted from December 2013 to April 2014 in TNAU Coimbatore to find out the effect of mulching on soil physical properties and soil moisture distribution pattern under drip irrigation. Soil physical properties such as bulk density, particle density, and porosity were determined. Bulk density decreased in mulched treatments compared to control. Initially, the bulk density of the soil was 1.37 g cc^{-1}. After the crop harvest, the bulk density of the mulched plots decreased to a range of 1.29–1.32 g cc^{-1}, whereas in control plots, there was not that much change in the bulk density of the soil compared to the mulching treatments. Porosity of the presowing and postharvest soil samples were measured. Porosity was less before transplanting (42.31%) but increased to greater extent in mulched treatments in postharvest observations. But the amount of increase was less in the control. More moisture was distributed in deeper soil layer below the emitter compared to top layer, and moisture distribution decreased with increase in radial distance from the emitter. It was also seen that at the beginning, water saturated the soil near the emitter and infiltration was slower, but at later stages, water penetrated to deeper layer without any loss as in the case of surface irrigation.

17.1 INTRODUCTION

Increasing the water supply in India is questionable. Policy to achieve water security and food security is to increase the water use efficiency (WUE) and water productivity, producing more with less water. In all water sectorial uses, particularly the agriculture sector receives nearly 85% of the available water resources, but with poor on farm water efficiency not exceeding 50%. Major efforts are directed towards the agriculture through increasing crop water productivity, reducing water losses, and raising the WUE. With the reduction in ground water levels, reduction in arable land, the increment in urbanization, and soil erosion due to deforestation, Tamil Nadu state may face an acute food crisis in the near future. Under such circumstances in the near future, it will become impossible to feed the entire growing population using conventional systems of agricultural production.[2] Technically, several approaches are now implemented for better water saving and uniform distribution in the irrigated agriculture; among them are the introduction of the new irrigation techniques such as surface and subsurface drip irrigation, sprinkler irrigation, and pivot systems.

Irrigation techniques such as border, check basin, and furrow have lower efficiency and nonuniform moisture distribution than drip and sprinkler irrigation methods. Many studies have indicated that the WUE of best drip fertigated treatment was 94% higher than the control furrow irrigated treatment presently being used by farmers.[3] Drip irrigation is the most effective way to supply water and nutrients to the plants not only to save water but also to increase the yield of fruit and vegetable crops. Deep percolation losses are also higher in conventional irrigation methods compared to microirrigation methods. Mulching reduced the bulk density of soil because mulches stop the impact of a rain drop and splash thereby prevents soil compaction, reduces surface runoff, and increases infiltration. The increased porosity and decreased compaction due to decreased soil bulk density in plastic film mulched plots may have enhanced aeration and microbial activities in the soil thus resulting in increased root penetration and cumulative feeding area leading to increased plant growth and yield.

This chapter focuses on moisture distribution patterns under drip irrigation and mulching.

17.2   MATERIALS AND METHODS

17.2.1   LOCATION OF STUDY AREA

The study area was selected in the farmer's field situated at Thithipalayam, which is 14 km from TNAU, Coimbatore. The field is located at $10°57'5.8''$ N to $76°52'29.6''$ E latitude, with a mean altitude of 465 m above the mean sea level. The topography of the experimental plot was uniform and leveled.

17.2.2   WEATHER AND CLIMATE

The mean annual rainfall of the study area is 612 mm. About 55% of annual rainfall is received during northeast monsoon season and 30% during the southwest monsoon. The annual maximum mean and annual minimum mean temperatures were 32.5°C and 20.1°C, respectively. The average relative humidity of the area is 56.8% and sunshine hours range from 3 to 10 h. The mean evaporation ranges from 3.5 to 7.6 mm per day.

17.2.3 SOIL PHYSICAL AND CHEMICAL PROPERTIES

The experimental field has soils with sandy clay loam texture with a pH of 7.7 and a good electrical conductivity of 0.6 dSm^{-1}. Table 17.1 shows physical and chemical properties of the initial soil.

TABLE 17.1 Initial Physical and Chemical Properties of Soil.

Soil properties	Property	Value
Physical characters	Bulk density, g cc^{-1}	1.37
	Particle density, g cc^{-1}	2.36
	Porosity, (%)	42.31
Chemical properties	Available N, kg ha^{-1}	180.3
	Available P, kg ha^{-1}	132.8
	Available K, kg ha^{-1}	360
	pH	7.7
	EC, dSm^{-1}	0.6

17.2.4 IRRIGATION SYSTEM

Irrigation water source was a nearby tube well from which water was pumped using a 10 HP vertical three stage submersible pump and conveyed through screen filters to the polyvinyl chloride (PVC) main line pipes of 75 mm diameter. PVC submain of 63 mm diameter was connected to the main line to which low-density polyethylene (LDPE) laterals of 16 mm diameter were connected. Each lateral was provided with individual taps for controlling irrigation. Along the laterals, inline drippers of 4 lph were installed at a spacing of 60 cm. Submains and laterals were plugged at the end with an end cap.

17.2.5 EXPERIMENTAL DESIGN AND TREATMENT DETAILS

The experiment was designed under *factorial randomized block design* (FRBD) with mulching thickness and fertilizer levels. Each treatment combination was replicated thrice. Two types of plastic mulching films of different thickness and one control without mulch were selected for the study (M_1—black plastic mulch of 25 µm thickness, M_2—black plastic mulch of 50 µm thickness, and M_3). The treatment details are given below:

T₁ Black plastic mulch of 25 µm thickness with 80% RDF.
T₂ Black plastic mulch of 25 µm thickness with 100% RDF.
T₃ Black plastic mulch of 25 µm thickness with 120% RDF.
T₄ Black plastic mulch of 50 µm thickness with 80% RDF.
T₅ Black plastic mulch of 50 µm thickness with 100% RDF.
T₆ Black plastic mulch of 50 µm thickness with 120% RDF.
T₇ No mulch with 80% RDF.
T₈ No mulch with 100% RDF.
T₉ No mulch with 120% RDF.

17.2.6   SOIL PHYSICAL PROPERTIES

The soil from each treatment before and after harvest was analyzed by standard procedures for physical characters such as bulk density, particle density, and porosity and the values were calculated by using eqs 17.1–17.3. The bulk density of soil quantifies soil compactness. Undisturbed soil core samples were collected from each plot before planting and after harvest. The core samples were used to determine the bulk density using core method.[6]

$$\text{Bulk density (BD)} = [\text{Dry weight of soil}]/ [\text{Soil volume including pore space}] \quad (17.1)$$

$$\text{Porosity} = [(1 - \text{BD/PD}) \times 100 \quad (17.2)$$

$$\text{Particle density (PD)} = \text{Weight of particle}/ [\text{Final volume} - \text{Initial volume}]) \quad (17.3)$$

17.2.7   SOIL MOISTURE DISTRIBUTION PATTERN

The wetting pattern of soil under different mulches was analyzed by taking moisture content at different horizontal distances and depths. In order to study the soil moisture distribution in soil, samples were collected at a distance of 0, 15, 30, and 45 cm from emitter along the horizontal direction and at the surface and at a depth of 10, 20, and 30 cm. The samples were collected immediately after irrigation, after 1 day, after 2 days of irrigation, and just before the next irrigation. Using gravimetric method, the soil moisture was calculated. Soil samples were taken using tube type soil augers and were kept in moisture boxes and covered immediately with lids. The samples were weighed along with the moisture box (W_2) and then placed in an oven at 105°C for 24 h until all moisture was driven off. It was weighed again and the weight (W_3) was noted. The soil moisture content was expressed as a percentage by weight on dry basis. Soil moisture contour maps were plotted by using the computer software package "Surfer" of the windows version.

The percentage of moisture content = $[(W_2 - W_1)/(W_3 - W_1)] \times 100$ (17.4)

where W_1 = weight of the empty container with lid (g); W_2 = weight of the container with lid and moist soil (g); and W_3 = weight of the container with lid and dry soil (g).

17.2.8 DIAMETER OF THE WETTING FRONT

Field observations were done to measure the horizontal and vertical movement of the wetting front over and below the surface of the field. The diameter of the wetting front was measured over different periods of time during emission and soil was cut vertically along the diagonal downward to record the vertical movement of wetting front. The vertical wetting front depth was measured exactly below the dripper position. The rate of horizontal wetting front advancement and vertical wetted zone depth from the emitter at different time interval were measured and the wetting front advance equation was developed.

17.3 RESULTS AND DISCUSSION

17.3.1 SOIL PHYSICAL PARAMETERS

Bulk density was decreased in mulched treatments compared to control. Initially, the bulk density of the soil was 1.37 g cc^{-1}. After the crop harvest, the bulk density of the mulched plots decreased to a range of

1.29–1.32 g cc^{-1}, whereas in control plots, there was not that much change in the bulk density of the soil compared to the mulching treatments (Table 17.2). Mulching reduced the bulk density of soil because mulches stop the impact of raindrops and splash thereby preventing soil compaction, reducing surface runoff, and increasing infiltration. The increased porosity and decreased compaction due to decreased soil bulk density in plastic film mulched plots may have enhanced aeration and microbial activities in the soil thus resulting in increased root penetration and cumulative feeding area leading to increased plant growth and yield. These results are in line with those by Mbah et al.[4]

TABLE 17.2 Soil Physical Parameters Before Transplanting and After Last Harvest of Tomato Crop.

Treatments	Bulk density (g cc^{-1})		Particle density (g cc^{-1})		Porosity (%)	
	Initial	After harvest	Initial	After harvest	Initial	After harvest
T$_1$	1.37	1.32	2.36	2.40	42.31	45.11
T$_2$	1.37	1.30	2.36	2.35	42.31	44.21
T$_3$	1.37	1.31	2.36	2.43	42.31	47.31
T$_4$	1.37	1.29	2.36	2.31	42.31	44.11
T$_5$	1.37	1.32	2.36	2.39	42.31	44.50
T$_6$	1.37	1.30	2.36	2.34	42.31	44.32
T$_7$	1.37	1.35	2.36	2.38	42.31	43.00
T$_8$	1.37	1.36	2.36	2.41	42.31	43.20
T$_9$	1.37	1.35	2.36	2.42	42.31	44.90

Porosity was less before transplanting (42.31%) but was increased to a greater extent in mulched treatments in postharvest observations. But the amount of increase was less in the control. Plastic film mulches increased the total porosity of the soil relative to the control, thus conforming to the findings of Jayaseeli,[1] who found that total porosity is directly related to aeration porosity whereas water holding capacity is inversely related. The amount of pore space was significantly higher in mulched treatments than nonmulched treatments.

17.3.2 SOIL MOISTURE DISTRIBUTION PATTERNS

Drip irrigation system is designed to apply precise amount of water near the plant with a certain degree of uniformity. The uniformity describes how evenly an irrigation system distributes water over a field. It is regarded as one of the important features for selection, design, and management of the irrigation system as uniform moisture distribution positively affect the crop yield. The soil moisture content at 0–10, 10–20, and 20–30 cm depths from the surface at different distances from the emitter were estimated just before irrigation, 2 h after irrigation, 1 day after irrigation, and 2 days after irrigation. In drip irrigated plot, it was observed that the moisture content was reduced as the distance from emitter was increased horizontally. The highest moisture content of 28.8% was observed below the emitter at 10–30 cm depth just before irrigation. The mean maximum soil moisture content (39%) was observed below the emitter at the depth of 0–20 cm immediately after 2 h of irrigation. It was also noted that recorded value moisture content was increased in all depths after irrigation.

17.3.3 SOIL MOISTURE CONTOUR MAPS

The soil moisture contents estimated at different depths and distances from emitter were plotted by using the computer software golden package "surfer" of windows version and are shown in Figures 17.1–17.4. The soil moisture contour maps show the moisture available at different depths in the vertical and horizontal direction in the experimental field, before and after irrigation. Moisture content lines were drawn in terms of percentage moisture content. It was observed that moisture lines are close to one another at 15–30 cm depth vertically and 15–30 cm away from the emitter horizontally. It indicates that more moisture was distributed in deeper soil layers below the emitter compared to the top layer and moisture distribution was decreased with increase in radial distance from the emitter. It was also seen that at the beginning, water saturated the soil near the emitter and infiltration was slower, but at later stages, water penetrated to deeper layer without any loss as in the case of surface irrigation.

Figures 17.2 and 17.3 show moisture distribution in drip irrigated plots before irrigation and immediately after irrigation. It was clear that moisture distribution followed the same trend as described above. It was also observed that soil moisture (28.8%) was less before irrigation compared

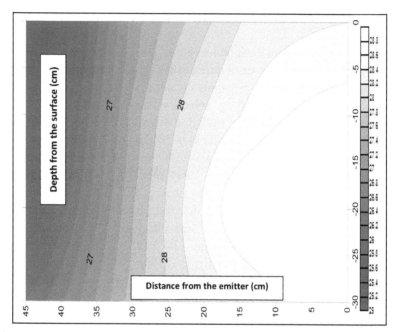

FIGURE 17.1 Soil moisture change after last harvest.

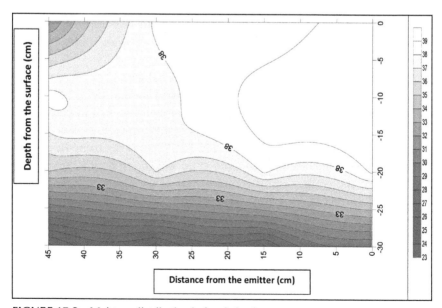

FIGURE 17.2 Moisture distribution before irrigation.

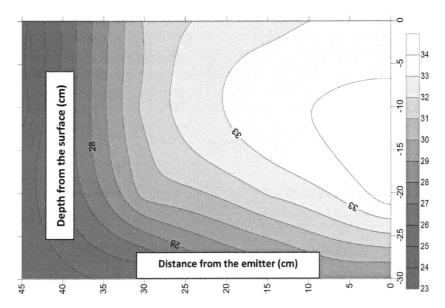

FIGURE 17.3   Moisture distribution 2 h after irrigation.

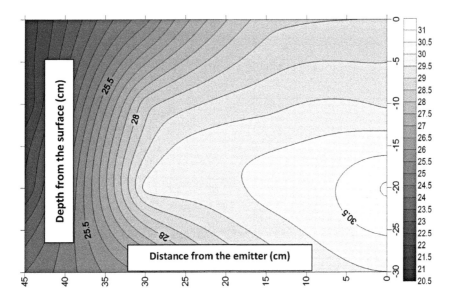

FIGURE 17.4   Moisture distribution 2 days after irrigation.

to after irrigation. After 2 h of irrigation, moisture content of 39% was observed from surface to 20 cm depth. In drip irrigation treatment, it was clear that more moisture was found below the emitters at greater depths (20–30 cm). The reason for higher moisture content in the lower horizons might be due to water stored in soil pores with minimum evaporation loss. Soil moisture content was lesser in the surface layer than in depths at different locations from the emitter. This might be due to more evaporation from the soil surface compared to lower layers.

17.3.4   DETERMINATION OF HORIZONTAL WETTING FRONT AND VERTICAL WETTING FRONT

The horizontal and vertical wetting front widths at different intervals were determined, and wetting front advancement equation was developed (Figs. 17.5 and 17.6).

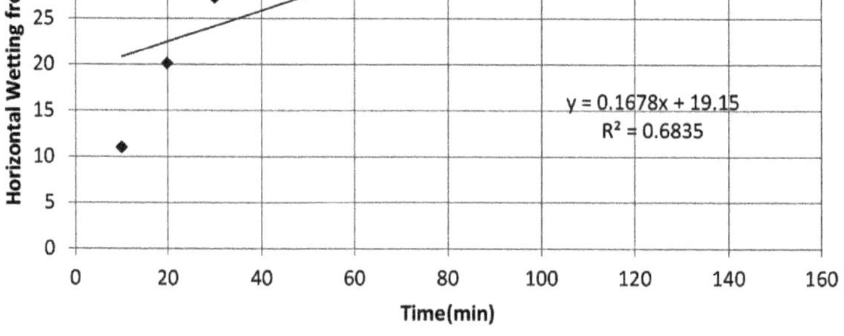

FIGURE 17.5   Horizontal wetting front advance.

The width of horizontal wetting front was increased from 11 (10 min) to 37 cm (135 min). A similar trend was noticed in case of vertical wetting front depth advancement. It was initially 5.92 cm (10 min) and reached a depth of 19.93 cm (60 min). The rate of advance in the wetted zone was found decreasing from 1.1 to 0.27 cm min^{-1} horizontally and 0.59 to 0.14 cm min^{-1} vertically. After 135 min, there was no further advancement in diameter of wetting front, it was stopped.

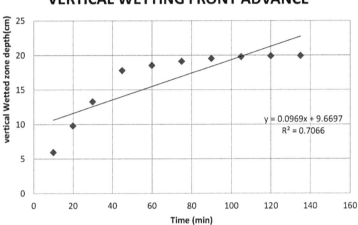

FIGURE 17.6   Vertical wetting front advance.

Table 17.3 indicates that water moved at a greater distance from emitter laterally than vertical under observation. This is probably associated with gravity and deep percolation. This behavior of horizontal wetted diameter

TABLE 17.3   Advancement of Horizontal Wetting Front and Vertical Wetted Zone Depth.

Elapsed time, (min)	Horizontal wetted zone radius, (cm)	Rate of advance in horizontal wetted zone, (cm min^{-1})	Vertical wetted zone depth, (cm)	Rate of advance in vertical wetted zone, (cm min^{-1})
10	11.0	1.10	5.92	0.59
20	20.1	1.00	9.79	0.48
30	27.3	0.91	13.27	0.44
45	32.8	0.78	17.78	0.39
60	34.2	0.57	18.53	0.30
75	35.3	0.47	19.13	0.25
90	36.1	0.40	19.53	0.21
105	36.6	0.34	19.78	0.18
120	36.9	0.30	19.89	0.16
135	37.0	0.27	19.93	0.14

versus time and vertical moment of water versus time confirmed the findings of Shrivastava et al.[5] who observed both the vertical and lateral spread of water in the soil profile in different irrigation levels, and found that in all irrigation levels lateral moment was more than the vertical moment; and as the irrigation level increased, magnitude of lateral spread was found to be more than the vertical spread.

17.4 SUMMARY

Plastic film mulched plots may have enhanced aeration and microbial activities in the soil thus resulting in increased root penetration and cumulative feeding area leading to increased plant growth and yield. It lowers fertilizer costs by holding surface runoff and deep percolation (leaching) to a minimum. It increases net returns by increasing crop yields and crop quality. Uniform moisture distribution reduces the water stress, in turn, increases the yield. Mulching along with drip irrigation is more economical and increases the yield and quality.

KEYWORDS

- bulk density
- drip irrigation
- horizontal wetting front
- leaching
- microbial activities
- porosity
- vertical wetting front
- water stress

REFERENCES

1. Jayaseeli, D.; Raj, S. P. Physical Characteristics as a Function of Its Particle Size to be Used as Soilless Medium. *Am. Eurasian J. Agric. Sci.* **2010,** *8*(4), 431–437.

2. Joseph, A.; Muthuchamy, I. Productivity, Quality and Economics of Tomato (*Lycopersicon esculentum Mill.*) Cultivation in Aggregate Hydroponics—A Case Study from Coimbatore Region of Tamil Nadu. *Indian J. Sci. Technol.* **2014,** *7*(8), 1078–1086.

3. Kumari, R.; Kaushal, A.; Singh, K. G. Water Use Efficiency of Drip Fertigated Sweet Pepper Under the Influence of Different Kinds and Levels of Fertilizers. *Indian J. Sci. Technol.* **2014,** *7*(10), 1538–1543.

4. Mbah, C. N.; Mbagwu, J. S. C.; Onyia, V. N. Effects of Application of Bio-fertilizers on Soil Densification, Total Porosity, Aggregate Stability and Maize Grain Yield in a Dystricleptoso. *J. Sci. Tech.* **2004,** *10*, 74–85.

5. Shrivastava, G. R.; Nayak, S. Soil Moisture Distribution as Influenced by Drip Irrigation Supply and Planting Pattern in Heavy Soils of Madhya Pradesh. *J. Agric. Technol.* **2011,** *7*(4), 1177–1186.

6. Yahya, Z. A.; Husin, J.; Talib, J.; Othman, O. H. Machinery Compaction Effects on Physical Properties of Bernam Series in an Oil Palm Plantation. *Am. J. Appl. Sci.* **2009,** *6*(12), 2006–2009.

CHAPTER 18

EFFECTS OF DIFFERENT COLOR PLASTIC MULCHES ON GROWTH AND YIELD OF BANANA

V. D. PARADKAR*, N. N. FIRAKE, and SUNIL D. GORANTIWAR

Department of Irrigation and Drainage Engineering, Dr. Annasaheb Shinde College of Agricultural Engineering (Dr. ASCAE), Mahatma Phule Agricultural University (MPKV), Rahuri 413722, India

Corresponding author. E-mail: paradkarvd@gmail.com

CONTENTS

ABSTRACT

The research was conducted to evaluate the effect of color plastic mulches on growth, yield, and quality of banana at MPKV, Rahuri, Maharashtra, India during 2014–15. Six color mulches used were yellow, blue, silver, white, red, and pervious plastic mulches. Six mulch treatments and seventh control treatment were given irrigation with 48% of pan evaporation. Eighth control treatment was given irrigation with 100% of crop evapotranspiration and ninth treatment with surface irrigation at 1 IW/CPE ratio. The results indicated that the highest plant height, stem girth, and number of functional leaves were obtained under treatment T_3 (silver black plastic mulch). The highest yield (84.45 t/ha) was obtained in T_3 due to improved microclimate both above and below mulch. Significantly less quantity of water (1015 mm) required for plants in T_3 due to reduced total duration of crop. The highest water use efficiency (81.83 kg/ha mm) was obtained in treatment T_3 due to less water requirement and comparatively higher yield. The quality of banana was not significantly influenced by different color mulches. The highest NDVI value (0.8943) was observed in treatment T_6 (pervious plastic mulch), followed by that (0.8918) in T_3 at 300 DAT and minimum (0.8104) in T_9. The research indicated that the adoption of silver black plastic mulch with daily drip irrigation at 48% pan evaporation resulted in 31.64% increase in yield over no mulch treatment.

18.1 INTRODUCTION

Mulch is any material laid on the soil surface to provide a favorable environment for the growth of plant and conservation of moisture. Some materials may be more beneficial than others.[1] Greater marketable yield is observed with the use of plastic mulches (up to 24–65% increase) compared to bare soil, because of conservation of moisture, improved microclimate both beneath and above the soil surface, light reflection, and reductions in weed population.[2] Because of imperviousness of plastic mulches, it prevents direct evaporation of moisture from the soil and thus limits the water losses and soil erosion over the surface.

The use of black polyethylene mulch (BPM) in vegetable production has been reported to control the weed incidence, reduce nutrient losses, and improve the hydrothermal regimes of soil.[3] The silver black and black plastic mulch may reduce weeds by 95–98%.[10] Different color mulches like yellow, gray, and blue reflect different radiation patterns back into the plant canopy, and affect plant growth and development. Some colors attract certain insects, for example, mulches might be used in a field to grow "catch crops" to pull insects away from other crops.[5] Thus, different crops respond differently to a specific color of mulch. Hence, there is a need to adopt specific color mulch for a particular crop.

Banana (*Musa paradisiaca* L.) is the fourth most important food crop in the world after rice, wheat, and maize with a world production of around 80 million t in 2006. In the world, India is the largest producer of banana with an annual production of 23.205 million t from an area of 0.647 million ha. The cycle time of Grand Naine variety of banana is slightly shorter, bunches are slightly heavier and fingers slightly longer. These advantages add up to a higher annual yield of the extra-large fruit of Grand Naine variety.[9]

Therefore, the research study in this chapter was undertaken to study the effect of different color plastic mulches on the performance of banana under drip irrigation.

18.2 MATERIALS AND METHODS

The field experiment was conducted at the Research cum Demonstration Farm of Precision Farming Development Centre (PFDC) of Dr. Anna-saheb Shinde College of Agricultural Engineering, Mahatma Phule Krishi Vidyapeeth (Agricultural University), Rahuri, India, during December 2014–December 2015. The field experiment was laid out in a randomized block design consisting of seven treatments with three replications. The size of each plot was 7 m × 3.5 m.

The color mulches for different treatments were 30 μm in thickness, 7.5 m in length, and 2.1 m in width. The soil at the site was clay type. The field capacity, permanent wilting point, and bulk density were 40.09%, 17.37%, and 1.24 g-cm^{-3}, respectively. The treatments were as follows:

- T_1 = Yellow black plastic mulch with daily drip irrigation at 48% pan evaporation.
- T_2 = Blue black plastic mulch with daily drip irrigation at 48% pan evaporation.
- T_3 = Silver black plastic mulch with daily drip irrigation at 48% pan evaporation.

- T_4 = White black plastic mulch with daily drip irrigation at 48% pan evaporation.
- T_5 = Red black plastic mulch with daily drip irrigation at 48% pan evaporation.
- T_6 = Pervious black plastic mulch with daily drip irrigation at 48% pan evaporation.
- T_7 = No mulch with daily drip irrigation at 48% pan evaporation.

All treatments were given with daily drip irrigation at 48% of pan evaporation. This was based on the recommendation of irrigation by PFDC, MPKV, Rahuri, for the banana crop as 60% of pan evaporation, which was for nonmulched treatment. Thus, considering average water saving of mulch as 20%, drip irrigation was scheduled as 48% of pan evaporation for mulched treatments.[4] The observations such as *absorbed photosynthetically active radiations* (APAR), photosynthesis rate, soil temperature, and biometric observations of banana were recorded periodically.

18.3   RESULTS AND DISCUSSION

18.3.1   *WATER REQUIREMENT OF BANANA CROP*

The plants under silver black plastic mulch recorded lowest seasonal water requirement because of reduced duration of the crop (Table 18.1). Similar results were observed by Salvin et al.[11]

TABLE 18.1   Water Requirement of Banana Under Different Treatments.

Treatment	Seasonal gross depth of water applied (mm)
T_1	1114
T_2	1130
T_3	1015
T_4	1087
T_5	1081
T_6	1060
T_7	1142

18.3.2 ABSORBED PHOTOSYNTHETICALLY ACTIVE RADIATIONS

The highest APAR were observed in plants under silver black plastic mulch at 240 days after transplanting (DAT), followed by plants under white black plastic mulch, and it was minimum in no mulch treatment (control plot). APAR was increased continuously up to 240 DAT and then was decreased.[6] Periodical data for APAR (μmol m^{-2} s^{-1}) of banana are shown in Table 18.2.

TABLE 18.2 Absorbed Photosynthetically Active Radiations of Banana for Different Treatments.

Treatments	Photosynthetically active radiation (μmol m^{-2} s^{-1}) at					
	60 DAT	120 DAT	180 DAT	240 DAT	300 DAT	Harvest
T$_1$ (YBPM, 48% Ep)	295.09	483.43	827.60	1025.27	946.32	891.35
T$_2$ (BBPM, 48% Ep)	196.72	328.85	699.75	931.80	932.56	846.65
T$_3$ (SBPM, 48% Ep)	307.67	573.25	882.50	1060.78	975.55	900.27
T$_4$ (WBPM, 48% Ep)	295.62	511.71	866.00	1040.47	972.37	895.17
T$_5$ (RBPM, 48% Ep)	216.08	353.77	727.73	926.17	935.39	864.61
T$_6$ (PPM, 48% Ep)	192.25	314.35	655.35	876.00	897.19	842.37
T$_7$ (NM, 48% Ep)	177.67	290.65	614.50	855.51	864.15	800.53
SEM±	12.36	12.63	18.89	24.60	23.41	21.42
CD at 5%	37.07	37.88	56.63	73.76	70.17	64.20

18.3.3 PHOTOSYNTHESIS RATE

The highest rate of photosynthesis (μmol CO_2 m^{-2} s^{-1}) was observed in plants under silver black plastic mulch at 300 DAT, followed by plants under white black plastic mulch and yellow black plastic mulch, and was minimum in no mulch treatment.[7] Periodical data for photosynthesis rate of banana are given in Table 18.3.

TABLE 18.3 Photosynthesis Rate of Banana Under Different Treatments.

Treatments	Photosynthesis rate (μmol CO_2 m^{-2} s^{-1}) at					
	60 DAT	120 DAT	180 DAT	240 DAT	300 DAT	Harvest
T_1 (YBPM, 48% Ep)	3.22	6.77	14.26	18.96	28.55	27.33
T_2 (BBPM, 48% Ep)	2.95	6.26	13.70	17.25	27.89	26.23
T_3 (SBPM, 48% Ep)	3.63	7.95	15.05	19.72	29.33	28.38
T_4 (WBPM, 48% Ep)	3.47	7.56	14.58	19.16	29.02	28.16
T_5 (RBPM, 48% Ep)	3.00	7.27	14.30	18.79	28.38	26.44
T_6 (PPM, 48% Ep)	3.16	6.19	12.48	17.13	25.90	24.08
T_7 (NM, 48% Ep)	2.40	5.58	12.11	16.39	25.25	23.87
SEM±	0.10	0.17	0.36	0.48	0.72	0.67
CD at 5%	0.31	0.51	1.07	1.43	2.14	2.02

18.3.4 SOIL TEMPERATURE

The highest monthly average daily soil temperature was observed under red black plastic mulch in the month of May, followed by blue black plastic mulch, and minimum in white black plastic mulch. This was because of absorption of more incoming radiations by dark color mulches and more reflection by light color mulches. A similar trend was observed in the case of seasonal average daily soil temperature. Periodical data for soil temperature are given in Table 18.4.

18.3.5 BIOMETRIC PARAMETERS

18.3.5.1 PLANT HEIGHT

Table 18.5 indicates plant height, stem girth, and number of functional leaves per plant under all treatments. The highest plant height at harvesting stage (320 DAT) was observed for plants under silver black plastic mulch, followed by plants under previous plastic mulch, and minimum in no mulch treatment.

TABLE 18.4 Monthly Average Daily Soil Temperatures Under Different Treatments.

Month	Treatments								
	T₁ (YBPM, 48% Ep)	T₂ (BBPM, 48% Ep)	T₃ (SBPM, 48% Ep)	T₄ (WBPM, 48% Ep)	T₅ (RBPM, 48% Ep)	T₆ (PPM, 48% Ep)	T₇ (NM, 48% Ep)	T₈ (NM, 100% ETc)	T₉ (SI, 1.00 IW/CPE)
January 15	22.63	23.00	22.25	21.75	24.00	22.38	22.50	22.50	22.93
February 15	24.00	23.58	22.75	22.50	24.00	23.00	23.50	23.50	24.00
March 15	24.38	25.50	24.13	23.75	25.75	25.00	25.13	24.25	24.80
April 15	26.25	27.38	25.63	25.13	27.00	26.25	26.45	25.88	26.50
May 15	29.13	30.50	28.00	27.63	30.63	28.63	29.50	29.34	29.75
June 15	27.50	28.75	26.63	26.13	28.63	27.25	27.38	27.00	27.63
July 15	25.63	26.50	24.88	24.00	26.00	25.38	25.88	25.38	25.95
August 15	27.13	27.50	26.00	25.63	27.38	26.88	27.00	26.63	27.13
September 15	25.38	26.50	24.75	25.45	25.88	25.25	26.00	25.38	26.00
October 15	25.75	26.50	25.25	25.25	26.25	26.13	26.38	25.88	26.25
November 15	25.25	25.58	24.25	23.75	25.88	25.00	25.75	25.00	25.20
December 15	24.00	23.75	25.75	22.25	25.15	24.75	25.00	25.50	26.00
Seasonal average	25.58	26.25	25.02	24.43	26.38	25.49	25.87	25.52	26.01

TABLE 18.5   Growth Characteristics of Banana Under Different Treatments.

Treatments	Plant height (cm)	Stem girth (cm)	No. of functional leaves per plant
T_1 (YBPM, 48% Ep)	243.78	59.79	13.33
T_2 (BBPM, 48% Ep)	217.54	56.93	12.08
T_3 (SBPM, 48% Ep)	266.74	63.99	14.08
T_4 (WBPM, 48% Ep)	237.19	59.07	12.92
T_5 (RBPM, 48% Ep)	236.32	58.77	12.50
T_6 (PPM, 48% Ep)	252.59	63.50	13.42
T_7 (NM, 48% Ep)	202.70	52.89	11.50
SEM. ±	7.00	1.75	1.67
CD at 5%	20.99	5.25	7.72

18.3.5.2   STEM GIRTH

Similar to plant height, the highest stem girth at harvesting stage (320 DAT) was observed for plants under silver black plastic mulch, followed by plants under previous plastic mulch, and minimum in no mulch treatment.

18.3.5.3   NUMBER OF FUNCTIONAL LEAVES PER PLANT

The maximum number of functional leaves per plant at harvesting (320 DAT) was observed for plants under silver black plastic mulch and minimum in no mulch treatment.

18.3.5.4   DURATION PARAMETERS OF BANANA

18.3.5.4.1   Days Required for Flowering

The data on duration parameters of banana are given in Table 18.6. The plants under silver black plastic mulch required a minimum number of days (220 days) to flowering. This earliness in the flowering over other treatments could be attributed to better microclimate above and below the mulch with less competition resulting in shortening of its vegetative phase and causing earlier flowering. The maximum number of days (264 days) required for flowering were observed in plants under no mulch treatments.

TABLE 18.6 Duration Parameters of Banana Under Different Treatments.

Treatments	Mean days to flower	Total duration (days)
T_1 (YBPM, 48% Ep)	230.33	347.00
T_2 (BBPM, 48% Ep)	254.67	367.67
T_3 (SBPM, 48% Ep)	220.33	323.33
T_4 (WBPM, 48% Ep)	238.67	352.67
T_5 (RBPM, 48% Ep)	248.67	362.00
T_6 (PPM, 48% Ep)	225.67	339.33
T_7 (NM, 48% Ep)	263.67	373.67
SEM ±	6.73	9.35
CD at 5%	20.17	28.03

18.3.5.4.2 Total Duration

A similar trend, as days required to flower, was observed in the case of total duration of banana crop. The plants under silver black plastic mulch recorded a minimum total duration of 323 days. The maximum total duration (374 days) was observed with plants under no mulch treatments.

18.3.6 YIELD PARAMETERS OF BANANA

18.3.6.1 NUMBER OF HANDS PER BUNCH

The yield attributes are shown in Table 18.7. Significantly, superior number of hands per bunch was observed in plants under silver black plastic mulch due to better air–water balance in soil and improved microclimate beneath the crop canopy caused due to reflectance from mulch.

TABLE 18.7   Banana Yield and Yield Attributes Under Different Treatments.

Treatments	No. of hands per bunch	No. of fingers per bunch	Bunch weight	Average fruit weight	Yield	Length of fruit	Girth of fruit
			kg	g	t/ha	cm	cm
T₁ (YBPM, 48% Ep)	8.00	120.52	21.57	185.67	70.67	18.08	12.67
T₂ (BBPM, 48% Ep)	6.92	114.67	19.19	154.00	63.61	17.08	12.17
T₃ (SBPM, 48% Ep)	9.25	130.15	25.86	200.33	84.45	20.08	14.03
T₄ (WBPM, 48% Ep)	7.53	117.57	20.75	174.67	68.10	17.50	12.53
T₅ (RBPM, 48% Ep)	7.36	115.67	19.84	161.67	65.80	17.36	12.37
T₆ (PPM, 48% Ep)	8.42	126.50	24.75	193.67	81.67	18.39	13.28
T₇ (NM, 48% Ep)	6.17	108.67	17.75	150.50	57.73	16.50	11.58
SEM±	0.23	4.15	0.66	5.02	2.04	0.50	0.38
CD at 5%	0.68	12.43	1.99	15.04	6.11	1.49	1.14

18.3.6.2   *NUMBER OF FINGERS PER BUNCH*

The maximum number of fingers per bunch was recorded in plants under silver black plastic mulch followed by pervious mulch and yellow black plastic mulch. The minimum number of fingers per bunch was recorded in plants under no mulch treatment.

18.3.6.3   *BUNCH WEIGHT*

Highest bunch weight (25.86 kg) of banana was observed in plants under silver black plastic mulch and lowest (17.75 kg) in no mulch treatment.

18.3.6.4   *AVERAGE WEIGHT OF FRUIT*

Maximum fruit weight (200.33 g) was recorded in plants under silver black plastic mulch, followed by plants under pervious plastic mulch, and minimum (150.50 g) in no mulch treatment.

18.3.6.5   YIELD (TONS/HA)

As the yield is a function of bunch weight, it also exhibited similar trend as bunch weight. Maximum yield (84.45 t/ha) was recorded in plants under silver black plastic mulch, followed by that under pervious plastic mulch, and minimum (57.73 t/ha) in no mulch treatment.[8]

18.3.6.6   LENGTH OF FRUIT

The highest and significantly superior fruit length (20.08 cm) was recorded in plants under silver black plastic mulch followed by that in pervious mulch and yellow black plastic mulch. The minimum length of fruit (16.50 cm) was observed in plants under no mulch treatment.

18.3.6.7   GIRTH OF FRUIT

Similar to length, the girth of fruit also exhibited the same trend. The maximum girth of fruit (14.03 cm) was recorded in plants under silver black plastic mulch, followed by pervious plastic mulch, and minimum (11.58 cm) in no mulch treatments.

18.3.6.8   TSS OF BANANA FRUIT

No significant differences were observed in TSS of banana fruit pulp under different treatments. TSS ranged from 18.00 to 18.47°Brix.

18.3.6.9   ACIDITY OF BANANA FRUIT

No significant differences were observed in acidity of banana fruit pulp under different treatments. Acidity ranged from 0.25% to 0.28%.

18.4   SUMMARY

The research was conducted to study the effect of six color plastic mulches along with no mulch treatment on growth, yield, and quality of banana at MPKV, Rahuri, Maharashtra, India. The plastic mulches were yellow, blue, silver, white, red, and pervious plastic mulches. All treatments were given irrigation with 48% of pan evaporation. Results indicated that highest plant height, stem girth, and number of functional leaves were obtained under silver black plastic mulch. The highest yield (84.45 t/ha) was obtained in silver black plastic mulch due to improved microclimate both above and below mulch. Significantly less quantity of water was required for plants under silver black plastic mulch due to the reduced total duration of the crop. The quality of banana was not significantly influenced by different color mulches. The research indicated that adoption of silver black plastic mulch with daily drip irrigation at 48% pan evaporation resulted in 31.64% increase in yield compared to no mulch treatment.

KEYWORDS

- banana
- growth
- yield
- mulch
- irrigation

REFERENCES

1. Agrawal, N.; Agrawal, S. Effect of Drip Irrigation and Mulches on the Growth and Yield of Banana cv. Dwarf Cavendish. *Indian J. Hortic.* **2005,** *62*(3), 238–240.
2. Ahmed, B. A.; Mohammed, A. A.; Ihsan, M. I. *Effect of Drip Irrigation System and Fertigation on Growth, Yield and Quality of Banana;* RUFORUM: Uganda, 2010; pp 20–24.
3. Ashrafuzzaman, M.; Halim, M. A. Effect of Plastic Mulch on Growth and Yield of Chilli (*Capsicum annuum* L.). *Braz. Arch. Technol.* **2011,** *54*, 321–330.
4. Berad, S. M. Use of Liquid Fertilizers through Drip Irrigation to Banana. Unpublished Ph.D. (Agri.) Thesis, Mahatma Phule Krishi Vidyapeeth, Rahuri (MH), India, 1996.
5. Csizinszky, A. A.; Schuster, D. J.; Kring, J. B. Color Mulches Influence Yield and Insect Pest Population in Tomatoes. *HortScience* **1995,** *120*(5), 778–784.

6. Daughtry, C. S.; Gallo, K. P.; Goward, S. N.; Prince, S. D.; Kustas, W. P. Spectral Estimates of Absorbed Radiation and Phytomass Production in Corn and Soybean Canopies. *Remote Sens. Environ.* **1992,** *39*, 141–152.

7. Decoteau, D. R.; Kasperbauer, M. J.; Hunt, P. G. Bell Pepper Plant Development over Mulches of Diverse Colors. *HortScience* **1990,** *25*(4), 460–462.

8. Goenaga, R.; Irizarry, H. Yield of Banana Grown with Supplemental Drip on Ultisol. *Expt. Agric.* **1998,** *34*, 439–448.

9. Jagandi, S. Effect of Surface Mulches and Different Depth of Irrigation Levels in Banana. *Indian J. Appl. Res.* **2012,** *1*(4), 1–2.

10. Rajablariani, H.; Rafezi, R.; Hassankhan, F. In *Using Colored Plastic Mulches in Tomato (Lycopersicon esculentum* L.) *Production,* 4th International Conference on Agriculture and Animal Science, IPCBEE, 2012, Vol. 47.

11. Salvin, S.; Baruah, K.; Bordoloi, S. K. Drip Irrigation Studies in Banana. *Crop Res. (Hisar)* **2000,** *20*(3), 489–493.

CHAPTER 19

PERFORMANCE OF DRIP-IRRIGATED MANGO, SAPOTA, AND CASHEW NUT

ASHOK R. MHASKE[*]

Agricultural Engineering Division, College of Agriculture at Nagpur, Dr. Panjabrao Deshmukh Krishi Vidyapeeth (State Government Agricultural University), Nagpur 440001, Maharashtra, India

[*]*Corresponding author. E-mail: mhaskear@gmail.com*

CONTENTS

ABSTRACT

The micro irrigation is one of the water saving method of irrigation being employed at large scale in various countries. Drip irrigation is a novel irrigation method in India. A field experiment was conducted to evaluate the performance of drip irrigation systems on horticulture plantation at Zonal Agricultural Research Station, Sindewahi, District Chandrapur (Maharashtra State). In this study, the performance of drip irrigation on growth and development of horticultural crops, namely, mango, sapota, and cashew nut at Ranwadi watershed during the year 2005–06 to 2006–07 was evaluated. The result revealed the effect of drip irrigation treatment on growth and development of mango, sapota, and cashew nut. It was found that the treatment with 40 L water per day per plant through drip was found superior than all other treatment in respect of height (31.71 cm), canopy (1193 cm^2), and diameter (2.71 cm) of the stem of mango plant. In respect of sapota plant, treatment with 60 L water per day per plant was found satisfactorily superior in respect of height (26.2 cm), canopy (706 cm^2), and diameter (2.41 cm) stem of sapota plant. In respect of cashew nut plant, the treatment with 48 L of water alternate day per plant was found statistically significant and was at par in case of height (46.70 cm), canopy (1906 cm^2), and diameter (2.94 cm) stem of cashew nut plant.

19.1 INTRODUCTION

The micro irrigation is one of the water-saving methods of irrigation being employed at large scale in various countries. The government of Maharashtra has given more emphasis to micro irrigation system and has been adopting them on a large scale to save water.[12] Drip irrigation is a novel irrigation method adopted in India and abroad. In drip irrigation, liquid fertilizers can be added in the irrigation water.[6] It also diminishes leaching of nutrients.[5] The installation costs are too high for the production of most annual crops but the production of high-value perennial crops is economically profitable.

The advent of increasing water scarcity in this century will observe less increase in irrigated land availability for food production than in the past. Novel irrigation technologies need to be tested under local environments and particularly in agricultural production systems of developing countries. Kanannavar et al.[9] stated that irrigation can benefit yields and enhance water use efficiency (WUE) in water-limited environments; the potential for full irrigation is decreasing with increased competition from the domestic and industrial sectors. Thus, the main challenge confronting both rainfed and irrigated agriculture is to improve WUE and sustainable water use for agriculture.[2]

India is facing a tremendous challenge in meeting the food needs of the rapidly growing population. There are small-, medium-, and large-scale irrigation systems. To this end, both irrigated and dry land cropping areas will have to be developed or improved in the future. However, these tasks will not be easy, as the cost of developing large-scale and medium-scale-level irrigation is now skyrocketing. Therefore, efficient utilization of water resources and development of small-scale irrigation schemes at the family level is crucial for countries like India that has a huge water resource, yet the population is chronically food insecure. Aujla[3] stated that micro irrigation system resulted in 30–70% water savings in various orchard crops and vegetables along with 10–60% increase in yield as compared to conventional methods of irrigation. Shock et al.[14] stated that it is prudent to make efficient use of water and bring more area under irrigation through available water resources. Zaman[16] stated that this can be achieved by introducing advanced methods of irrigation and improved water management practices.

Mango, sapota, and cashew nut are dominant horticultural fruit crops in India, and are best suited for drip irrigation. However, not sufficient research has been done to study the effects of drip irrigation on the growth of horticultural plantation in Eastern Vidarbha zone of Maharashtra state having sandy clay loam soil in a monoculture paddy cropping system. The present study was planned to evaluate the effects of the amount of water through drip irrigation on growth and development of horticultural crop at Ranwadi watershed (district Chandrapur) in Maharashtra State. The aim of the study was to evaluate the growth response of mango (*Mangifera indica*), sapota (*Manilkara zapota*), and cashew nut (*Anacardium occidentale*) under drip irrigation system by measuring the height, canopy, and diameter of the stem in sandy clay loam soil of Eastern Vidarbha zone.

19.2   MATERIALS AND METHODS

19.2.1   EXPERIMENTAL SITE

A field study was carried out at the research farm of Dr. Punjabrao Deshmukh Krishi Vidyapeeth, Akola situated in Eestern Vidarbha zone at a latitude of 79°39′ E and longitude of 20°17′ N at mean sea level of 222 m. The horticultural crops (mango, sapota, and cashew nut) were planted at Ranwadi micro watershed, Zonal Agricultural Research Station, Sindewahi, District Chandrapur (Maharashtra State) in 2000. Plant to plant and row to row spacing was 15 ft × 15 ft for mango and sapota; it was 20 ft × 20 ft for cashew nut. The study on this plantation of mango, sapota, and cashew nut was initiated in the year 2002–03 and continued till 2006–07.

A drip irrigation system was fitted to the above horticultural plantation. Seven main treatments were as follows:

T_1—Basin irrigation (control method) at field capacity of 10 days interval based on the evaporation and considering the 5 cm depth and 5 days irrigation scheduling.[10]

T_2—40 L water per alternate day per plant.

T_3—40 L water per day per plant.

T_4—48 L water per alternate day per plant.

T_5—48 L water per day per plant.

T_6—60 L water per alternate day per plant.

T_7—60 L water per day per plant as stated by the author.[9]

Fertilizer doses (Table 19.1) and other package and practices recommended by Dr. Panjabrao Deshmukh Krishi Vidyapeeth, Akola were used in all treatments in June–July, September–October, and January–February.

TABLE 19.1   N–P–K Doses (Grams).

Crop	Fertilizer doses (g)		
	N	P	K
Mango	720	900	300
Sapota	900	450	400
Cashew nut	900	450	400

19.2.2   EXPERIMENTAL DESIGN AND MATERIALS

A randomized block design was applied and placement of drippers and irrigation duration for different water applications were as follows:

- For 40 L application—Dripper per plant = 8 LPH (2 no) + 4 LPH (1 no)
  - Irrigation hours—2 h.
- For 48 L application—Dripper per plant = 8 LPH (3 no)
  - Irrigation hours—2 h.
- For 60 L application—Dripper per plant = 16 LPH (1 no) + 4 LPH (1 no)
  - Irrigation hours—3 h.

19.2.3   IRRIGATION SYSTEM

The water source was a borewell and it was dug out near the embankment pond, which was constructed for recharging of the borewell. Four plants, each of mango, sapota, and cashew nut for each treatment were selected for the study. Water was applied to the plant through drip irrigation system by doing the calibration of the drippers. Each dripper discharged 4, 8, and 16 L of water per hour as stated above and the water through drip was applied in each treatment. Plants were irrigated from October to March for the year 2005–06 to 2006–07. The drippers were attached to dripline according to water requirement, and for every row, separate line was used for providing irrigation to the plant. The plants were irrigated from the month of October to the start of the rainy season, but the observation on the growth of the plant was recorded from October to March (during good growth period).

19.2.4   DATA COLLECTION

Monthly observation of height, canopy, and diameter of the plant was recorded. All the four plants per treatment were considered for observations on development and growth of the plants.

19.3   RESULT AND DISCUSSION

19.3.1   EFFECTS OF DRIP IRRIGATION ON GROWTH AND DEVELOPMENT OF MANGO

The data pertaining to the growth and development of the mango plant are reported in Table 19.2. The treatment differences in case of mean height, mean canopy, and mean diameter of the stem for mango plant were significant. From Table 19.2, it is observed that the treatment T_3 (40 L of water per day per plant through drip) was statistically superior than all other treatments in respect of height (31.77 cm), canopy area (1193 cm²), and diameter (2.71 cm) of the stem of mango plants. The results are in agreement with those by Hanson et al.[8]

TABLE 19.2   Effect of Drip Irrigation Treatments on Height, Canopy, and Diameter of the Stem of Mango Plants (2006–07).

Treatment	Mean height (cm)		Mean canopy (cm²)		Mean diameter of stem (cm)		Survival (%)
	Mean	% increase over control	Mean	% increase over control	Mean	% increase over control	
T_1	17.63	–	509	–	1.98	–	100
T_2	20.56	16.62	604	18.66	2.12	7.07	100
T_3	31.77	80.20	1193	134.38	2.71	36.86	100
T_4	24.04	36.40	822	61.49	2.34	18.18	100
T_5	29.4	66.72	1058	100.78	2.58	30.30	100
T_6	26.58	50.76	918	80.35	2.47	24.75	100
T_7	22.82	29.43	686	34.77	2.26	14.14	100
"F" test	Sig.	–	Sig.	–	Sig.	–	–
SEM±	2.56	–	173	–	0.34	–	–
CD	7.19	–	487	–	0.95	–	–
CV%	19.75	–	36.64	–	24.09	–	–

There was 80.20% increase in mean height, 134.38% increase in mean canopy area, and 36.86% increase in mean diameter of the stem compared to control treatment. To establish the correlation between the diameter of stem and canopy of the mango plant, power correlation between diameter and canopy of mango plant was used (Fig. 19.1). A fairly good value of R^2 of 0.987 shows that data variation has been explained by the model and shows good fit to data.

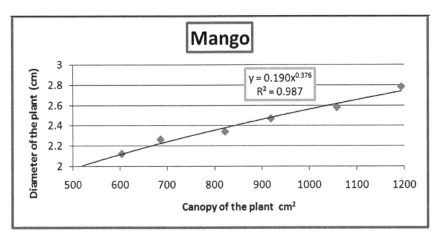

FIGURE 19.1   Correlation between the diameter and canopy of the mango plant.

19.3.2   EFFECT OF DRIP IRRIGATION ON GROWTH AND DEVELOPMENT OF SAPOTA PLANT

The data on the growth and development of sapota plant are reported in Table 19.3. It is observed that the treatment T_6 (60 L of water per alternate day plant) was statistically superior than all other treatments in respect of height (26.2 cm), canopy area (706 cm^2), and diameter (2.41 cm) of the stem of sapota plant. The treatment differences in respect of height, canopy, and diameter of the stem of the sapota plant were found to be significant.[4] There was 67.41% increase in height, 108.90% increase in the canopy, and 38.51% increase in diameter of the stem compared to the control of the sapota plants.[15] Power correlation was developed to establish the correlation between the diameter of the stem and canopy of sapota plant (Fig. 19.2). Regression coefficients with a fairly good value equal to 0.957 show that data variation has been explained by the model successfully with good fit to the data.

TABLE 19.3 Effects of Drip Irrigation Treatments on Height, Canopy, and Diameter of Stem of Sapota Plants (2006–07).

Treatments	Mean height (cm)	% increase over control	Mean canopy (cm²)	% increase over control	Mean diameter of stem (cm)	% increase over control	Survival (%)
T_1	15.65	–	338	–	1.74	–	100
T_2	16.80	7.35	389	15.09	1.94	7.14	100
T_3	25.45	56.23	630	86.40	2.39	37.50	100
T_4	21.04	34.44	497	47.04	2.15	23.56	100
T_5	23.15	47.92	555	64.20	2.24	28.74	100
T_6	**26.2**	**67.41**	**706**	**108.9**	**2.41**	**38.51**	100
T_7	18.75	19.80	427	26.33	2.06	18.39	100
"F" test	Sig.	–	Sig.	–	Sig.	–	–
SEM±	1.70	–	67.44	–	0.04	–	–
CD	4.79	–	189.42	–	0.13	–	–
CV%	17.25	–	30.94	–	4.70	–	–

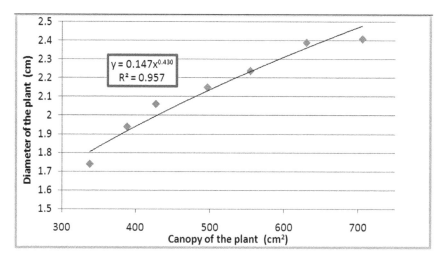

FIGURE 19.2 Correlation between the diameter and canopy of the sapota plants.

19.3.3   EFFECTS OF DRIP IRRIGATION TREATMENTS ON GROWTH AND DEVELOPMENT OF CASHEW NUT PLANT

The data on the development and growth of cashew nut plant in different treatments of drip irrigation are reported in Table 19.4. The treatment differences regarding height, canopy, and diameter of the stem were significant. In this study, it was observed that plants irrigated with treatment T_4 (48 L water per alternate day plant) were found statistically significant and superior in case of height (46.7 cm), canopy (1906 cm^2), and diameter (2.94 cm) of the stem.

TABLE 19.4   Effects of Drip Irrigation Treatments on Height, Canopy, and Diameter of the Stem of Cashew Nut Plants (2006–07).

Treatments	Mean height (cm)	% increase over control	Mean canopy (cm²)	% increase over control	Diameter of stem (cm)	% increase over control	Survival (%)
T_1	22.39	–	686	–	2.08	–	100
T_2	29.86	33.36	828	20.70	2.19	5.28	100
T_3	39.39	75.92	1420	106.99	2.78	27.90	100
T_4	**46.70**	**108.57**	**1906**	**177.84**	**2.94**	**41.35**	**100**
T_5	36.00	60.78	1186	72.88	2.65	17.79	100
T_6	42.90	91.60	1631	137.75	2.81	35.01	100
T_7	33.70	50.51	1030	50.14	2.36	13.47	100
"F" test	Sig.	–	Sig.	–	Sig.	–	–
SEM±	4.45	–	121	–	0.32	–	–
CD	12.49	–	326	–	0.90	–	–
CV%	21.05	–	21.94	–	22.33	–	–

The increase was observed in drip irrigation treatment T_4 in height, canopy, and diameter of the stem up to 108.57%, 177.84%, and 41.35%, respectively compared to control plots. Power model was developed to establish the correlation between the diameter and canopy of cashew nut plant (Fig. 19.3). The developed model had R^2 of 0.99, thus indicating that data variation had been explained by the model with a good fit.

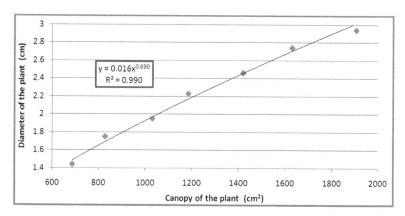

FIGURE 19.3   Correlation between the diameter and height of the cashew nut plants.

19.3.4   EFFECT OF DRIP IRRIGATION ON THE HEIGHT OF MANGO, SAPOTA, AND CASHEW NUT PLANTS

Effects of drip irrigation treatment including control on mean height of mango, sapota, and cashew nut plants are depicted in Figure 19.4. From the data, it can be concluded that height of mango was superior in T_3 (40 L amount of water per day per plant) followed by T_5 which was at par. The height of sapota was observed highest in T_6 (60 L water per alternate day per plant) followed by T_3 and T_5. The height of the cashew nut plant was observed higher for T_4 (48 L water per alternate day per plant) followed by T_6 and T_3.

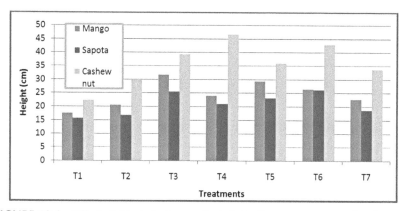

FIGURE 19.4   Effect of drip irrigation on the height of mango, sapota, and cashew nut plants.

Results indicate that the treatment T_3 for mango, treatment T_6 for sapota, and treatment T_4 for cashew nut were found to be superior in terms of height. Further, the height of cashew nut plant was higher followed by mango plant and sapota plant. Similar results have been reported by Sankaranarayanan.[13]

19.3.5   EFFECT OF DRIP IRRIGATION ON CANOPY OF MANGO, SAPOTA, AND CASHEW NUT PLANTS

Effects of drip irrigation treatment including control on the mean canopy of mango, sapota, and cashew nut plants are depicted in Figure 19.5. The data indicate that canopy of mango was superior in T_3 (40 L of water per day per plant) followed by T_5 which was at par. The canopy of sapota was highest in T_6 (60 L water per alternate day per plant) followed by T_3 and T_5. Further, the canopy of the cashew nut plant was observed higher for T_4 (48 L water per alternate day per plant) followed by T_6 and T_3. The treatment T_3 for mango, treatment T_6 for sapota, and treatment T_4 for cashew nut were found to be superior in terms of the canopy and similar results have been reported by Abbey et al.[1] Also the canopy of the cashew nut plant was more followed by mango and sapota plant.

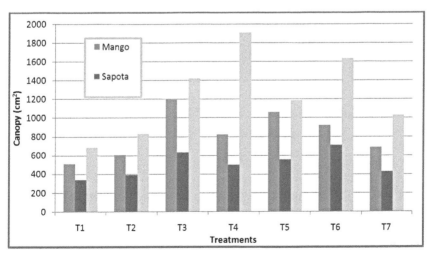

FIGURE 19.5   Effect of drip irrigation on canopy of mango, sapota, and cashew nut plants.

19.3.6   EFFECT OF DRIP IRRIGATION ON STEM DIAMETER OF MANGO, SAPOTA, AND CASHEW NUT PLANTS

Effects of drip irrigation on mean stem diameter of mango, sapota, and cashew nut plants are depicted in Figure 19.6. The data indicated that diameter of the stem of mango plant was superior in T_3 (40 L water per day per plant) followed by T_5 which was at par. The diameter of the stem of sapota was observed highest in T_6 (60 L water per alternate day per plant) followed by T_3 and T_5. Further, the diameter of the stem of the cashew nut plant was more for T_4 (48 L water per alternate day per plant) followed by T_6 and T_3. The treatment T_3 for mango, treatment T_6 for sapota, and treatment T_4 for cashew nut were found superior in terms of the diameter of the stem. Also, the diameter of the stem of the cashew nut plant was observed to be more followed by mango and sapota plant.

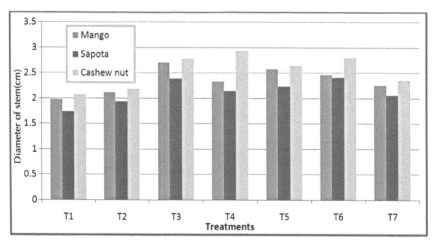

FIGURE 19.6   Effect of drip irrigation on the diameter of mango, sapota, and cashew nut plants.

19.4   FUTURE PROSPECTIVES AND RESEARCH OPPORTUNITIES

In the Indian context, water scarcity has compelled the farmers to adopt drip irrigation to use the water more efficiently for the high-valued crop. To achieve the target, more research is needed on thrust area, such as:

- To reduce the cost of system and problem free system.
- Package of practice for various crops with drip irrigation.
- Improve the performance of drip irrigation especially on clogging.
- The thickness of the lateral should be minimum to reduce the cost.
- Design and evaluation of the drip irrigation for horticultural plantation and row crops in the different agroecological region.
- Design and evaluation of pressurized irrigation as adjunct canal irrigation system.

19.5 CONCLUSIONS

The author conducted experiments at Zonal Agricultural Research Station, Sindewahi, district Chandrapur, to study the performance of drip irrigation on growth and development of horticultural crops (mango, sapota, and cashew nut) at Ranwadi watershed during the year 2005–06 and 2006–07. The results revealed that effect of drip irrigation treatment in growth and development of mango, sapota, and cashew nut was different under different treatments. It was found that the treatment with 40 L water per day per plant through drip irrigation was found superior than all other treatments in respect of height (31.70 cm), canopy (1193 cm²), and diameter (2.71 cm) of the stem of mango plant. For sapota plant, treatment with 60 L water per alternate day per plant was found satisfactorily superior in respect of height (26.2 cm), canopy (706 cm²), and diameter (2.41 cm). In case of cashew nut plant, the treatment with 48 L water alternate day per plant was found statistically significant and was at par in case of height (46.70), canopy (1906), and stem diameter (2.94) of cashew nut plant.

19.6 SUMMARY

There was a significant difference in the growth and development of the horticultural plants (mango, sapota, and cashew nut) in terms of height, canopy, and diameter of stem due to drip irrigation. Cashew nut plant showed better response with 48 L water per alternate day per plant compared to control; followed by mango crop with 40 L water per day per plant and sapota with 60 L water per alternate day per plant compared to control treatment. Drip irrigation method facilitates the water and air proportion suitable for the growth of the plant and is based on the evapotranspiration. Drip irrigation provides desired quantity of water to the plant.

KEYWORDS

- **calibration**
- **canopy**
- **drip irrigation**
- **evapotranspiration**
- **field capacity**
- **horticultural crop**
- **perennial crops**
- **scheduling**
- **stem diameter**
- **watershed**

REFERENCES

1. Abbey, L.; Joyce, D. C. Water-deficit Stress and Soil Type Effects on Spring Onion Growth. *J. Veg. Crop Prod.* **2004,** *10*(2), 5–18.
2. Anonymous. *Annual Report on Management of Water Resources Through Efficient Utilisation of Water and Ground Water Recharge*; Punjab Agricultural University: Ludhiana, India, 2007; pp 16–18.
3. Aujla, M. S.; Thind, H. S.; Butter, G. S. Cotton Yield and Water Use Efficiency at Various Levels of Water and N Through Drip Irrigation Under Two Method of Planting. *Agric. Water Manag.* **2005,** *71*(2), 167–179.
4. Bagali, A. N.; Patil, H. B.; Guled, M. B.; Patil, R. V. Effect of Scheduling of Drip Irrigation on Growth, Yield and Water Use Efficiency of Onion (*Allium cepa L.*). *Karnataka J. Agric. Sci.* **2012,** *25*(1), 116–119.
5. Bryla, D. R.; Thomas, J. T.; James, E. A.; Johnson, R. S. Growth and Production of Young Peach Trees Irrigation by Furrow, Micro Jet, Surface Drip or Subsurface Drip Systems. *HortScience* **2003,** *38*, 1112–1116.
6. Gethe, R. M.; Pawar, V. S.; Pathan, S. H.; Sonawane, D. A.; Kadlag, A. D. Influence of Planting Layouts, Irrigation Regimes and Fertilizer Levels on Growth and Yield of Onion Under Microsprinkler. *J. Maharashtra Agric. Univ.* **2006,** *31*(3), 272–274.

7. Gunduz, M.; Korkmaz, N.; Asik, S.; Unal, H. B.; Avci, M. Effects of Various Irrigation Regimes on Soil Water Balance, Yield and Fruit Quality of Drip-irrigated Peach Trees. *J. Irrig. Drain. Eng.* **2011,** *137,* 426–434.

8. Hanson, B. R.; May, D. M.; Chwankl, L. J. Effect of Irrigation Frequency on Subsurface Drip Irrigated Vegetables. *HortTechnology* **2003,** *13*(1), 115–120.

9. Kanannavar, P. S.; Kumathe, S. S.; Premanand, B. D.; Kawale, N. *Water Saving and Economics of Banana Production Under Drip Irrigation in North Eastern Dry Zone of Karnataka*; Birasa Agricultural University: Ranchi (Jharkhand) India, 2009; p 60.

10. Komilov, B.; Ibragimov, N.; Esanbekov, Y.; Evett, S.; Lee, H. In *Irrigation Scheduling Study of Drip Irrigated Cotton by Use of Soil Moisture Neutron Probe*, Proceeding of the National Workshop on Developing Cotton and Winter Wheat Agrotechnologies, UNCGRI, Tashkent, Uzbekistan, Dec 24–25, 2002.

11. Nalayini, P.; Raja, R.; Kumar, A. A. Evapotranspiration Based Scheduling of Irrigation Through Drip for Cotton (*Gossipium hirsuitum*). *Indian J. Agron.* **2006,** *51*(3), 232–235.

12. Pawar, B. R.; Landge, V. V.; Deshmukh, D. S.; Yeware, P. P. Economics of Banana Production in Drip Irrigated and Flood Irrigated Gardens. *Int. J. Commer. Bus. Manag.* **2010,** *3*(1), 88–91.

13. Sankaranarayanan, K.; Nalayini, P.; Sabesh, M.; Usha-Rani, S.; Nachane, R. P.; Gopalakrishnan, N. *Low Cost Drip-cost Effective and Precision Irrigation Tool in Cotton*; Techni. Bull. 1/2011; Central Institute for Cotton Research, Regional Station: Coimbatore (Tamil Nadu), India, 2011; p 30.

14. Shock, C. C.; Flock, R.; Feibert, E.; Shock, C. A.; Jensen, L.; Klauzer, J. *Drip Irrigation Guide for Onion Growers in the Treasure Valley*; Oregon State University Agricultural Experiment Station Special Report 1062, 2005; pp 173–176. http://www.cropinfo.net/AnnualReports/2004/ranger%20umatilla%20compare 04.php

15. Siag, M.; Kaushal, M. P.; Buttar, G. S. Impact of Drip Line Spacing on Cotton Growth and Yield. *J. Agric. Eng.* **2010,** *47*(4), 47–50.

16. Zaman, W. U.; Arshad, M.; Saleem, A. Distribution of Nitrate-nitrogen in the Soil Profile Under Different Irrigation Methods. *Int. J. Agric. Biol.* **2001,** *3*(2), 208–209.

APPENDIX
GLOSSARY OF TECHNICAL TERMS

Calibration is the process of comparison of measurement values delivered by a device under test with a calibration standard of known accuracy.

Canopy is the uppermost branches of the trees in a forest, forming a more or less continuous layer of foliage.

Correlation is a statistical technique that is used to measure and describe the strength and direction of the relationship between two variables. Correlation requires two scores from the same individuals.

Drip irrigation saves water and fertilizer by allowing water to drip slowly to the roots of many different plants, either onto the soil surface or directly onto the root zone, through a network of valves, pipes, tubing, and emitters.

Evapotranspiration is the process by which water is transferred from the land to the atmosphere by evaporation from the soil and other surfaces and by transpiration from plants.

Irrigation scheduling is the process used by irrigation system managers to determine the correct frequency and duration of watering.

Perennial crop or simply *perennial* is a plant that lives for more than 2 years. The term is often used to differentiate a plant from shorter lived annuals and biennials. The term is also widely used to distinguish *plants* with little or no woody growth from trees and shrubs, which are also technically *perennials*.

Survival percentages are obtained by following a group of individuals over time. This gives us the finite survival rate which is the ratio of a number of individuals alive at end of time period to the number of individuals alive at the start of the time period.

Water requirement is the amount of water needed by the various crops to grow optimally.

INDEX